Methods in Molecular Biology

Series Editor
John M. Walker
School of Life and Medical Sciences
University of Hertfordshire
Hatfield, Hertfordshire, UK

For further volumes:
http://www.springer.com/series/7651

For over 35 years, biological scientists have come to rely on the research protocols and methodologies in the critically acclaimed *Methods in Molecular Biology* series. The series was the first to introduce the step-by-step protocols approach that has become the standard in all biomedical protocol publishing. Each protocol is provided in readily-reproducible step-by-step fashion, opening with an introductory overview, a list of the materials and reagents needed to complete the experiment, and followed by a detailed procedure that is supported with a helpful notes section offering tips and tricks of the trade as well as troubleshooting advice. These hallmark features were introduced by series editor Dr. John Walker and constitute the key ingredient in each and every volume of the *Methods in Molecular Biology* series. Tested and trusted, comprehensive and reliable, all protocols from the series are indexed in PubMed.

Immunometabolism

Methods and Protocols

Edited by

Suresh Mishra

Faculty of Health Sciences, Department of Internal Medicine, University of Manitoba, Winnipeg, MB, Canada; Faculty of Health Sciences, Department of Physiology and Pathophysiology, University of Manitoba, Winnipeg, MB, Canada

Humana Press

Editor
Suresh Mishra
Faculty of Health Sciences
Department of Internal Medicine
University of Manitoba
Winnipeg, MB, Canada

Faculty of Health Sciences
Department of Physiology and Pathophysiology
University of Manitoba
Winnipeg, MB, Canada

ISSN 1064-3745 ISSN 1940-6029 (electronic)
Methods in Molecular Biology
ISBN 978-1-0716-0801-2 ISBN 978-1-0716-0802-9 (eBook)
https://doi.org/10.1007/978-1-0716-0802-9

Cover caption: Immunofluorescence staining of membrane (red) and mitochondrial (green) markers in macrophage.

This Humana imprint is published by the registered company Springer Science+Business Media, LLC, part of Springer Nature.
The registered company address is: 1 New York Plaza, New York, NY 10004, U.S.A.

Preface

Immunometabolism is an emerging field of biomedical investigation at the interface of the historically distinct disciplines of immunology and metabolism [1]; it incorporates both the role of immune cells in metabolic homeostasis in the body and the impact of interconnected metabolic pathways on immune cell functions [2, 3]. The birth of this new research frontier can be traced back to the understanding that obesity affects the immune system and promotes inflammation, also known as meta-inflammation (or alternatively chronic low-grade inflammation). For more than 50 years, physicians and scientists have observed a close association between metabolic disorders and systemic inflammation [4]. Studies dating from the 1960s found that individuals with type 2 diabetes mellitus have higher circulating concentrations of active complement and acute-phase reactants [5, 6], both of which are classical markers of an inflammatory state. However, the site of inflammation and their pathological implications remained obscure until about 25 years ago [4]. In the 1990s, it was reported that the obesity-induced expression of tumor necrosis factor-α (TNF-α) exists in the adipose tissue of both rodents and humans, and it has been proposed that TNF-α mediates obesity-related insulin resistance [7]. Subsequently, the transcriptional evidence for the presence of macrophages in adipose tissue was found [8, 9]. Later, it was shown that the macrophage levels correlated positively with adiposity, and most of the TNF-α and other inflammatory molecules were derived from adipose tissue macrophages [8, 9].

Although the initial immunometabolism studies focused on adipose tissue, it is now evident that the metabolic activation of the immune system is not limited to obesity [4]. Growing evidence suggests that interacting metabolic pathways in immune cells play a central role in their functional plasticity and have been the focus of intense interest as therapeutic targets to harness the full potential of the immune system [2, 3]. The scientific community does not yet fully understand how and why immune cells commit to a particular metabolic fate, or the immunological consequences of reaching a metabolic endpoint by one pathway versus another. The multilevel interactions between the metabolic and immune systems suggest pathogenic mechanisms that may underlie many metabolic and immune diseases and offer substantial therapeutic promise [1]. Research on immunotherapy has been conducted for over a century; nevertheless, the last decade has seen an increase in interest in studying immunotherapy. However, a number of challenges remain due to its limited effectiveness and treatment-related adverse effects. It is anticipated that incorporating immunometabolism and manipulating immune cell functions will provide some much-needed ways to improve the effectiveness of promising immunotherapy, and reduce the unintended side effects, as well as improve the treatment and prevention of a wide variety of pathologies and chronic diseases. Thus, the molecular underpinning of immunometabolism has become a priority to maximize the therapeutic efficacy of immunotherapy.

This book is dedicated to showcasing the tremendous effort and progress made over the last few decades in developing techniques and protocols, and in utilizing recent technological advances for probing and manipulating adipose and immune cells, and subsequently their functions and immunometabolic consequences. All chapters are written by experts in their particular fields and cover a wide range of topics related to the study of immunometabolism using different experimental approaches in combination with new tools and techniques.

Many chapters in this protocols book are written using macrophages as a model immune cell type (including murine and human cell lines, as well as primary cells) because macrophages are the most prominent cells of the innate immune system that regulate a variety of inflammatory, host defense, and wound repair processes, and as such have been studied extensively for cell differentiation, gene regulation, and signal transduction. In addition, well-established procedures exist to isolate, culture, and activate mouse and human macrophages. Importantly, an extraordinary plasticity of macrophages to their surrounding microenvironment makes them a unique therapeutic target for a variety of immunometabolic diseases. Moreover, protocols using adipocytes, dendritic cells, and T cells as model cell lines, as well as measurement of glucose metabolism at the systemic level, have also been included, as it relates to immunometabolism.

The single-cell RNA sequencing (scRNA-seq) allows an unbiased approach for uncovering a new level of cellular heterogeneity and dynamics of a diverse biological system, including the immune system, as it enables a comprehensive analysis of the transcriptome of individual cells by next-generation sequencing. Optimization of the technical procedures performed prior to RNA-seq analysis is imperative to the success of a scRNA-seq experiment. Janilyn Arsenio describes three major experimental procedures: (1) the isolation of immune $CD8a^{+}$ T cells from primary murine tissue; (2) the generation of single-cell cDNA libraries using the 10x Genomics Chromium Controller and the Chromium Single Cell 3′ Solution; and (3) cDNA library quality control. In this protocol, $CD8a^{+}$ T cells are isolated from murine spleen tissue, but any cell type of interest can be enriched and used for the single-cell cDNA library generation and subsequent RNA-seq experiments.

Cellular metabolism has emerged as a major player in the regulation of the functional plasticity of different immune cell types. For example, the production of lactate by macrophages has been associated with their polarization and function. Baeza-Lehnert et al. describe imaging protocols to characterize the metabolism of cultured human macrophages using a genetically encoded fluorescent sensor specific for lactate. This protocol allows determining the kinetic parameters of monocarboxylate transporter 4 and lactate production at the single-cell level. The authors have also provided practical advice regarding sensor expression, imaging, and data analysis. Importantly, the spatiotemporal resolution of this technique is amenable to the study of fast events at the single-cell level in different immune cells and other cell types.

In addition to the measurement of the transcriptome and cellular metabolism at a single-cell level, recent developments have enabled a parallel analysis of both the transcript and protein at a single-cell level by using antibodies conjugated to barcoded oligonucleotides. These antibodies allow the "i" of protein levels to be presented in nucleotide format, permitting a sequencing-based detection of both modalities at a single-cell level. Tapio Lönnberg and colleagues present a simple and reliable method for the conjugation of oligonucleotides with antibodies and a protocol for their use in single-cell transcriptome sequencing. This protocol addresses the significant challenges associated with the biological and functional interpretation of newly identified cell populations using scRNA-seq.

The stable isotope labeling of metabolites is a technique employed to investigate the movement of a metabolite through a cell's enzymatic machinery. The information obtained from this process allows for the determining of the relative fluxes of metabolites through a biochemical pathway, and the contribution of specific metabolites to the total metabolite pool. For instance, the incorporation of stable isotope labels into specific fatty acids allows for the discrimination of newly synthesized fatty acids from those that are in preexisting pools, or fatty acids that have been imported from an extracellular source. Kevin Williams and Steven Bensinger describe a workflow for a total cellular fatty acid analysis in macrophages, which combines a fatty acid methyl ester analysis (by gas chromatography–mass spectrometry) with isotopic labeling. This approach can elucidate the synthetic pathways being engaged by the cells and the relative contribution of synthesis and import to maintain lipid content, which is an important component of cellular metabolism in immune cells.

Discovery-based quantitative proteomics is a useful method to unravel complex protein networks and protein-protein interactions. Saiful Chowdhury and coworkers describe protocols for the proteomics network analysis of polarized macrophages in response to pro- and anti-inflammatory agents. They provide detailed protocols, a quantitative proteomic analysis by mass spectrometry data, a protein network analysis by bioinformatics, and a validation of targets through biochemical methods (e.g., immunocytochemistry, immunoblotting, gene silencing, and real-time PCR).

The intercellular communication or cross talk between different cell types, including intra-organ and interorgan cross talk, engaged in metabolic and immune regulation (e.g., adipocytes, hepatocytes, macrophages, dendritic cells, lymphocytes) plays a crucial role in immunometabolism at the systemic level and their dysregulation in the development of a number of metabolic and immune diseases, including different types of cancer. In this context, exosomes have been identified as a crucial player in the intercellular cross talk in health and disease [10]. Thus, it is crucial to develop protocols to investigate the intercellular cross talk between metabolic and immune cells, including the role of exosomes in this process. To this end, Fridman et al. describe methods for the isolation and polarization of mouse peritoneal macrophages, the purification of exosomes from the conditioned media of the polarized macrophages, and the characterization of the resulting exosomes. In addition, they provide protocols to study exosome-based communication between two cell types using macrophages and pancreatic cancer cells as an example, thus mimicking a tumor microenvironment. This protocol may be used to study exosome-based communication in other experimental settings as well. In addition, Inbar Azoulay-Alfaguter and Adam Mor provide protocols for the isolation and characterization of T lymphocyte-derived exosomes using mass spectrometry. Particularly, they describe a centrifugation approach, combined with mass spectrometry characterization, as a means to study exosomes derived from primary human T lymphocytes. As mass spectrometry is a very sensitive method, this protocol can be applied when limited samples are available.

Three-dimensional cultures are better able to reflect the tumor microenvironment than two-dimensional monolayer cultures, by facilitating cell-cell interactions in the appropriate spatial dimensions. Tanya N. Augustine describes the isolation and co-culture of immune cells with tumor cell lines in a three-dimensional system in a biologically relevant scaffold facilitated by a basement membrane extract. This protocol allows for the assessment of immune-tumor cell interactions in spatial dimensions that reflect the in vivo tumor microenvironment. This protocol may be adapted for different cell types, and for determining a response to therapeutic agents.

Continuing on the theme of intercellular communication, Monk et al. describe methodologies for the co-culture of mature adipocytes (differentiated 3T3-L1 pre-adipocyte cell line) with primary immune cell subsets purified from mouse splenic mononuclear cells using magnetic MicroBead positive selection. MicroBead-based positive selection may be used to purify multiple immune cell populations sequentially from a single mouse spleen, thereby providing diversity in the types of immune cells that can be co-cultured with adipocytes. Additionally, the authors provide the experimental procedures for co-culturing adipocytes and immune cells in two different co-culture systems, including a cell contact-dependent co-culture system wherein the cells are in direct physical contact, as well as a cell contact-independent, soluble mediator-driven co-culture system, wherein a transwell semipermeable membrane physically separates the cells.

Ghia and colleagues have elaborated methods for the isolation of macrophages from a variety of murine sources, including peritoneal, bone marrow-derived, and alveolar macrophages, which are extensively used to explore both the immunobiology and pathophysiology of several diseases. In addition, the authors describe the phenotypic characterization of polarized human monocytic THP-1-derived macrophages and murine RAW264.7 cells (a macrophage cell line). Ikeogu et al. describe methods for isolation and preparation of bone marrow-derived immune cells for metabolic analysis, including macrophages, dendritic cells, and neutrophils from mice.

The posttranslational modifications by ADP-ribosylation and phosphorylation are important regulators of cellular pathways. While mass spectrometry-based methods for the study of protein phosphorylation are well developed, protein ADP-ribosylation methodologies are still in development. Nita-Lazar and colleagues describe an immobilized metal affinity chromatography: a phosphoenrichment matrix-based method to enrich ADP-ribosylated peptides, which have been cleaved down to their phosphoribose attachment sites by a phosphodiesterase, thus isolating the ADP-ribosylated and phosphorylated proteomes simultaneously for their quantitative analysis. Importantly, this protocol allows the achievement of a robust and relative quantification of changes in the posttranslational modification using dimethyl labeling, a straightforward and economical choice, which can then be used on lysate from any cell type, including primary tissue. The protocol has been optimized to work in ADP-ribosylation-compatible buffers and with a protease-laden lysate from the macrophage cells.

As mentioned above, cellular metabolism plays a central role in the activation and effector functions of macrophages. Intracellular pathogens subvert the immune functions of macrophages to establish an infection by modulating the metabolism of the macrophages. Cumming et al. describe how the Seahorse Extracellular Flux analyzer (XF) can be used to study changes in the bioenergetic metabolism of the macrophages induced by infection with mycobacteria. The XF simultaneously measures the oxygen consumption and extracellular acidification of the macrophages noninvasively in real time, and together with the addition of metabolic modulators, substrates, and inhibitors enables measurements of the rates of oxidative phosphorylation, glycolysis, fatty acid oxidation, and ATP production.

Another important immune cell type is dendritic cells (DCs), which serve as the bridge between innate and adaptive immunity, and which are promising therapeutic targets for cancer and immune-mediated disorders. This can be achieved by differentiating them into either immunogenic or tolerogenic DCs by modulating their metabolic pathways (including glycolysis, oxidative phosphorylation, and fatty acid metabolism) to orchestrate their desired function. Thus, understanding the metabolic regulation of DC subsets and functions not only will improve our understanding of DC biology and immune regulation, but can also open up opportunities for treating immune-mediated ailments and cancer by adjusting endogenous T-cell responses through DC-based immunotherapies. Wei et al. describe a method to analyze this dichotomous metabolic reprogramming of DCs for generating a reliable and effective DC cell therapy product. Particularly, by using a pharmacological nuclear factor (Nrf2) activator as an example, they illustrate the metabolic profile of tolerogenic DCs.

The mitochondrial membrane potential ($\Delta\psi$) is an established indicator of the functional metabolic status of mitochondria, which accounts for approximately 90% of all available ATP for cellular activities. There are several experimental approaches to measure $\Delta\psi$ levels, ranging from fluorometric evaluations to electrochemical probes. Teodoro et al. describe the evaluation of the mitochondrial membrane potential using fluorescent dyes or a membrane-permeable cation (TPP+) electrode in isolated mitochondria and intact cells. Moreover, the authors discuss the advantages and disadvantages of several of these methods, ranging from one method that is dependent on the movement of a particular ion, tetraphenylphosphonium (TPP^+) with a selective electrode, to the selection of a fluorescent dye from various types to achieve the same goal. These methods are highly sensitive, fast, accurate, and a simple mode of evaluation of $\Delta\psi$ levels in respiring mitochondria, either isolated or still inside the cell.

Apart from cellular metabolism, mitochondrial dynamics (i.e., mitochondrial fission and fusion) coincide with effectors and memory T-cell differentiation, resulting in metabolic reprogramming. In general, freshly collected immune cells are preferred for such measurements, as frozen cells are not considered ideal for immunometabolic analyses. However, the use of freshly collected clinical samples is not always possible due to the logistic difficulties of having to complete analyses within a few hours of blood collection. Clovis Palmer and coworkers describe methods for the multiparametric analysis of mitochondrial dynamics in T cells from cryopreserved peripheral blood mononuclear cells. They have optimized and validated a simple cryopreservation protocol for peripheral blood mononuclear cells, yielding an astonishing >95% cellular viability, and preserved metabolic and immunologic properties. By combining fluorescent dyes with cell surface antibodies, the authors demonstrate how to analyze mitochondrial density, membrane potential, and reactive oxygen species production in CD4 and CD8 T cells from cryopreserved clinical samples.

Finally, Xu et al. describe methods to measure glucose homeostasis at the systemic level (which is an integral component of immunometabolism) by measuring blood glucose disposal and insulin sensitivity utilizing glucose tolerance and insulin tolerance tests. The authors also provide valuable tips for consideration while performing these tests, as well as data presentation and interpretation.

In addition to the previously discussed protocol chapters, opinion and review chapters have been included within this book, relating to fundamental aspects in the field of immunometabolism and their implications. The first chapter is titled "Immunometabolism and Its Potential to Improve Current Limitations of Immunotherapy" and is authored by Andrew Sheppard and Joanne Lysaght. The authors have provided an excellent account of the promising discoveries made in this field, and future directions in the field can move in to enhance therapeutic effectiveness. The second chapter is authored by Mishra et al., in which the authors have highlighted the need to advance our understanding of sex differences in this field, hence the title "Sex Differences in Immunometabolism: An Unexplored Area." It is anticipated that these two thought-provoking opinion chapters, along with a variety of experimental protocols, will provide a valuable source of information and motivation for researchers in this emerging and promising field of immunometabolism.

I am indebted to all of the authors for spending their valuable time to contribute to this book. Importantly, I would like to thank John Walker, the series editor of *Methods in Molecular Biology*, for the opportunity, as well as his guidance and help during the whole process. I would also like to thank my laboratory members for their help and contribution to this project. Finally, I would like to thank Nivida Mishra for her contributions in copyediting the manuscript. Her expertise in rectifying the book was invaluable, and the final product was made better by her involvement.

Winnipeg, MB, Canada ***Suresh Mishra***

References

1. Mathis D (2011) Immunometabolism: an emerging frontier. Nat Rev Immunol 11:81–83
2. Pearce EL, Pearce EJ (2013) Metabolic pathways in immune cell activation and quiescence. Immunity 38:633–643
3. Puleston DJ, Villa M, Pearce E (2017) Ancillary activity: beyond core metabolism in immune cells. Cell Metab 26:131–141
4. Ferrante AW Jr (2013) Macrophages, fat, and the emergence of immunometabolism. J Clin Invest 123:4992–4993
5. Ganrot PO, Gydell K, Ekelund H (1967) Serum concentration of α-2-macroglobulin, haptoglobin and α-1-antitrypsin in diabetes mellitus. Acta Endocrinol (Copenh) 55:537–544
6. Powell ED, Field RA (1966) Studies on salicylates and complement in diabetes. Diabetes 15:730–733
7. Hotamisligil GS, Shargill NS, Spiegelman BM (1993) Adipose expression of tumor necrosis factor-α: direct role in obesity-linked insulin resistance. Science 259:87–91
8. Weisberg SP, McCann D, Desai M, Rosenbaum M, Leibel RL, Ferrante AW Jr (2003) Obesity is associated with macrophage accumulation in adipose tissue. J Clin Invest 112:1796–1808
9. Xu H, et al (2003) Chronic inflammation in fat plays a crucial role in the development of obesity-related insulin resistance. J Clin Invest 112:1821–1830
10. Lee YS, Wollam J, Olefsky JM (2018) An integrated view of immunometabolism. Cell 172:22–40

Contents

Preface v
Contributors xiii

1 Single-Cell Transcriptomics of Immune Cells: Cell Isolation and cDNA Library Generation for scRNA-Seq 1
Janilyn Arsenio

2 Monitoring Lactate Dynamics in Individual Macrophages with a Genetically Encoded Probe 19
Felipe Baeza-Lehnert, Carlos A. Flores, Anita Guequén, and L. Felipe Barros

3 The Conjugation of Antibodies for the Simultaneous Detection of Surface Proteins and Transcriptome Analysis at a Single-Cell Level 31
Iivari Kleino, Eliisa Kekäläinen, and Tapio Lönnberg

4 Cellular Fatty Acid Analysis in Macrophage Using Stable Isotope Labeling 47
Kevin J. Williams and Steven J. Bensinger

5 Proteomics Network Analysis of Polarized Macrophages 61
Jayanta K. Chakrabarty, Abu Hena Mostafa Kamal, A. D. A. Shahinuzzaman, and Saiful M. Chowdhury

6 The Purification and Characterization of Exosomes from Macrophages 77
Eran Fridman, Lana Ginini, Ziv Gil, and Neta Milman

7 Isolation and Characterization of T Lymphocyte-Exosomes Using Mass Spectrometry 91
Inbar Azoulay-Alfaguter and Adam Mor

8 Analysis of Immune-Tumor Cell Interactions Using a 3D Co-culture Model 103
Tanya N. Augustine

9 Studying Adipocyte and Immune Cell Cross Talk Using a Co-culture System 111
Jennifer M. Monk, Danyelle M. Liddle, Amber L. Hutchinson Lindsay E. Robinson

10 A Gene Expression Analysis of M1 and M2 Polarized Macrophages 131
Nour Eissa, Hayam Hussein, and Jean-Eric Ghia

11 Simultaneous, Quantitative Characterization of Protein ADP-Ribosylation and Protein Phosphorylation in Macrophages 145
Casey M. Daniels, Arthur Nuccio, Pauline R. Kaplan Aleksandra Nita-Lazar

12 The Analysis of *Mycobacterium tuberculosis*-Induced Bioenergetic Changes in Infected Macrophages Using an Extracellular Flux Analyzer 161
Bridgette M. Cumming, Vineel P. Reddy, and Adrie J. C. Steyn

13 Analyzing the Metabolic Phenotype of Bone Marrow-Derived Dendritic Cells by Assessing Their Oxygen Consumption and Extracellular Acidification 185
Hsi-Ju Wei, John J. Letterio, and Tej K. Pareek

14 The Evaluation of Mitochondrial Membrane Potential Using Fluorescent Dyes or a Membrane-Permeable Cation (TPP^+) Electrode in Isolated Mitochondria and Intact Cells 197
João S. Teodoro, Ivo F. Machado, Ana C. Castela, Anabela P. Rolo, and Carlos M. Palmeira

15 The Multiparametric Analysis of Mitochondrial Dynamics in T Cells from Cryopreserved Peripheral Blood Mononuclear Cells (PBMCs) 215
Jesse J. R. Masson, Matias Ostrowski, Gabriel Duette, Man K. S. Lee, Andrew J. Murphy, Suzanne M. Crowe, and Clovis S. Palmer

16 The Measurement of Whole-Body Glucose Homeostasis in Mice 225
Yang Xin Zi Xu, Sudharsana R. Ande, and Suresh Mishra

17 Immunometabolism and Its Potential to Improve the Current Limitations of Immunotherapy 233
Andrew D. Sheppard and Joanne Lysaght

18 Sex Differences in Immunometabolism: An Unexplored Area 265
Suresh Mishra, Geetika Bassi, and Yang Xin Zi Xu

19 Isolation and Preparation of Bone Marrow-Derived Immune Cells for Metabolic Analysis 273
Nnamdi M. Ikeogu, Chidalu A. Edechi, Gloria N. Akaluka, Aida Feiz-Barazandeh, and Jude E. Uzonna

Index *281*

Contributors

GLORIA N. AKALUKA • *Department of Immunology, Max Rady College of Medicine, University of Manitoba, Winnipeg, MB, Canada*

SUDHARSANA R. ANDE • *Faculty of Health Sciences, Department of Internal Medicine, University of Manitoba, Winnipeg, MB, Canada*

JANILYN ARSENIO • *Department of Internal Medicine, University of Manitoba, Winnipeg, MB, Canada; Manitoba Centre for Proteomics and Systems Biology, Winnipeg, MB, Canada; Department of Immunology, University of Manitoba, Winnipeg, MB, Canada*

TANYA N. AUGUSTINE • *School of Anatomical Sciences, Faculty of Health Sciences, University of the Witwatersrand, Johannesburg, South Africa*

INBAR AZOULAY-ALFAGUTER • *Iovance Biotherapeutics, Tampa, FL, USA*

FELIPE BAEZA-LEHNERT • *Centro de Estudios Científicos—CECs, Valdivia, Chile; Universidad Austral de Chile, Valdivia, Chile*

L. FELIPE BARROS • *Centro de Estudios Científicos—CECs, Valdivia, Chile*

GEETIKA BASSI • *Faculty of Health Sciences, Department of Physiology and Pathophysiology, College of Medicine, University of Manitoba, Winnipeg, MB, Canada*

STEVEN J. BENSINGER • *UCLA Lipidomics Laboratory, Los Angeles, CA, USA; Department of Microbiology, Immunology and Molecular Genetics, David Geffen School of Medicine, University of California, Los Angeles, Los Angeles, CA, USA*

ANA C. CASTELA • *Department of Life Sciences, University of Coimbra, Coimbra, Portugal; Center for Neurosciences and Cell Biology, University of Coimbra, Coimbra, Portugal*

JAYANTA K. CHAKRABARTY • *Department of Chemistry and Biochemistry, University of Texas at Arlington, Arlington, TX, USA*

SAIFUL M. CHOWDHURY • *Department of Chemistry and Biochemistry, University of Texas at Arlington, Arlington, TX, USA*

SUZANNE M. CROWE • *Life Sciences, Burnet Institute, Melbourne, VIC, Australia; Department of Infectious Diseases, Monash University, Melbourne, VIC, Australia*

BRIDGETTE M. CUMMING • *Africa Health Research Institute, Durban, South Africa*

CASEY M. DANIELS • *Functional Cellular Networks Section, Laboratory of Immune System Biology, National Institute of Allergy and Infectious Diseases, National Institutes of Health, Bethesda, MD, USA*

GABRIEL DUETTE • *INBIRS, Facultad de Medicina, Buenos Aires, Argentina*

CHIDALU A. EDECHI • *Department of Pathology, Max Rady College of Medicine, University of Manitoba, Winnipeg, MB, Canada*

NOUR EISSA • *Department of Immunology, Max Rady College of Medicine, Rady Faculty of Health Science, University of Manitoba, Winnipeg, MB, Canada; Children's Hospital Research Institute of Manitoba, University of Manitoba, Winnipeg, MB, Canada; Department of Internal Medicine, Section of Gastroenterology, Max Rady College of Medicine, Rady Faculty of Health Sciences, University of Manitoba, Winnipeg, MB, Canada; University of Manitoba IBD Clinical and Research Centre, University of Manitoba, Winnipeg, MB, Canada*

AIDA FEIZ-BARAZANDEH • *Department of Immunology, Max Rady College of Medicine, University of Manitoba, Winnipeg, MB, Canada*

CARLOS A. FLORES • *Centro de Estudios Científicos—CECs, Valdivia, Chile*

ERAN FRIDMAN • *The Laboratory for Applied Cancer Research, Department of Otolaryngology, Head and Neck Surgery, The Head and Neck Center, Rambam Healthcare Campus, The Technion, Israel Institute of Technology, Haifa, Israel*

JEAN-ERIC GHIA • *Department of Immunology, Max Rady College of Medicine, Rady Faculty of Health Science, University of Manitoba, Winnipeg, MB, Canada; Children's Hospital Research Institute of Manitoba, University of Manitoba, Winnipeg, MB, Canada; Department of Internal Medicine, Section of Gastroenterology, Max Rady College of Medicine, Rady Faculty of Health Sciences, University of Manitoba, Winnipeg, MB, Canada; University of Manitoba IBD Clinical and Research Centre, University of Manitoba, Winnipeg, MB, Canada*

ZIV GIL • *The Laboratory for Applied Cancer Research, Department of Otolaryngology, Head and Neck Surgery, The Head and Neck Center, Rambam Healthcare Campus, The Technion, Israel Institute of Technology, Haifa, Israel*

LANA GININI • *The Laboratory for Applied Cancer Research, Department of Otolaryngology, Head and Neck Surgery, The Head and Neck Center, Rambam Healthcare Campus, The Technion, Israel Institute of Technology, Haifa, Israel*

ANITA GUEQUÉN • *Centro de Estudios Científicos—CECs, Valdivia, Chile*

HAYAM HUSSEIN • *Department of Immunology, Max Rady College of Medicine, Rady Faculty of Health Science, University of Manitoba, Winnipeg, MB, Canada; Department of Parasitology and Animal Diseases, Veterinary Research Division, National Research Centre, Giza, Egypt*

AMBER L. HUTCHINSON • *Department of Human Health and Nutritional Sciences, University of Guelph, Guelph, ON, Canada*

NNAMDI M. IKEOGU • *Department of Immunology, Max Rady College of Medicine, University of Manitoba, Winnipeg, MB, Canada*

ABU HENA MOSTAFA KAMAL • *Department of Chemistry and Biochemistry, University of Texas at Arlington, Arlington, TX, USA*

PAULINE R. KAPLAN • *Functional Cellular Networks Section, Laboratory of Immune System Biology, National Institute of Allergy and Infectious Diseases, National Institutes of Health, Bethesda, MD, USA*

ELIISA KEKÄLÄINEN • *Translational Immunology Research Program, University of Helsinki, Helsinki, Finland; Department of Bacteriology and Immunology, University of Helsinki and Helsinki University Hospital, Helsinki, Finland*

IIVARI KLEINO • *Translational Immunology Research Program, University of Helsinki, Helsinki, Finland; Department of Bacteriology and Immunology, University of Helsinki and Helsinki University Hospital, Helsinki, Finland*

TAPIO LÖNNBERG • *Turku Bioscience Centre, University of Turku and Åbo Akademi University, Turku, Finland*

MAN K. S. LEE • *Division of Immunometabolism, Haematopoiesis and Leukocyte Biology Laboratory, Baker Heart and Diabetes Institute, Melbourne, VIC, Australia*

JOHN J. LETTERIO • *The Case Comprehensive Cancer Center, Case Western Reserve University, Cleveland, OH, USA; Division of Pediatric Hematology/Oncology, Department of Pediatrics, Case Western Reserve University, Cleveland, OH, USA; Angie Fowler Cancer Institute, Rainbow Babies and Children's Hospital, University Hospitals, Cleveland, OH, USA; Celloram Inc., Cleveland, OH, USA*

DANYELLE M. LIDDLE • *Department of Human Health and Nutritional Sciences, University of Guelph, Guelph, ON, Canada*

JOANNE LYSAGHT • *Cancer Immunology and Immunotherapy Group, Trinity Translational Medicine Institute, St. James's Hospital, Dublin, Ireland*
IVO F. MACHADO • *Department of Life Sciences, University of Coimbra, Coimbra, Portugal; Center for Neurosciences and Cell Biology, University of Coimbra, Coimbra, Portugal*
JESSE J. R. MASSON • *Life Sciences, Burnet Institute, Melbourne, VIC, Australia*
NETA MILMAN • *The Laboratory for Applied Cancer Research, Department of Otolaryngology, Head and Neck Surgery, The Head and Neck Center, Rambam Healthcare Campus, The Technion, Israel Institute of Technology, Haifa, Israel*
SURESH MISHRA • *Faculty of Health Sciences, Department of Internal Medicine, University of Manitoba, Winnipeg, MB, Canada; Faculty of Health Sciences, Department of Physiology and Pathophysiology, University of Manitoba, Winnipeg, MB, Canada*
JENNIFER M. MONK • *Department of Human Health and Nutritional Sciences, University of Guelph, Guelph, ON, Canada*
ADAM MOR • *Columbia Center for Translational Immunology, Columbia University Irving Medical Center, New York, NY, USA*
ANDREW J. MURPHY • *Division of Immunometabolism, Haematopoiesis and Leukocyte Biology Laboratory, Baker Heart and Diabetes Institute, Melbourne, VIC, Australia*
ALEKSANDRA NITA-LAZAR • *Functional Cellular Networks Section, Laboratory of Immune System Biology, National Institute of Allergy and Infectious Diseases, National Institutes of Health, Bethesda, MD, USA*
ARTHUR NUCCIO • *Functional Cellular Networks Section, Laboratory of Immune System Biology, National Institute of Allergy and Infectious Diseases, National Institutes of Health, Bethesda, MD, USA*
MATIAS OSTROWSKI • *INBIRS, Facultad de Medicina, Buenos Aires, Argentina*
CARLOS M. PALMEIRA • *Department of Life Sciences, University of Coimbra, Coimbra, Portugal; Center for Neurosciences and Cell Biology, University of Coimbra, Coimbra, Portugal*
CLOVIS S. PALMER • *Life Sciences, Burnet Institute, Melbourne, VIC, Australia; Department of Infectious Diseases, Monash University, Melbourne, VIC, Australia; Department of Microbiology and Immunology, University of Melbourne, Melbourne, VIC, Australia*
TEJ K. PAREEK • *The Case Comprehensive Cancer Center, Case Western Reserve University, Cleveland, OH, USA; Division of Pediatric Hematology/Oncology, Department of Pediatrics, Case Western Reserve University, Cleveland, OH, USA; Angie Fowler Cancer Institute, Rainbow Babies and Children's Hospital, University Hospitals, Cleveland, OH, USA; Celloram Inc., Cleveland, OH, USA*
VINEEL P. REDDY • *Department of Microbiology, University of Alabama at Birmingham, Birmingham, AL, USA*
LINDSAY E. ROBINSON • *Department of Human Health and Nutritional Sciences, University of Guelph, Guelph, ON, Canada*
ANABELA P. ROLO • *Department of Life Sciences, University of Coimbra, Coimbra, Portugal; Center for Neurosciences and Cell Biology, University of Coimbra, Coimbra, Portugal*
A. D. A. SHAHINUZZAMAN • *Department of Chemistry and Biochemistry, University of Texas at Arlington, Arlington, TX, USA*
ANDREW D. SHEPPARD • *Cancer Immunology and Immunotherapy Group, Trinity Translational Medicine Institute, St. James's Hospital, Dublin, Ireland*
ADRIE J. C. STEYN • *Africa Health Research Institute, Durban, South Africa; Department of Microbiology, University of Alabama at Birmingham, Birmingham, AL, USA; UAB*

Centers for AIDS Research and Free Radical Biology, University of Alabama at Birmingham, Birmingham, AL, USA
JOÃO S. TEODORO • *Department of Life Sciences, University of Coimbra, Coimbra, Portugal; Center for Neurosciences and Cell Biology, University of Coimbra, Coimbra, Portugal*
JUDE E. UZONNA • *Department of Immunology, Max Rady College of Medicine, University of Manitoba, Winnipeg, MB, Canada*
HSI-JU WEI • *Department of Biochemistry, School of Medicine, Case Western Reserve University, Cleveland, OH, USA; The Case Comprehensive Cancer Center, Case Western Reserve University, Cleveland, OH, USA*
KEVIN J. WILLIAMS • *Department of Biological Chemistry, David Geffen School of Medicine, University of California, Los Angeles, Los Angeles, CA, USA; UCLA Lipidomics Laboratory, Los Angeles, CA, USA*
YANG XIN ZI XU • *Faculty of Health Sciences, Department of Physiology and Pathophysiology, College of Medicine, University of Manitoba, Winnipeg, MB, Canada*

Chapter 1

Single-Cell Transcriptomics of Immune Cells: Cell Isolation and cDNA Library Generation for scRNA-Seq

Janilyn Arsenio

Abstract

Single-cell RNA-sequencing (scRNA-seq) enables a comprehensive analysis of the transcriptome of individual cells by next-generation sequencing. ScRNA-seq offers an unbiased approach to investigate the cellular heterogeneity and dynamics of diverse biological systems, including the immune system. Optimization of the technical procedures performed prior to RNA-seq analysis is imperative to the success of a scRNA-seq experiment. Here, three major experimental procedures are described: (1) the isolation of immune CD8a^{+} T cells from primary murine tissue, (2) the generation of single-cell cDNA libraries using the 10× Genomics Chromium Controller and the Chromium Single Cell 3′ Solution, and (3) cDNA library quality control. In this protocol, CD8a^{+} T cells are isolated from murine spleen tissue, but any cell type of interest can be enriched and used for single-cell cDNA library generation and subsequent RNA-seq experiments.

Key words Single-cell suspension, CD8a^{+} T cells, Single-cell cDNA libraries, 10× Genomics Chromium™, Single-cell RNA-seq

1 Introduction

The application of single-cell gene expression profiling technologies to immunological studies has revolutionized our molecular understanding of cell development, differentiation, and role of the immune system in health and disease. Single-cell RNA-sequencing (scRNA-seq) has provided a high-resolution investigation of the whole transcriptome of single cells, permitting the identification of novel regulators of immune cell differentiation and the interrogation of immune cellular heterogeneity [1–4]. In 2009, Tang et al. first published data analyses on whole-transcriptome sequencing of a single mouse cell, revealing novel insights into the complexity of the transcriptome at the single-cell level [5]. Over the past decade, numerous scRNA-seq platforms have been developed, including plate-based technologies such as STRT-seq [6], SMART-seq [7], SMART-seq2 [8], MARS-seq [9], CEL-seq [10],

Suresh Mishra (ed.), *Immunometabolism: Methods and Protocols*, Methods in Molecular Biology, vol. 2184,
https://doi.org/10.1007/978-1-0716-0802-9_1, © Springer Science+Business Media, LLC, part of Springer Nature 2020

commercial microfluidics platform Fluidigm C1, and droplet-based and microwell technologies, including Drop-seq [11], InDrop [12], Seq-well [13], Microwell-seq [14], and commercial 10× Genomics Chromium Controller [15]. The technical details of these technologies have been extensively reviewed [16, 17]. Of importance, the success of a scRNA-seq experiment, irrespective of the type of platform used, is highly dependent on the integrity of the single-cell suspension obtained for analysis and quality of the single-cell cDNA libraries for sequencing, which are generated from these cells.

Here, the isolation of CD8a$^+$ T-cell suspensions from murine tissue, followed by single-cell cDNA library generation and library quality control procedures, are overviewed. In this protocol, cDNA library generation is performed using the 10× Genomics Chromium Controller and the Chromium Single Cell 3′ Solution Reagents Kit. The 10× Genomics Chromium Controller is a microfluidics platform, which enables gene expression profiling of 500–10,000 single cells per sample. GemCode Technology incorporates 10× barcodes to index the transcriptome of individual cells. Input cells are partitioned into gel bead-in-emulsions (GEMs), enabling the simultaneous generation of sequence-ready cDNA libraries of ≥500 cells. To produce GEMs, the 10× barcoded gel beads are combined with the 10× genomics master mix, sorted cells of interest (CD8a$^+$ T cells, as detailed in this chapter), and partitioning oil; then added onto the 10× Genomics Chromium Chip B, as designated in the Chromium Single Cell 3′ Solution Reagents Kits v3 user guide; and placed into the Chromium Controller. The incubation of the GEMs with 10× genomics reverse transcription reagents generates full-length cDNA from the polyadenylated mRNAs of one cell, all of which share the same 10× barcode and contain a unique molecular identifier (UMI). Following incubation, the GEMs are broken and first-strand cDNA is purified using Dynabeads MyOne SILANE magnetic beads. A PCR amplification of the barcoded, full-length cDNA is then performed to produce a sufficient cDNA yield for library generation. The construction of cDNA libraries involves fragmentation and size selection of the cDNA amplicons, end repair, A-tailing, adaptor ligation, and PCR to incorporate a sample index (the P5 and P7 primers, which are compatible with Illumina sequencing technology). cDNA quantification and cDNA library quality control are assessed using the Agilent 2100 Bioanalyzer and Qubit Fluorometer. The resulting single-cell cDNA libraries in this protocol can then be sequenced on Illumina Sequencers (MiSeq, NextSeq 500/550, HiSeq 2500, HiSeq 3000/4000, and NovaSeq).

2 Materials

Recommendations for working with associated materials in these experiments are as follows. Prepare all working solutions in a regulated biosafety cabinet to keep sterile. Store cell solutions and all other reagents at 4 °C, or as indicated. For cDNA library generation and quality control assessment, ensure that the laboratory workspace is clean and free of debris. Wipe down the workspace and all pipettes used with RNase cleaner prior to the start of the experiment. Use nuclease-free filtered pipette tips and gloves at all times.

2.1 The Isolation of Immune CD8a$^+$ T Cells from Murine Tissue

1. Cell media: Hanks balanced salt solution (HBSS) supplemented with 1% fetal bovine serum (FBS).
2. Red blood cell lysis buffer: Store at room temperature.
3. Cell isolation buffer: 1× Phosphate buffer saline, 0.5% FBS or 0.5% BSA, and 2 mM EDTA, pH 7.2–8.0.
4. Biotin-labeled antibodies against non-CD8a$^+$ T cells (MACS Miltenyi Biotec mouse CD8a$^+$ T cell isolation kit).
5. Anti-biotin microbeads.
6. MACS Miltenyi Biotec LS cell separation column: Store at room temperature in the dark.
7. MACS Miltenyi Biotec Manual Separator for magnetic bead isolation.
8. FACS Buffer: 1× Phosphate buffer saline, 5% FBS, 0.1% NaN_3 sodium azide, pH 7.2–8.0.
9. FITC anti-mouse CD3ε and APC anti-mouse CD8a$^+$ T-cell antibodies.
10. 1× PBS: Store at room temperature.
11. 15 and 50 mL conical tubes.
12. 1.5 mL Microcentrifuge tubes.
13. 70 μm Cell strainer.
14. Sterile 3 mL syringe.
15. 60 mm (diameter) × 15 mm (height) petri dish.
16. 5 and 10 mL serological pipettes.
17. P10, P200, P1000 pipettes and corresponding pipette tips.
18. Corning Falcon Test Tube with Cell Strainer Snap Cap.
19. Flow cytometer.

2.2 The Generation of Single-Cell cDNA Libraries

1. PCR tubes (0.2 mL 8-tube strips and caps).
2. 1.5 mL Nuclease-free microcentrifuge tubes.
3. Nuclease-free water.
4. Buffer EB: 10 mM Tris–Cl, pH 8.5, store at room temperature.

5. Pure ethanol (200 Proof, anhydrous): Store at room temperature in appropriate flammable-regulated storage area.
6. SPRIselect Reagent Kit (Beckman Coulter): Store at room temperature.
7. 50% Glycerol.
8. Vortex mixer.
9. Minicentrifuge with PCR 8-tube strip adapter.
10. Thermal cycler.
11. 10× Genomics Chromium Single Cell 3′ Reagents Kit: Store reagents at either room temperature, 4 °C, −20 °C, or −80 °C as indicated in the 10× Genomics Chromium Single Cell 3′ Reagents v3 user guide.
12. 10× Genomics Chromium Chip B and gasket.
13. 10× Genomics Chromium Controller.
14. 10× Genomics magnet.
15. Dynabeads MyOne SILANE (Thermo Fisher Scientific): Store at 4 °C.
16. Agilent 2100 Bioanalyzer instrument.
17. Agilent Bioanalyzer High Sensitivity DNA kit: Store reagents at 4 °C in the dark and Agilent Bioanalyzer High Sensitivity Chips at room temperature.

2.3 cDNA Library Quality Control

1. Agilent Bioanalyzer High Sensitivity DNA chip and kit reagents.
2. Agilent 2100 Bioanalyzer instrument.
3. Qubit dsDNA HS (High Sensitivity) Assay Kit (Invitrogen): Store kit reagents at 4 °C.
4. Qubit Fluorometer (Invitrogen).

3 Methods

3.1 The Isolation of Immune CD8a⁺ T Cells from Murine Tissue

1. In a biosafety cabinet, mash spleen tissue in a 60 mm (diameter) × 15 mm (height) petri dish in 5 mL of cell media with the flat end of a sterile 3 mL syringe. Using a 5 mL serological pipet, pipet up the tissue suspension and pass through a 70 μm cell strainer placed atop a 50 mL conical tube. Add 2 mL of cell media to rinse the petri dish to attain as much of the tissue suspension and pass the excess tissue suspension through the same 70 μm strainer. Centrifuge the 50 mL conical tube at 300 × *g* for 5 min at 4 °C. Discard supernatant, leaving behind a red cell pellet.

2. To the red cell pellet, add 1 mL of red blood cell lysis buffer, vortex quickly to mix, and incubate at room temperature for 4 min. Following the 4-min incubation, add 3 mL of cell media to the conical tube and centrifuge it at 300 × *g* for 1 min at 4 °C. Remove and discard supernatant. Resuspend the cell pellet in 1 mL of cell isolation buffer. Count cells.
3. Perform enrichment of $CD8a^+$ T cells by incubating the spleen-derived cell suspension with a combination of biotin-conjugated antibodies against CD4, CD11b, CD11c, CD19, CD45R (B220), CD49b, anti-MHC-class II, Ter-119, and TCR γ/δ (*see* **Note 1**). This antibody cocktail is available from MACS Miltenyi Biotec as the mouse $CD8a^+$ T-cell isolation kit. Isolate $CD8a^+$ T cells by the depletion of magnetically labeled non-$CD8a^+$ T cells. Resuspend the cell pellet at 10^7 cells per 40 μL of cell isolation buffer (*see* **Note 2**) and transfer the cell suspension to a 1.5 mL microcentrifuge tube. Add 10 μL of the antibody cocktail to this cell suspension. Gently pipette mix the cell suspension mixture five times. Incubate the cells and the antibody mixture on ice for 10 min. Centrifuge the cell suspension at 400 × *g* for 1 min and discard the supernatant. Wash the cell pellet with 1 mL of cell isolation buffer, gently pipetting the cells to mix. Centrifuge the cells at 300 × *g* for 1 min and discard the supernatant.
4. Resuspend the cells in 30 μL of cell isolation buffer per 10^7 cells. Add 20 μl of an anti-biotin microbeads per 10^7 cells. Gently pipette the cell mixture five times to mix. Incubate on ice for 10 min. Centrifuge the cell suspension at 400 × *g* for 1 min and discard the supernatant. Wash the cell pellet with 1 mL of cell isolation buffer, gently pipetting the cells to mix. Centrifuge the cells at 300 × *g* for 1 min and discard the supernatant. Resuspend the cell pellet in 500 μL of cell isolation buffer and keep on ice.
5. Place an LS cell separation column in the magnetic field of the appropriate MACS Separator (MACS Miltenyi Biotec) and an empty 15 mL conical tube below the column for sample collection. Wash the column with 3 mL cell isolation buffer (*see* **Note 3**). Add 500 μL cell suspension to the column and collect the flow-through in the 15 mL conical tube. This is the negative fraction, containing the unlabeled $CD8a^+$ T cells (*see* **Note 4**). Wash the column three times, each with 3 mL of cell isolation buffer.
6. Centrifuge the 15 mL conical tube containing the unlabeled $CD8a^+$ T cells at 300 *g* for 5 min at 4 °C. Aspirate the supernatant carefully, leaving behind approximately 0.5 mL of supernatant. Gently resuspend the cell pellet in the remaining 0.5 mL and transfer single-cell suspension to a new 1.5 mL

microcentrifuge tube. Centrifuge the microcentrifuge tube at 300 × *g* for 1 min at 4 °C. Remove supernatant carefully so as not to disrupt the cell pellet. Resuspend the cell pellet in 1 mL cell media and count the cells.

7. Use flow cytometry to confirm the purity of the enriched CD8a$^+$ T cells and to sort the purified CD8a$^+$ T-cell suspension for downstream library preparation and RNA sequencing. To stain the enriched CD8a$^+$ T cells for FACS sorting, first resuspend the cells at a concentration of 1–5 million cells/mL in 100 μL ice-cold staining buffer in a 1.5 mL microcentrifuge tube (*see* **Note 5**). Add 1 μL of FITC anti-mouse CD3ε and 1 μL APC anti-mouse CD8a$^+$ T-cell antibody to the cell suspension at a 1:100 ratio (*see* **Note 6**). Gently pipette up and down the cell suspension and antibody mixture to mix well. Incubate the cells and antibody mixture for at least 15 min on ice in the dark.
8. Following the 15-min incubation on ice, centrifuge the tube containing the cell suspension and antibody mixture at 300 × *g* for 1 min at 4 °C in a microcentrifuge. Discard the supernatant and add 0.5 mL ice-cold 1× PBS to the cells to wash. Centrifuge the cells at 400 × *g* for 1 min at 4 °C, and repeat with a second wash. Resuspend the cells in ~300 μL cell media. Filter the single-cell suspension using a Corning Falcon Test Tube with Cell Strainer Snap Cap (*see* **Note 7**).
9. Place the sample on ice and sort CD8a$^+$ T-cell-positive, CD3-ε-positive cells into cell media. Exclude dead cells and doublets during cell sorting. Record cell viability percentage and cell count of sorted purified CD8a$^+$ T cells. Keep sorted cells on ice.

3.2 Generation of Single-Cell cDNA Libraries

The generation of single-cell cDNA libraries for subsequent RNA-sequencing is performed according to the 10× Genomics Chromium Single Cell 3′ Reagents Kits v3 user guide, using the Chromium Controller and Chromium Single Cell Gene Expression Solution. It is recommended that the user follow the extensively detailed protocol in the Chromium Single Cell 3′ Reagents Kits v3 user guide, particularly for 10× Genomics master mix compositions as indicated in the procedures below. All steps are to be performed on the laboratory bench.

1. Centrifuge the sorted CD8a$^+$ T cells at 400 × *g* for 1 min at 4 °C and remove cell media. Wash the cell pellet in 200 μL of 1× PBS + 0.04% BSA (*see* **Note 8**). Spin down the cells at 400 × *g* for 1 min at 4 °C in the microcentrifuge and repeat with a second wash with 200 μL of 1× PBS + 0.04% BSA.
2. Prepare cells for cell capture in the 10× Genomics Chromium system according to the Chromium Single Cell 3′ Reagents Kits v3 user guide. Based on the cell suspension volume

calculator table of the Chromium Single Cell 3′ Reagents Kits v3 protocol and cell count of the sorted CD8a$^+$ T cells, calculate the volume of 1× PBS + 0.04% BSA required to suspend the sorted CD8a$^+$ T cells to form a cell stock concentration of 1200 cells/μL. This cell stock concentration is used for a targeted cell recovery of 6000 cells (*see* **Note 9**). Keep cells on ice.

3. For GEM Generation and Barcoding, prepare the 10× Genomics master mix (RT Reagent, Template Switch Oligo, Reducing Reagent B, RT Enzyme C) on ice, using nuclease-free 1.5 microcentrifuge tubes and nuclease-free, filtered pipette tips. Per sample, add 33.4 μL of master mix into one tube of a PCR 8-tube strip on ice.
4. Prepare the cell suspension. For a targeted cell recovery of 6000 cells, add 8.0 μL of cells from the cell stock concentration of 1200 cells/μL with 38.6 μL of nuclease-free water, according to the volume calculator table of the Chromium Single Cell 3′ Reagents Kits v3 protocol. Keep cells on ice.
5. Carefully place the Chromium Chip B in the 10× Genomics Chip Holder (*see* **Note 10**). Add the appropriate volume of 50% glycerol solution into the chip wells that will not be used. For instance, if only one well of the 8-well chip will be used for one cell sample, fill the remaining 7 wells with 50% glycerol solution.
6. Gently pipette mix the cell suspension and add it to the tube of the PCR 8-tube strip containing the master mix. Gently pipette mix the cell suspension and master mix and carefully load 75 μL of this mixture into the bottom center of the first well in the row labeled 1 on the Chromium Chip B (*see* **Note 11**).
7. Place the 10× gel bead strip into the 10× vortex adapter and vortex for 30 s. To recover the gel beads after vortexing, flick the gel bead strip in a quick, downward motion, and ensure that the liquid levels in each tube of the gel bead strip look equal. Pierce open the foil seal of the gel bead strip and slowly pipette up 40 μL of gel beads. Gently dispense the gel beads into the first well in row labeled 2 on the Chromium Chip B (*see* **Note 12**). Add 140 μL of 10× genomics partitioning oil into the first well of row labeled 3 on the Chromium Chip B. Repeat with a second aliquot of 140 μL of 10× genomics partitioning oil for a total volume of 280 μL per well (*see* **Note 13**).
8. To attach the 10× gasket on top of the Chromium Chip B, align the top-left notch of the gasket to the top-left corner of the 10× chip holder. Ensure that the gasket is fastened onto the 10× chip holder and confirm that the gasket holes align with the wells (*see* **Note 14**).

9. On the Chromium Controller, press the eject button on its screen to eject the tray. Carefully place the 10× chip holder containing the Chromium Chip B and gasket on the tray, aligning the top-left notch of the chip holder with the controller tray. Confirm that the controller lists Chromium Single Cell B program on its screen and then press the play button on the controller to start the program.
10. Following the run, press the eject button on the controller and carefully remove the 10× chip holder and place it on the lab bench workspace. Discard the gasket. To open the chip holder, fold the lid backwards slowly until it clicks and the chip holder is positioned at 45°. The wells of the Chromium Chip B should now be exposed. Take caution to avoid any spillover of the 10× partitioning oil into other wells while opening the 10× chip holder (*see* **Note 15**).
11. To transfer the GEMs from the Chromium Chip B into a PCR tube, slowly pipette up 100 μL GEMs by placing the pipette tip to the lowest point of the recovery well in the top row of the Chromium Chip B. Remove the pipette tip from the well and visually analyze the GEMs in the pipette tip. The GEMs should appear opaque and uniform among all channels (*see* **Note 16**). Slowly dispense the GEMs into a clean tube of a PCR 8-tube strip on ice. Incubate the GEMs in a thermal cycler for reverse transcription (RT) with the program as follows:
    (a) Lid temperature: 53 °C; reaction volume of 125 μL; step 1: 53 °C for 45 min; step 2: 85 °C for 5 min; step 3: 4 °C hold.
    (b) Samples can be stored at 4 °C for 72 h or at −20 °C for 1 week.
12. To clean up the GEM-RT reaction, add 125 μL 10× genomics recovery agent to the sample at room temperature and let sit for 1 min without mixing. A biphasic mixture will appear, containing the 10× genomics recovery agent/partitioning oil at the bottom of the PCR tube (pink in color), and a clear aqueous phase on top.
13. Slowly remove 125 μL 10× genomics recovery agent from the bottom of the tube without removing any of the aqueous phase of the sample. Prepare the Dynabeads MyOne SILANE cleanup mix (10× Genomics Cleanup Buffer and Reducing Agent B, Dynabeads MyOne SILANE, and nuclease-free water) according to the Chromium Single Cell 3′ Reagents Kits v3 user guide on page 30.
14. Add 200 μL Dynabeads MyOne SILANE cleanup mix to each sample and pipette mix ten times. Incubate at room temperature for 10 min (*see* **Note 17**). Halfway through the incubation time, pipette mix the GEM-RT and Dynabeads mixture to resuspend the SILANE beads.

15. After the 10-min incubation, place the PCR tube on the 10× magnet-high position until the solution becomes clear. Remove the supernatant and add 300 μL of 80% ethanol to the pellet while the PCR tube is still on the magnet (*see* **Note 18**). Let sit for 30 s, and then remove the ethanol. Repeat the ethanol wash with 200 μL of 80% ethanol. Let sit for 30 s, and then remove the ethanol. Centrifuge the PCR tube briefly and place it on the 10× magnet-low position. Remove the excess ethanol and air-dry for 1 min.
16. Remove the PCR tube from the magnet and add 35.5 μL 10× elution solution I (buffer EB, 10% Tween 20, 10× reducing reagent B) (*see* **Note 19**). Pipette mix and incubate at room temperature for 2 min. Place the PCR tube on the 10× magnet-low position until the solution clears. Transfer 35 μL sample to a new PCR tube.
17. For cDNA amplification, prepare the 10× genomics cDNA amplification reaction mix (Amp mix and cDNA primers) on ice. Add 65 μL of cDNA amplification reaction mix to 35 μL of the eluted sample. Pipette mix and centrifuge briefly. Incubate in a thermal cycler with the program as follows:
    (a) Lid temperature: 105 °C; reaction volume: 100 μL.
    (b) Step 1: 98 °C for 3 min; step 2: 98 °C for 15 s; step 3: 63 °C for 20 s; step 4: 72 °C for 1 min; step 5: go to step 2. The number of cycles depends on the cell load. Based on the 10× genomics cycle number optimization table on page 32 of the v3 user guide, a cell load of 500–6000 cells requires 12 cycles; step 6: 72 °C for 1 min; step 7: 4 °C hold.
18. For cDNA cleanup, the SPRIselect reagent is used. Vortex the SPRIselect reagent to resuspend any settled beads. Add 60 μL (0.6× the volume of cDNA amplification reaction volume) of SPRIselect reagent to the cDNA sample in the PCR tube and pipette mix 15 times. Incubate at room temperature for 5 min.
19. Place the PCR tube on the 10× magnet-high position until the solution clears. Discard the supernatant. Add 200 μL of 80% ethanol to the pellet and let sit for 30 s. Remove the ethanol while the PCR tube is still on the magnet. Repeat with a second wash of 200 μL of 80% ethanol added to the pellet. Incubate at room temperature for 30 s before removing the ethanol. Centrifuge the PCR tube quickly and place back onto the 10× magnet-low position. Remove the remaining ethanol and let air-dry for 2 min (*see* **Note 20**).
20. Remove the PCR tube from the 10× magnet, add 40.5 μL Buffer EB to the PCR tube, and gently pipette mix. Incubate at room temperature for 2 min. Place the PCR tube on the 10× magnet-high position until the solution clears. Transfer 40 μL of the sample to a new PCR tube. Samples can be stored at 4 °C for 72 h or at −20 °C for 4 weeks.

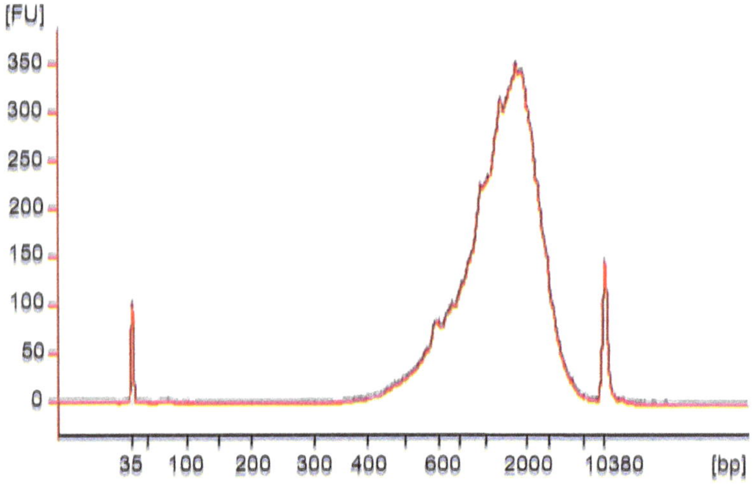

Fig. 1 Representative Agilent Bioanalyzer Trace of CD8a⁺ T-cell cDNA. Shown is an electropherogram of 1 μL diluted (1:10) cDNA. The cDNA content range is between ~400 and ~9000 bp

21. To assess the cDNA quality, use the Agilent Bioanalyzer High Sensitivity DNA chip on an Agilent 2100 Bioanalyzer instrument (*see* **Note 21**). Dilute the cDNA sample 1:10 and run 1 μL of cDNA on an Agilent Bioanalyzer High Sensitivity chip (*see* **Note 22**). Evaluate the electropherogram of the sample (Fig. 1).
22. To obtain the concentration of the cDNA sample, in the Agilent 2100 Bioanalyzer software, manually select the region on the electropherogram between ~200 and ~9000 bp. The cDNA concentration in this region will be shown in pg/μL in the software. To calculate the cDNA total yield in ng, use the following formula:

$$\begin{aligned}\text{Total cDNA yield (ng)} = {} & \text{concentration of cDNA (pg/L)} \\ & \times \text{Elution volume (e.g.40 L)} \\ & \times \text{Dilution Factor (e.g.10)}/1000\ \text{(pg/ng)}\end{aligned}$$

23. According to the 10× Genomics Chromium Single Cell 3′ Reagents Kits v3 user guide, use 25% of total cDNA yield to generate the cDNA libraries (0.25 × total cDNA yield (ng)). Transfer 10 μL (25%) of cDNA sample to a PCR tube on ice. Store the remaining cDNA sample at −20 °C for up to 4 weeks.
24. Generation of single-cell cDNA libraries first involves fragmentation, end repair, and A-tailing steps. Prepare the 10× genomics fragmentation mix (fragmentation buffer, fragmentation enzyme) on ice, as on page 37 of the v3 user guide. To the 10 μL of cDNA, add 25 μL buffer EB and 15 μL 10× genomics fragmentation mix. Pipette mix the contents

15 times on ice, and then briefly centrifuge the PCR tube (*see* **Note 23**). Place the PCR tube in a pre-cooled thermal cycler and incubate the reaction with the program as follows:

(a) Lid temperature: 65 °C; reaction volume: 50 μL; fragmentation step: 32 °C for 5 min; end repair and A-tailing step: 65 °C for 30 min; step 3: 4 °C hold.

25. Following the thermal cycler program, vortex to mix SPRIselect reagent (*see* **Note 24**). Add 30 μL (0.6×) of the SPRIselect reagent to the sample, pipette mix, and incubate at room temperature for 5 min. Place the PCR tube on the 10× magnet-high position until the solution becomes clear. Transfer 75 μL of the supernatant to a new PCR tube.

26. Add 10 μL of the SPRIselect reagent (0.8×) to the sample, pipette mix, and incubate at room temperature for 5 min. Place the tube on the 10× genomics magnet-high position until the solution becomes clear. Discard 80 μL of the supernatant without disrupting any of the beads. Add 125 μL of the 80% ethanol to the PCR tube while still on the 10× genomics magnet. Incubate for 30 s. Carefully remove the ethanol and repeat to perform a second 80% ethanol wash. Briefly centrifuge the PCR tube and place the tube on the 10× genomics magnet-low position until the solution becomes clear. Carefully remove any excess ethanol (*see* **Note 25**).

27. Remove the PCR tube from the magnet, add 50.5 μL of buffer EB, and pipette mix to fully resuspend the pellet. Incubate at room temperature for 2 min. Place the PCR tube on the 10× genomics magnet-high position until the solution becomes clear. Transfer 50 μL of the sample to a new PCR tube, being careful not to aspirate any SPRIselect reagent into the sample.

28. To complete the generation of single-cell cDNA libraries for subsequent RNA-seq involves adaptor ligation and sample index PCR steps. Prepare the 10× genomics adaptor ligation mix (ligation buffer, DNA ligase, adaptor oligos) as on page 39 of the v3 user guide, on ice. Add 50 μL of the 10× genomics adaptor ligation mix to 50 μL sample from the previous step. Pipette mix 15 times and briefly centrifuge the mixture. Incubate the sample in a thermal cycler with the program as follows:

    Lid temperature: 30 °C; reaction volume: 100 μL; step 1: 20 °C for 15 min, step 2: 4 °C hold.

29. Add 80 μL (0.8×) of the SPRIselect reagent to the sample, pipette mix, and incubate at room temperature for 5 min. Place the PCR tube on the 10× genomics magnet-high position until the solution becomes clear, and then carefully remove the supernatant. Add 200 μL of the 80% ethanol to

the pellet and incubate for 30 s. Remove the ethanol and repeat for an additional wash. Briefly centrifuge the PCR tube and place it on the 10× genomics magnet-low position. Carefully remove any excess ethanol and air-dry for precisely 2 min.

30. Remove the PCR tube from the magnet, add 30.5 μL of buffer EB, and pipette mix. Incubate at room temperature for 2 min. Place the PCR tube on the 10× magnet-low position until the solution becomes clear.
31. Transfer 30 μL of sample to a new tube, being careful not to aspirate any SPRIselect reagent.
32. For the sample index PCR, record the 10× genomics sample index well ID on the 10× Genomics Chromium i7 Sample Index Plate that is used for each sample (*see* **Note 26**). Prepare the 10× genomics sample index PCR mix (10× Amp mix, SI primer) as on page 41 of the v3 user guide on ice. Pipette mix 60 μL of sample index PCR mix with 30 μL of sample, and add 10 μL of one 10× Genomics Chromium i7 Sample Index to the sample + sample index PCR mixture. Pipette mix and briefly centrifuge the contents. Incubate the mixture in a thermal cycler with the program as follows:
    (a) Lid temperature: 105 °C; reaction volume: 100 μL; step 1: 98 °C for 45 s; step 2: 98 °C for 20 s; step 3: 54 °C for 30 s; step 4: 72 °C for 20 s; step 5: go to step #2 for # cycles (*see* **Note 27**); step 6: 72 °C for 1 min; step 7: 4 °C hold.
33. After the thermal cycler program is complete, add 60 μL (0.6×) of the SPRIselect reagent to the sample, pipette mix, and incubate at room temperature for 5 min. Place the PCR tube on the 10× genomics magnet-high position until the solution becomes clear. Carefully transfer 150 μL of the supernatant into a new PCR tube.
34. Add 20 μL of the SPRIselect reagent to the sample, pipette mix, and incubate at room temperature for 5 min. Place the PCR tube on the 10× magnet-high position until the solution becomes clear. Remove 165 μL of the supernatant. To the pellet, add 200 μL of the 80% ethanol and let sit for 30 s. Remove the ethanol and repeat the ethanol wash once more. Briefly centrifuge the PCR tube and place it on the 10× genomics magnet-low position. Remove any excess ethanol.
35. Remove the PCR tube from the magnet, add 35.5 μL of buffer EB, pipette mix, and incubate at room temperature for 2 min. Place the PCR tube on the 10× genomics magnet-low position until the solution becomes clear. Transfer 35 μL of supernatant into a new PCR tube, labeled cDNA library. Store the cDNA library at −20 °C for a long term.

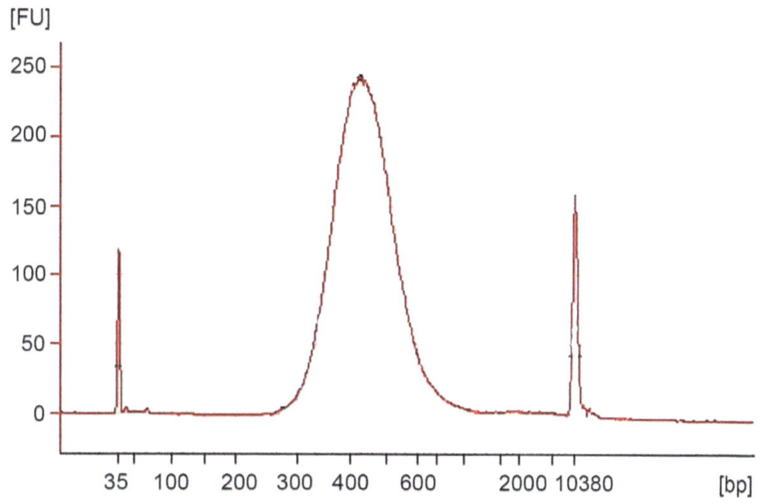

Fig. 2 Representative Agilent Bioanalyzer Trace of single-cell cDNA libraries. Shown is an electropherogram of 1 μL diluted (1:10) cDNA library with peak of the fragment distribution curve at ~450 bp

3.3 cDNA Library Quality Control (QC)

1. For cDNA library QC, use the Agilent Bioanalyzer High Sensitivity DNA Chip and Agilent 2100 Bioanalyzer instrument. Run 1 μL of 1:10 diluted cDNA library on an Agilent Bioanalyzer High Sensitivity Chip.
2. The average fragment size of the single-cell cDNA libraries can be determined from the Agilent Bioanalyzer trace (Fig. 2). Manually select the region between ~35 and ~10,000 bp for each cDNA library sample. It is expected that the majority of 10× Genomics Chromium library inserts will be between 400 and 600 bp in size (*see* **Notes 28** and **29**).
3. Determine the concentration of the cDNA library (ng/μL) using the Qubit dsDNA HS (High Sensitivity) Assay Kit and Qubit Fluorometer (Invitrogen). The concentration of the cDNA libraries for each sample is required to calculate the library input concentration (pooled or non-pooled samples) in downstream sequencing experiments on an Illumina sequencer. Calculate the cDNA library concentration as follows:

$$\begin{aligned}&\text{cDNA library concentration (nM)}\\&\quad = \text{cDNA library concentration (ng/L)}/660\ \text{g/mol}\\&\qquad \times \text{average fragment size of cDNA library (bp)} \times 10^6.\end{aligned}$$

4 Notes

1. The procedure outlined describes the enrichment of CD8a$^+$ T cells by negative selection. An alternative method of cell enrichment is positive selection, in which the cells of interest are

bound to biotin-labeled antibodies. In positive selection, the negative fraction containing the untargeted cells is eluted through the LS column into the collection tube, while the positive fraction containing the cells of interest is flushed out of the LS column by a plunger that is supplied with the column (MACS Miltenyi Biotec).

2. Keep the cell isolation buffer on ice in between steps of the cell enrichment procedure.
3. Do not let the LS column become dry while on the magnetic separator in between the first wash of the column to sample elution. If the cells are not ready to load onto the column immediately after the first wash of the column, remove the column from the magnetic separator until ready for use. This is to maintain the integrity of the column for efficient cell isolation by magnetic bead enrichment.
4. When adding the cell sample to the LS column, ensure that the cells have not settled and remain fully suspended in the buffer. Gently pipette the cell suspension up and down prior to loading onto the LS column.
5. For FACS sorting, cells can be incubated with fluorophores in either one of 1× PBS, cell media, or FACS buffer as the staining buffer. Cells can be stained in 1.5 mL centrifuge tubes or 96-well round-bottom plates on ice.
6. In general, 0.1–10 μg/mL of a fluorophore-conjugated antibody is used for flow cytometry staining, but it is important to predetermine the optimal concentration of each fluorophore-conjugated antibody prior to any experiment. This can be achieved by titrating the antibody in FACS buffer, followed by incubating the cells of interest with different concentrations of antibody. Differences in signal intensities as a result of the titrations can be analyzed on the flow cytometer at the time of sample acquisition. For multi-fluorophore flow cytometry experiments, it is essential to include controls. Negative (unstained cells) and single-stain controls (cells stained with only one of the fluorophore-conjugated antibodies) should be included in the flow cytometry panel design for compensation purposes on the cytometer, and to identify the positive expression of proteins of interest on the cells analyzed. Alternative controls include the use of isotype control antibodies and FMO (fluorescent-minus-one) controls.
7. Filtering the cell suspension prior to FACS sorting will remove any potential cell clumps, which is important to help minimize the chance of cell clogs in the cytometer during sample acquisition.
8. Store 1× PBS + 0.04% BSA at 4 °C. This solution is used for washing the cells prior to cell capture on the 10× genomics

chromium to minimize cellular aggregates and dead cells to maintain good integrity of the live cells.

9. Selection of the targeted cell recovery number can vary per experiment and will depend on the nature of the experimental design (e.g., how many single cells are desired to be analyzed and sequenced per experiment) and limitations of cell number availability.
10. Hold the 10× Genomics Chromium Chip B horizontal and avoid touching the barcode on the chip. It is important to fill all unused input wells on the chip (of all rows labeled 1, 2, and 3) with 50% glycerol solution prior to loading the wells that will be used. Do not add 50% glycerol to the recovery wells on the chip.
11. It is important to slowly load the cell suspension and the master mix into the center of the appropriate well on the Chromium Chip B to avoid introducing any air bubbles. The presence of air bubbles will interfere with the microfluidics in the chromium controller and may negatively affect the efficiency of cell capture and GEM generation. Visually assess each well containing the master mix and the cell suspension for any air bubbles. Hold a pipette tip by hand to gently pop and remove any air bubbles present.
12. To pierce open the foil of the 10× genomics gel bead strip, take a pipette tip in one hand and puncture the center of the foil seal of the gel bead strip to create a small hole. Using the same pipette tip, gently widen the size of the hole on the foil seal in order to fully open the seal. To aspirate the gel beads for loading onto the Chromium Chip B, place the pipette tip to the bottom of the well of the gel bead strip and slowly pipette up the beads without introducing any air bubbles during pipetting.
13. Do not forget to add the 10× partitioning oil, as failure to add this reagent at this step can cause chromium controller damage.
14. When attaching the 10× gasket onto the 10× chip holder, keep the gasket horizontal and do not touch the surface of the gasket that aligns with the Chromium Chip B wells. Do not press on the gasket when it is fastened to avoid any movement of the Chromium Chip B and spillover of the reagents in the wells.
15. Analyze the volumes in all rows labeled 1 through 3. An unequally high volume in any well can indicate a clog occurred during the controller run.
16. When assessing the quality of the GEMs, if there is partitioning oil in the pipette tip containing the GEMs, a clog may have occurred during the controller run.

17. After addition of the Dynabeads MyOne SILANE mix to each sample and during the incubation step, do not close the PCR tubes to avoid any overflow and loss of reagents.
18. Prepare a fresh solution of 80% ethanol every time for the post-GEM-RT cleanup. Prepare 80% ethanol using ethanol, pure (200 proof) and nuclease-free water. Keep the time duration for each 80% ethanol wash (30 s) and air-dry step (1 min) precise.
19. Prepare the 10× elution solution I before the Dynabeads MyOne SILANE cleanup mix to ensure that the elution solution is ready for its immediate addition to the PCR tube after the ethanol washes and air-drying step.
20. For this air-drying step, it is important to not exceed the 2 min of incubation at room temperature, as this will decrease the cDNA elution efficiency.
21. Equilibrate reagents of the Agilent Bioanalyzer High Sensitivity to room temperature in the dark approximately 30 min prior to using the Agilent 2100 Bioanalyzer instrument. Ensure that all sample wells contain marker reagent, including unused wells. Do not forget to load the ladder in the appropriate well on the chip. When loading the marker reagent and sample by pipetting, do not introduce air bubbles into the wells of the chip.
22. For cells with low-RNA material (less than 1 pg RNA/cell), 1 μL of undiluted cDNA can be run on the Agilent Bioanalyzer High Sensitivity Chip.
23. Prepare the 10× genomics fragmentation mix on ice and pre-cool the thermal cycler to 4 °C until ready for use.
24. Vortex the SPRIselect reagent to fully resuspend it before any sample addition as the SPRIselect reagent has a tendency to settle. Do not discard the supernatant at this step.
25. It is critical to perform these steps as quickly as possible to not let the sample overdry. Overdrying the sample will decrease the elution efficiency.
26. In multiplexed sequencing experiments, i.e., when the cDNA libraries from more than one sample or experimental condition will be sequenced together in the same run, it is critical to ensure that different (i.e., nonoverlapping) sample indices from the 10× Genomics Chromium i7 Sample Index Plate are used in order to identify each sample during the RNA-seq data analysis.
27. The number of cycles will depend on the concentration (ng) of the cDNA input (25% of total cDNA which was used in library construction) as determined in **steps 18** and **19**. The recommended number of index PCR cycles has been previously

determined as indicated on page 41 of the 10× Genomics Chromium Single Cell 3′ Reagents Kits v3 user guide.

28. Shown in Fig. 2 is a representative Agilent Bioanalyzer trace of one cDNA library sample with the majority of insert sizes (peak of the fragment distribution curve) at 500 bp. If the peak of the cDNA libraries fragment distribution curve on the Bioanalyzer trace ranges from 400 to 1000 bp, this indicates the presence of larger fragments in the libraries, which can still be submitted for sequencing.

29. Unexpected peaks in the electropherogram of cDNA libraries might appear at peak sizes <100 bp (indicates primer or primer dimer) or ~120 to ~125 bp (indicates adapter dimer). It is highly recommended to remove potential dimers in the cDNA libraries prior to RNA-seq. To remove the potential dimers, perform an additional SPRIselect cleanup step using 0.8× the total volume of library sample. Run 1 μL of undiluted cDNA library on an Agilent Bioanalyzer High Sensitivity DNA Chip and analyze the trace. The unexpected peaks at <125 bp should no longer be present.

Acknowledgments

This work was supported by the Canada Research Chairs Program, Canadian Foundation for Innovation John R. Evans Leaders Fund, Research Manitoba, Health Sciences Centre Foundation, Natural Sciences and Engineering Council of Canada Discovery Grant, and the Department of Internal Medicine, Max Rady College of Medicine, Rady Faculty of Health Sciences, University of Manitoba.

References

1. Kakaradov B, Arsenio J, Widjaja CE et al (2017) Early transcriptional and epigenetic regulation of CD8+ T cell differentiation revealed by single-cell RNA sequencing. Nat Immunol 18:422–432
2. Gaublomme JT, Yosef N, Lee Y et al (2015) Single-cell genomics unveils critical regulators of Th17 cell pathogenicity. Cell 163:1400–1412
3. Karmaus PWF, Chen X, Lim SA et al (2019) Metabolic heterogeneity underlies reciprocal fates of TH17 cell stemness and plasticity. Nature 565:101–105
4. Satija R, Shalek AK (2014) Heterogeneity in immune responses: from populations to single cells. Trends Immunol 35:219–229
5. Tang F, Barbacioru C, Wang Y et al (2009) mRNA-Seq whole-transcriptome analysis of a single cell. Nat Methods 6:377–382
6. Islam S, Kjallquist U, Moliner A et al (2011) Characterization of the single-cell transcriptional landscape by highly multiplex RNA-seq. Genome Res 21:1160–1167
7. Ramskold D, Luo S, Wang YC et al (2012) Full-length mRNA-Seq from single-cell levels of RNA and individual circulating tumor cells. Nat Biotechnol 30:777–782
8. Picelli S, Faridani OR et al (2014) Full-length RNA-seq from single cells using Smart-seq2. Nat Protoc 9:171–181
9. Jaitin DA, Kenigsberg E, Keren-Shaul H et al (2014) Massively parallel single-cell RNA-seq for marker-free decomposition of tissues into cell types. Science 343:776–779
10. Hashimshony T, Wagner F, Sher N et al (2012) CEL-Seq: single-cell RNA-Seq by multiplexed linear amplification. Cell Rep 2:666–673

11. Macosko EZ, Basu A, Satija R et al (2015) Highly parallel genome-wide expression profiling of individual cells using Nanoliter droplets. Cell 161:1202–1214
12. Klein AM, Mazutis L, Akartuna I et al (2015) Droplet barcoding for single-cell transcriptomics applied to embryonic stem cells. Cell 161:1187–1201
13. Gierahn TM, Wadsworth MH 2nd, Hughes TK et al (2017) Seq-well: portable, low-cost RNA sequencing of single cells at high throughput. Nat Methods 14:395–398
14. Han X, Wang R, Zhou Y et al (2018) Mapping the mouse cell atlas by microwell-Seq. Cell 172 (1091–1107):e1017
15. Zheng GX, Terry JM, Belgrader P et al (2017) Massively parallel digital transcriptional profiling of single cells. Nat Commun 8:14049
16. Papalexi E, Satija R (2018) Single-cell RNA sequencing to explore immune cell heterogeneity. Nat Rev Immunol 18:35–45
17. Proserpio V, Lonnberg T (2016) Single-cell technologies are revolutionizing the approach to rare cells. Immunol Cell Biol 94:225–229

Chapter 2

Monitoring Lactate Dynamics in Individual Macrophages with a Genetically Encoded Probe

Felipe Baeza-Lehnert, Carlos A. Flores, Anita Guequén, and L. Felipe Barros

Abstract

Lactate, the product of aerobic glycolysis, plays a dual role as fuel and intercellular signal in inflammation, immune evasion, and tumor progression. The production of lactate by macrophages has been associated with their polarization and function. Here we describe imaging protocols to characterize the metabolism of cultured human macrophages using a genetically encoded fluorescent sensor-specific for lactate. By superfusing cultures with increasing lactate concentrations and pharmacological inhibitors, it is possible to estimate the kinetic parameters of monocarboxylate transporter 4 (MCT4) and lactate production. Practical advice is given regarding sensor expression, imaging, and data analysis. The spatiotemporal resolution of this technique is amenable to the study of fast events at the single-cell level in different immune and other cell types.

Key words Laconic, FRET microscopy, Aerobic glycolysis, Warburg effect, MCT4

1 Introduction

The Warburg effect, a form of aerobic glycolysis [1], plays important roles in the function of macrophages. The hallmark of aerobic glycolysis is the production of lactate, a metabolic *cul-de-sac* located at the interphase between glycolysis and mitochondrial respiration. In addition to being an efficient fuel for oxidative cells such as neurons and red skeletal muscle fibers [2–4], lactate is a pleiotropic intercellular signal in brain and peripheral tissues [5–7], which is involved in various aspects of the immune response including inflammation, phagocytosis, immune evasion, and tumor progression [8] (Fig. 1a).

Fluorescence microscopy can resolve individual cells and monitor physiological parameters with high temporal resolution. In recent years, genetically encoded fluorescent sensors have become available for the detection of a number of metabolites [9]. Current protocols for the study of macrophages are based on the Förster

Suresh Mishra (ed.), *Immunometabolism: Methods and Protocols*, Methods in Molecular Biology, vol. 2184,
https://doi.org/10.1007/978-1-0716-0802-9_2,

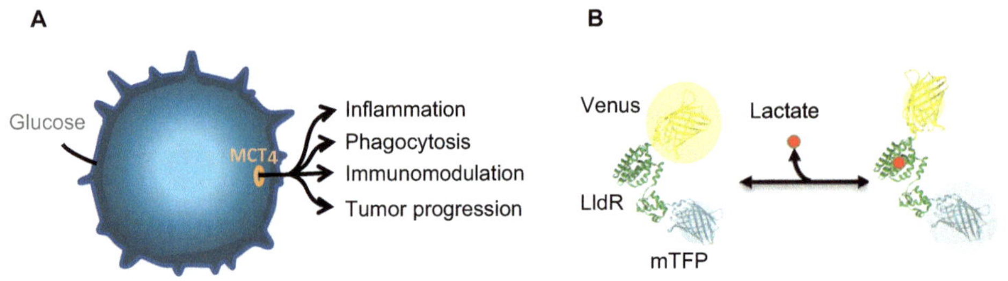

Fig. 1 Macrophage lactate and FRET-based lactate sensor Laconic. (**a**) Lactate produced by macrophages by the Warburg effect, a form of aerobic glycolysis, is involved in various aspects of the immune response. (**b**) FRET-based lactate sensor Laconic. The transcriptional regulator LldR from *Escherichia coli* is sandwiched between the fluorescent proteins mTFP and Venus. Upon lactate binding the efficiency in energy transfer from mTFP to Venus decreases

resonance energy transfer (FRET) lactate sensor Laconic [10] (Fig. 1b). The rationale of this method is as follows: First, the sensor is expressed in human macrophages cultured on glass coverslips, using viral vectors or plasmid transfection. Coverslips are then mounted under the objective of a fluorescence microscope capable of separating the emission of the FRET partners. Next, lactate permeability is computed as the speed of change in fluorescence ratio upon exposure of the cells to a lactate pulse, and kinetic parameters of the macrophage monocarboxylate transporter 4 (MCT4) are estimated by monitoring the response to pulses at increasing lactate concentrations. Finally, aerobic glycolysis is demonstrated as intracellular lactate accumulation after exposure of the cells to an MCT4 blocker.

2 Materials

2.1 Macrophage Culture and Polarization

1. THP-1 monocytes (ATCC® TIB-202™) or human monocytes isolated from blood sample.
2. Polylysine-coated 25 mm N°1 glass coverslips.
3. 35 mm Petri dishes.
4. RPMI 1640 cell culture medium supplemented with 10% fetal bovine serum (FBS).
5. Interferon gamma (INF-γ).
6. Lipopolysaccharide (LPS).

2.2 Expression of Laconic

1. Plasmid encoding the lactate sensor Laconic lactate. Available from Addgene (a nonprofit plasmid repository) (https://www.addgene.org/44238/) (*see* **Note 1**).
2. Plasmid containing dead Laconic, a version of the sensor that does not respond to lactate (as a negative control). Also available from Addgene (https://www.addgene.org/118627/).

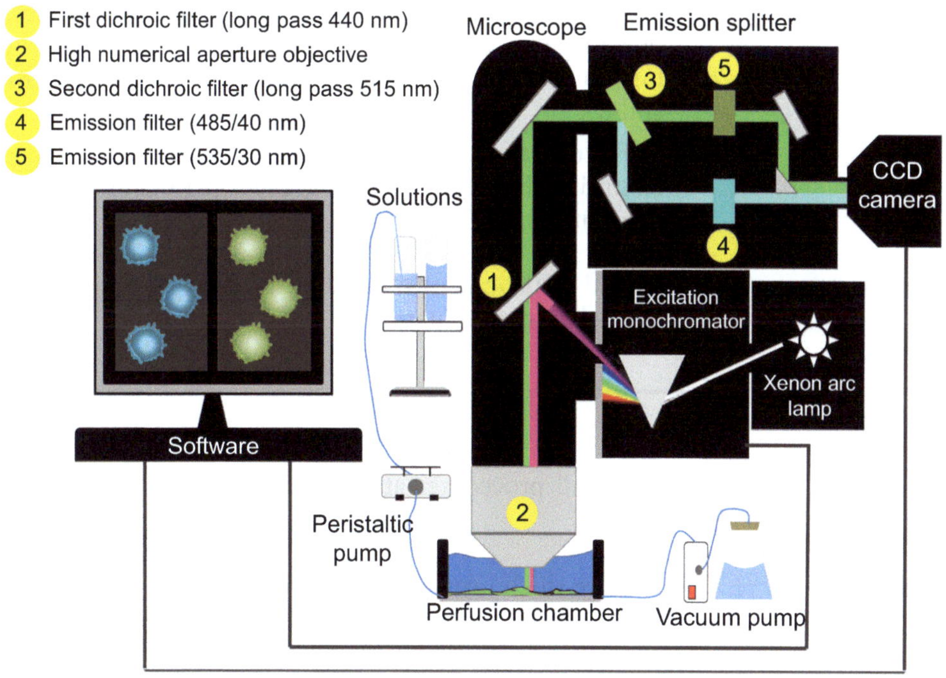

Fig. 2 Schematic representation of epifluorescence FRET setup. See text for more details

2.3 Expression of Laconic in Macrophages

1. A custom-made adenoviral vector containing Laconic (*see* **Note 2**).

2.4 Imaging

1. A wide-field epifluorescence microscope (e.g., Olympus BX51 as shown in Fig. 2) or any other equivalent microscope with appropriate filters to separate the mTFP and Venus emissions (*see* **Note 3**).
2. An Olympus F1000 laser scanning confocal equipped with a BX61WI upright microscope and a 20× 1.0 NA water immersion objective, as an alternative (*see* **Note 4**).

2.5 Perfusion System

1. 1.6 mm diameter plastic tubing for assembling the perfusion system.
2. A peristaltic pump (*see* **Note 5**).
3. A recording/perfusion chamber (e.g., Warner chambers or alternatively 3D printed chamber) and a vacuum pump.

2.6 Experimental Buffer for Perfusion

1. Bicarbonate-free buffer: 136 mM NaCl, 5 mM KCl, 1.25 mM $CaCl_2$, 1.25 mM $MgCl_2$, 10 mM HEPES, pH 7.4 adjusted with 1 M NaOH.
2. Bicarbonate buffer: 112 NaCl, 5 KCl, 1.25 $CaCl_2$, 1.25 $MgCl_2$, 10 HEPES, 24 $NaHCO_3$, pH 7.2 adjusted with 1 M NaOH and bubbled with 95% air/5% CO_2 (*see* **Note 6**).

3. 1 M glucose, sodium lactate, and sodium pyruvate stocks are prepared in water. Use aliquots to supplement experimental buffers to obtain the desired concentrations (*see* **Notes 7** and **8**).
4. Buffer containing the MCT4 blocker diclofenac is prepared by dissolving diclofenac powder in the experimental solution to a concentration of 0.5 mM (*see* **Notes 8** and **9**).

2.7 Data Analysis

A standard statistical software (e.g., SigmaPlot or GraphPad) for data analysis and presentation.

3 Methods

Carry out all procedures at room temperature, unless otherwise specified.

3.1 Macrophage Culture and Polarization

1. Culture the human monocytic macrophages as per the ATCC protocols (https://www.atcc.org/products/all/TIB-202.aspx) at 37 °C in 5% CO_2/95% air. Seed cells on polylysine-coated 25 mm N°1 glass coverslips in 35 mm Petri dishes in RPMI 1640 medium supplemented with 10% FBS [11].
2. On day 9, polarize M0 macrophages to M1 using 100 ng/mL interferon gamma and 10 ng/mL lipopolysaccharide, separately.

3.2 Expression of Laconic in Macrophages

1. On day 7 of the cell culture, infect macrophages with a custom-made adenoviral vector encoding Laconic. Add 10^{11} PFU to the 2 mL culture media in the Petri dish (*see* **Note 10**). Return cultures to the 37 °C incubator. Change media after 24 h.

3.3 Experimental Buffer for Superfusion

1. Superfuse cells with saline buffer (*see* **Note 11**) that may or may not include bicarbonate, depending on the biological question in hand. Bicarbonate buffers demand continuous CO_2 gassing, which is more physiological but cumbersome.

3.4 Imaging

1. Mount coverslips with cells on the imaging chamber and transfer to the microscope stage.
2. Check visually for cell morphology and sensor expression and distribution using the FITC cube (*see* **Note 12** and **13**).
3. Connect inlet and outlet tubing of the chamber and adjust flow to 2 mL/min (*see* **Note 14**).
4. Select an appropriate microscope field and illumination setting. As a general rule, samples should be exposed to the minimum amount of light that generates a reasonable signal-to-noise ratio. Excess illumination may result in photobleaching and/or photodamage.

5. Excite the Laconic at 430–440 nm, with the appropriate filters deployed to separate the mTFP and Venus emissions, using wide-field microscopy or confocal epifluorescence microscopy. It also works with two-photon microscopes [12] and microplate readers [13].
6. For epifluorescence microscopes equipped with xenon arc lamps and modern CCD cameras, typical acquisition times are between 0.1 and 0.5 s. Monochromator slits must not be too open to avoid excitation light getting beyond the dichroic filter.
7. In a confocal setup, the 440 nm laser line is typically set at less than 1%. If higher laser power is needed, contamination by autofluorescence may be an issue (*see* **Note 15**). Adjust photomultiplier voltage to avoid saturated pixels. Gain adjustments are not recommended as they introduce nonlinearity. Keep the pinhole totally open to maximize light collection. Obtain images at 512 × 512 pixels, which help monitoring of cell shape and movement over the course of the experiment without storing unnecessary detail.
8. Obtain images every 10 s to monitor lactate dynamics within the turnover time of the metabolite, which is typically in the order of minutes. Monitor experiment by a live plot of the fluorescence ratio (mTFP/Venus) in a few selected cells, a useful feature available in most commercial software.
9. Save images as .tiff files to be processed after the experiment.

3.5 Data Processing

Images may be processed with the dedicated software (MetaFluor or Olympus Fluoview) or alternatively with ImageJ (https://imagej.nih.gov).

1. Draw regions of interest (ROIs) around cells and a background ROI in an area devoid of cells. By checking the entire stack of images in both channels, make sure that cells did not migrate outside their ROIs and that the background ROI remained empty. For unevenly illuminated fields (e.g., tissue explants), it is advisable to draw a background ROI near every cell.
2. Import average fluorescence intensity values for each ROI and the background into an Excel spreadsheet. It is advisable to plot "raw data" to check for bleaching (*see* **Note 16**), focal drift (*see* **Note 17**), and low signal-to-noise ratio (*see* **Note 18**), e.g., Fig. 3.
3. Calculate the fluorescence ratio (mTFP/Venus) from the background-corrected ROIs from each channel (*see* **Note 19**). The ratiometric measure of Laconic makes it insensitive to some common sources of artifact (Fig. 3). Macrophage

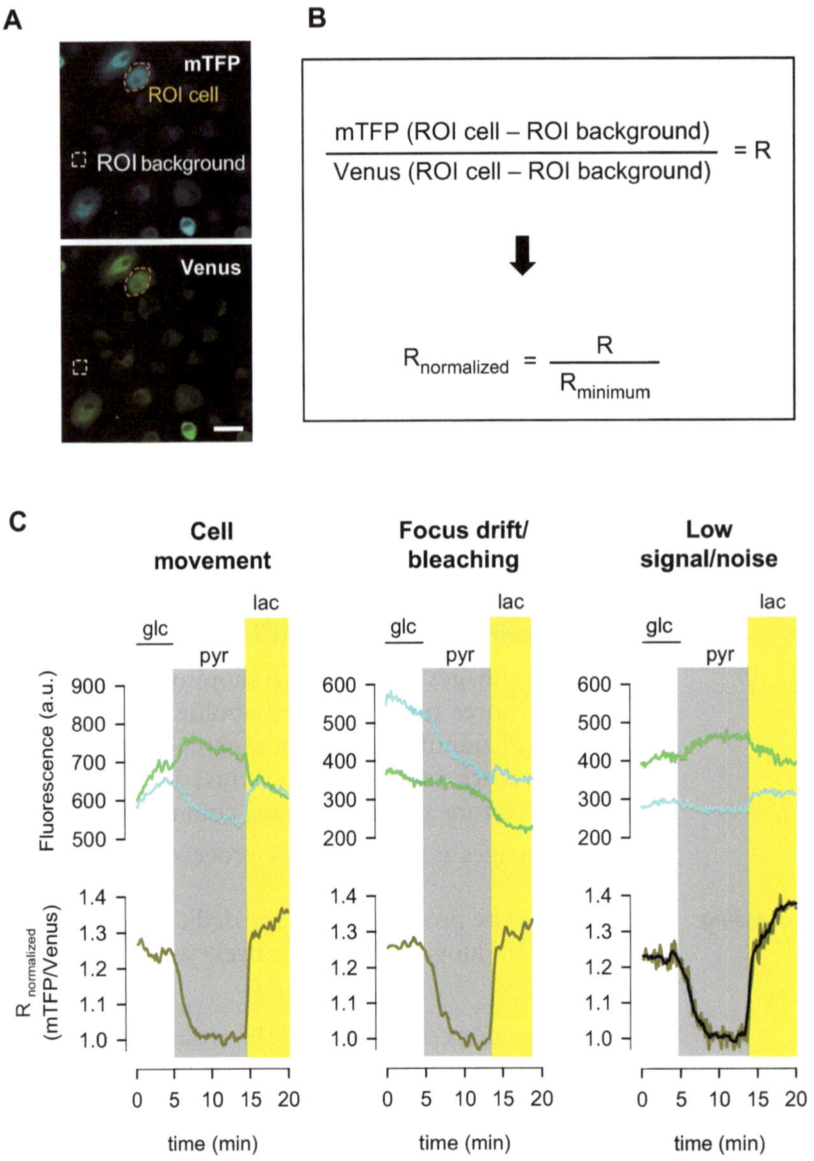

Fig. 3 Data processing of Laconic expressed in macrophages. (**a**) Separated mTFP (top) and Venus (bottom) emissions from an experimental stack of images are shown in pseudocolor. Bar is 30 μm. (**b**) The formulae explain how fluorescence intensity is first converted into ratio and then normalized. (**c**) Cells maintained in glucose (glc) were depleted of lactate by trans-acceleration with 10 mM pyruvate (pyr) and refilled with a pulse of 10 mM lactate (lac). Data from three different experiments are shown to illustrate how cell movement (left), focus drift/bleaching (center), and low signal/noise (right) affect the individual channels (top panels) but not the fluorescence ratio. The noisy trace (bottom-right) was corrected with a three-point running average (black trace). Notice how the Laconic ratio corrects flaws in imaging acquisition

lactate is high when the cell is exposed to glucose or lactate and low when exposed to pyruvate, which depletes cells from lactate through MCT trans-acceleration (*see* **Note 20**).

3.6 Assessment of Lactate Transport Capacity

1. First, deplete the lactate by superfusing the cell cultures with glucose- and lactate-free buffer for 10–20 min.
2. When the fluorescence ratio has stabilized, switch the perfusion to 10 mM lactate. A rapid increase in fluorescence ratio is expected, reaching a new steady state after 1–2 min (Fig. 4a).
3. The initial rate of fluorescence ratio increase is proportional to the maximum capacity of transport by MCT4. Estimate initial rates by linear regression.

3.7 Kinetic Parameters of Lactate Transport

1. Deplete the cells of the lactate by superfusing the cultures with glucose- and lactate-free buffer for 10–20 min.
2. Subsequently, superfuse cells with increasing lactate concentrations: 0.1, 0.5, 1, 5, 10, and 20 mM, with washes in between the exposures (Fig. 4b).
3. Use the plot of lactate concentration versus initial rates of fluorescence ratio increase to obtain the kinetic parameters of transport by adjusting a rectangular hyperbola to the data (Fig. 4b; *see* **Note 21**).
4. Note that the affinity of MCT4 for lactate measured directly with the lactate sensor was found to be much higher than that obtained by indirect pH measurements [14].

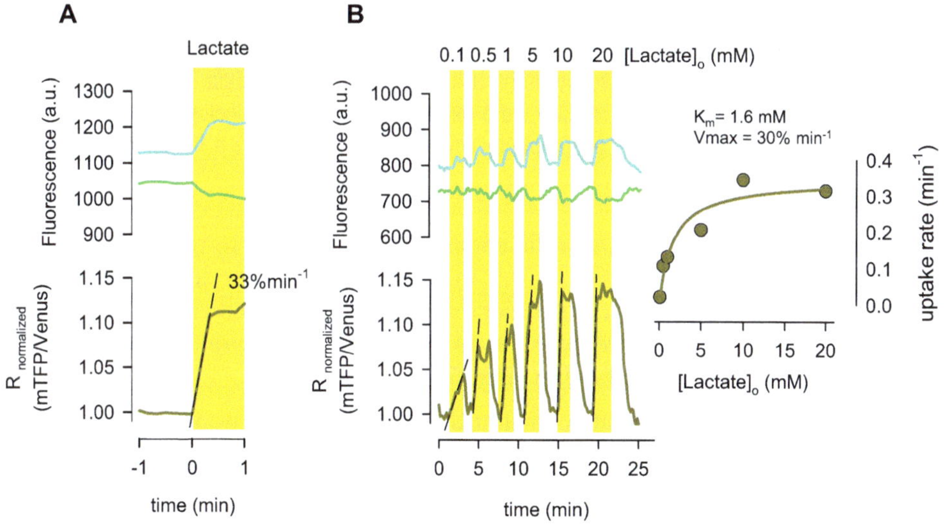

Fig. 4 MCT4 characterization in macrophages. (**a**) Lactate permeability was assessed with a short 10 mM lactate pulse. The rate of Laconic ratio change was computed by fitting a linear function to the data (dashed line). (**b**) Macrophages were exposed to increasing concentrations of lactate from 0.1 to 20 mM. The upper plot shows background-corrected mTFP and Venus signals. The bottom trace shows the ratio between the two proteins. A linear function was fitted to the initial segment of each lactate uptake (dashed lines). Saturation parameters were obtained by fitting a rectangular hyperbola to the initial uptake data (right)

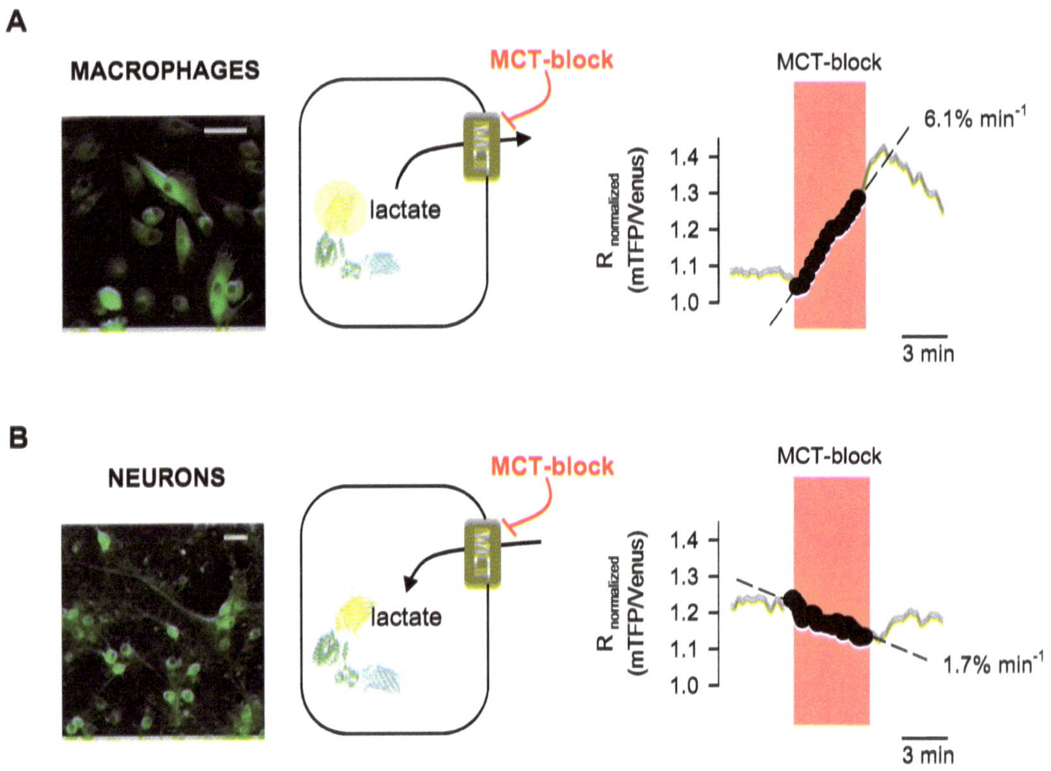

Fig. 5 Transport-stop protocol to monitor lactate production or consumption. (**a**) Macrophages incubated in 5 mM glucose were exposed to the MCT4 blocker diclofenac (0.5 mM) for 3 min. The rate of lactate accumulation (production) was computed by fitting a linear function to the data (dashed line). (**b**) Neurons incubated in 2 mM glucose and 1 mM lactate were exposed to the MCT2 blocker AR-C155858 (1 μM) for 3 min. The rate of lactate depletion (consumption) was computed by fitting a linear function to the data (dashed line)

3.8 Assessment of Lactate Production

1. Expose cells superfused with glucose to the MCT4 blocker diclofenac (0.5 mM). A slow increase in fluorescence ratio is expected as lactate accumulates within cells. The initial rate of fluorescence ratio increase informs on the speed of lactate production (Fig. 5a).
2. Lactate consumption may also be demonstrated using this type of approach, as exemplified by data from neurons (Fig. 5b).

3.9 Data Analysis

1. Import raw data into Excel spreadsheets for mathematical operations and data representation.

4 Notes

1. Laconic is a genetically encoded sensor for lactate [10]. It is based on LldR, a bacterial transcription regulator specific for lactate. In Laconic, LldR is sandwiched between the

fluorescent proteins mTFP and Venus, which undergo FRET due to their overlapping spectra. When lactate binds the ratio between mTFP and Venus emissions, intensities increase by about 40%. Laconic senses lactate between 1 μM and 10 mM and is insensitive to pH [12] and a host of other metabolites, including pyruvate [10].

2. It is possible to express Laconic using lipid transfection, but at low efficiency (<10%). Instead, use a custom-made adenoviral vector to obtain >90% of macrophages expressing the sensor.
3. Use an Olympus BX51 upright epifluorescence microscope (or any other equivalent microscope) equipped with a 20× 0.95 NA water immersion objective (Fig. 2), which combines a large field of view with efficient light harvesting. Achieve excitation with a xenon arc lamp and a CAIRN monochromator set at 430 nm. Excitation and emission are separated in the microscope fluorescence cube with a dichroic filter that passes wavelengths over 440 nm (Chroma 51017bs). A CAIRN Optosplit II equipped with a second dichroic mirror (515DCLP) and band-pass filters for mTFP emission (D485/40) and Venus emission (D535/30) juxtaposes the emissions on the chip of a Hamamatsu Orca CCD camera. Data acquisition is under the control of MetaFluor software.
4. Alternatively, image Laconic using an Olympus F1000 laser scanning confocal equipped with a BX61WI upright microscope and a 20× 1.0 NA water immersion objective. A 440 nm solid-state laser is used for excitation. Excitation and emission are separated in the microscope fluorescence cube with the 405–440/515 dichroic filter. Emission is split with a SDM510 dichroic filter and band-pass filters (465–495 nm for mTFP and 535–565 nm for Venus) and directed to two photomultiplier tubes. Data acquisition is under the control of the Olympus Fluoview software. If a 440 nm laser line is unavailable, the sensor may be imaged with the 405 nm laser (used for DAPI) that is standard in many confocal microscopes, but only with 50% efficiency.
5. Chambers may be superfused by gravity instead of using the peristaltic pump. The volume of superfusate trapped between the objective and coverslip is approx. 0.50 mL. Keep superperfusion at 2 mL/min so that the turnover time of the fluid in the chamber is about 15 s.
6. To achieve physiological pH in the bicarbonate buffer, 30 min of continuous bubbling is desirable.
7. Saline solutions supplemented with the required sugars and/or inhibitors are prepared on the experimental day. 1 M Stock solution for sugars can be kept at −20 °C for up to 6 months. For inhibitors 1000× stock is recommended. All the saline

solutions can be heated to physiological temperature and bubbled with 95% air/5% CO_2 gas, if needed.

8. Typically carry out experiments at room temperature (22–24 °C). At this temperature, experimental manipulations are easier and most metabolic events are present. However, when studying a new phenomenon, confirmatory experiments must be done at near-physiological temperature (35–36 °C). To avoid degassing, heat the solutions in a water bath and place an in-line heater before the chamber.
9. Avoid DMSO because it can interfere with cell metabolism even at low concentrations (0.1%, from 1000× stocks). Alternatively, control experiments should be run to discard artifacts.
10. After the addition of a viral vector to the Petri dish (usually a few microliters), make sure that the vector is homogenously distributed by repeated mixing with a 1 mL pipette.
11. The saline buffer is a solution containing physiological concentrations of relevant salts found in body fluids, but is lacking in amino acids. We have performed experiments as long as 3 h without any signs of cellular deterioration, so we recommend narrowing down recordings to this window of time.
12. The lactate sensor should distribute homogenously in the cytosol, with exclusion of nuclei (visible in any microscope) and organelles (visible under confocal microscopy). From our experience with astrocytes, neurons, and cell lines, when they are made to express too much, the sensor concentrates in bright spots that later coalesce into large lactate-insensitive vacuoles. Such cells should be discarded.
13. Intracellular lactate concentration ranges from hundreds of micromolar to millimolar, much higher than the typical concentrations of sensor expression (10–50 μM) [15]. Therefore, Laconic is not a significant lactate buffer.
14. If fast responses are of interest, accelerating the superfusion flow will imply a higher volume of experimental buffer. Another possibility is focusing on a microscope field close to the entrance of the experimental buffer. Expensive/scarce reagents may also be applied directly to the chamber using a pipette after stopping the superfusion.
15. We recommend in-house calibration of the laser power level. Take a cell culture without Laconic and increase the laser power until autofluorescence appears. Set the power as the maximum acceptable when recording cells.
16. The two fluorescent proteins used in Laconic are resistant to photobleaching. However, if illumination is too strong or

prolonged, some sort of bleaching may occur, and a decrease in the signal-to-noise ratio will be observed. Bleaching is minimized by reducing the frequency of acquisition. Revise expression protocol, since the need for high illumination is a consequence of low sensor expression in most cases.

17. Small acute changes in focus are accurately corrected by Laconic "ratiometric" measurements. In contrast, large acute changes may affect the fluorescence ratio in an unpredictable manner. Plotting raw data as a regular procedure is helpful in these cases. Check for changes in focus throughout the experiment. Cross-check the variations in the Laconic ratio with obvious changes in the raw data. Be careful with interpretations if big focus changes occur. The best recourse will be to do more experiments. There are commercially available devices that automatically correct focus drift. Typically, the source of signal drift is temperature change due to metal dilation on the microscope. Air conditioners or heaters in the room may prove to be useful.
18. If the signal-to-noise ratio is too low, we recommend using a running average of 3 or 5 measurements. Be aware that this procedure will reduce temporal resolution. This problem mostly arises from low sensor expression.
19. A common mistake is to process the data by first obtaining a ratio image (mTFP image divided by Venus image). This strategy introduces unnecessary noise by giving undue weight to regions with low sensor expression. Worse, small fluctuations in background noise in one of the channels may simulate drifts in fluorescence ratio. These ratio image problems may be minimized by applying a mask that selects only bright pixels (available in ImageJ). However, a disadvantage of the mask is that valuable information is lost.
20. To deplete cells of lactate we take advantage of a property of MCTs termed trans-acceleration. Pyruvate, an MCT substrate that is not recognized by Laconic, translocates to the cytosol, increasing the number of inward-facing transporters available for lactate efflux. Pyruvate trans-acceleration has been used to estimate the lactate level of brain cells in vivo [12].
21. As the absolute calibration of the Laconic FRET sensor is troublesome, express the initial velocity of lactate uptake and the Vmax for MCT4 as a percentage of the Laconic ratio change per minute (% min^{-1}). A new improved single fluorophore version of Laconic will be soon available in which absolute calibration will be feasible.

Acknowledgments

We thank Karen Everett for reading the manuscript. The Centro de Estudios Científicos (CECs) is funded by the Chilean Government through the Centers of Excellence Basal Financing Program of CONICYT.

References

1. Barros LF, Ruminot I, San Martín A, Lerchundi R, Fernández-Moncada I, Baeza-Lehnert F (2019) Aerobic glycolysis in the brain: Warburg and Crabtree contra Pasteur. Neurochem Res. https://doi.org/10.1007/s11064-020-02964-w
2. Brooks GA (2009) Cell-cell and intracellular lactate shuttles. J Physiol 587 (Pt 23):5591–5600
3. Magistretti PJ, Allaman I (2018) Lactate in the brain: from metabolic end-product to signalling molecule. Nat Rev Neurosci 19 (4):235–249
4. Barros LF, Weber B (2018) CrossTalk proposal: an important astrocyte-to-neuron lactate shuttle couples neuronal activity to glucose utilisation in the brain. J Physiol 596 (3):347–350
5. Ahmed K, Tunaru S, Tang C, Muller M, Gille A, Sassmann A et al (2010) An autocrine lactate loop mediates insulin-dependent inhibition of lipolysis through GPR81. Cell Metab 11(4):311–319
6. Barros LF (2013) Metabolic signaling by lactate in the brain. Trends Neurosci 36 (7):396–404
7. Liberti MV, Locasale JW (2016) Correction to: 'The Warburg effect: how does it benefit cancer cells?': [Trends in Biochemical Sciences, 41 (2016) 211]. Trends Biochem Sci 41 (3):287
8. Haas R, Cucchi D, Smith J, Pucino V, Macdougall CE, Mauro C (2016) Intermediates of metabolism: from bystanders to signalling molecules. Trends Biochem Sci 41 (5):460–471
9. Depaoli MR, Bischof H, Eroglu E, Burgstaller S, Ramadani-Muja J, Rauter T et al (2019) Live cell imaging of signaling and metabolic activities. Pharmacol Ther 202:98–119
10. San Martín A, Ceballo S, Ruminot I, Lerchundi R, Frommer WB, Barros LF (2013) A genetically encoded FRET lactate sensor and its use to detect the Warburg effect in single cancer cells. PLoS One 8(2):e57712
11. Xu R, Li C, Wu Y, Shen L, Ma J, Qian J, Ge J (2017) Role of KCa3.1 channels in macrophage polarization and its relevance in atherosclerotic plaque instability. Arterioscler Thromb Vasc Biol 37(2):226–236
12. Machler P, Wyss MT, Elsayed M, Stobart J, Gutierrez R, von Faber-Castell A et al (2016) In vivo evidence for a lactate gradient from astrocytes to neurons. Cell Metab 23 (1):94–102
13. Contreras-Baeza Y, Ceballo S, Arce-Molina R, Sandoval PY, Alegr K, Barros LF et al (2019a) MitoToxy assay: a novel cell-based method for the assessment of metabolic toxicity in a multiwell plate format using a lactate FRET nanosensor, laconic. bioRxiv:583096
14. Contreras-Baeza Y, Sandoval PY, Alarc R, Galaz A, Cort-Molina F, Alegr K et al (2019b) MCT4 is a high affinity transporter capable of exporting lactate in high-lactate environments. bioRxiv:586966
15. Miyawaki A, Griesbeck O, Heim R, Tsien RY (1999) Dynamic and quantitative Ca2+ measurements using improved cameleons. Proc Natl Acad Sci U S A 96(5):2135–2140

Chapter 3

The Conjugation of Antibodies for the Simultaneous Detection of Surface Proteins and Transcriptome Analysis at a Single-Cell Level

Iivari Kleino, Eliisa Kekäläinen, and Tapio Lönnberg

Abstract

Transcriptome analysis at a single-cell level with single-cell RNA sequencing (scRNA-seq) is a powerful method for detailed characterization of heterogeneous cell populations. Recent developments have enabled parallel analysis of both transcript and protein levels by using antibodies conjugated to barcoded oligonucleotides. These antibodies enable protein levels to be converted into nucleotide format, allowing the sequencing-based detection of both modalities at single-cell level. Here we present a simple and reliable method for conjugation of oligonucleotides with antibodies and a protocol for their use in single-cell transcriptome sequencing.

Key words scRNA-seq, CITE-seq, Transcriptome analysis, Transcriptomics, Proteomics

1 Introduction

Transcriptome analysis of cell populations at the level of individual cells by single-cell RNA-sequencing (scRNA-seq) allows data-driven characterization of cell type identities and regulatory states. This recent methodological advancement is revolutionizing many fields of biology, and the impact on immunology has been particularly high, due to the heterogeneity typically associated with immune cell populations [1–3]. Recent studies have revealed diverse novel subsets and cell states within T cells [4, 5], dendritic cells [6], and macrophages [7]. At the same time, the biological and functional interpretation of newly identified cell populations presents a significant challenge. A typical starting point for such efforts is the characterization of the expression of classical cell surface CD molecules that are used to define a cell type or state, and are backed by decades of research. However, in some cases these standard

Eliisa Kekäläinen and Tapio Lönnberg contributed equally to this work.

Suresh Mishra (ed.), *Immunometabolism: Methods and Protocols*, Methods in Molecular Biology, vol. 2184, https://doi.org/10.1007/978-1-0716-0802-9_3,

surface markers are not robustly detected by sequencing [8]. Moreover, mRNA abundances only partially predict the levels of their corresponding proteins [9, 10]. Finally, certain markers, such as CD45, are regulated at the level of splicing, and analysis of such markers is inherently beyond the capabilities of 3′-scRNA-seq.

These limitations have been addressed by recent methodological advances in the field of single-cell sequencing by adding modalities to the analysis, including the detection of proteins [8, 11]. In these applications, antibodies (or other feature recognizing molecules) are conjugated with oligonucleotides containing barcode sequences and A-nucleotide stretch (or other sequences compatible with the scRNA-seq chemistry). Following single-cell capture and lysis, the sequences are converted into cell-barcoded, double-stranded cDNA by the reverse transcriptase in parallel with the cellular mRNAs. The subsequent processing steps result in DNA sequencing libraries containing information on cellular transcriptomes and levels of antigens recognized by the antibody-oligo conjugates. A bioinformatic analysis of the sequenced libraries resolves this information into single-cell transcript and protein profiles for biological analyses.

While commercial oligo-conjugated antibodies are already available for many clones, they are expensive and may not be available for the clones of interest, or oligo labeling may not be optimal for the chosen scRNA-seq application. Here, we describe a simple and reliable method for the conjugation of oligonucleotides with antibodies and for their use, along with single-cell transcriptome sequencing (Fig. 1). In this protocol, dibenzocyclooctyne-PEG_4-N-hydroxysuccinimidyl ester (DBCO-PEG_4-NHS) molecules are first covalently attached to the antibody via amine groups. These activated antibodies are then conjugated to 5′-azide-modified oligos. Finally, the antibody-oligo conjugates are cleaned up of the remaining free oligo.

2 Materials

1. Phosphate-buffered saline (PBS).
2. Target antibody in a suitable format.
3. Dibenzocyclooctyne-PEG_4-N-hydroxysuccinimidyl ester (DBCO-PEG_4-NHS).
4. HPLC-purified 5′-azide-DNA-oligo (available from oligo manufacturers).
5. Dialysis membrane cups: Thermo's Slide-A-Lyzer MINI Dialysis Units 3500 MWCO or equivalent.
6. Optional: Dry polyethylene glycol (PEG), with MW 10000 or higher.

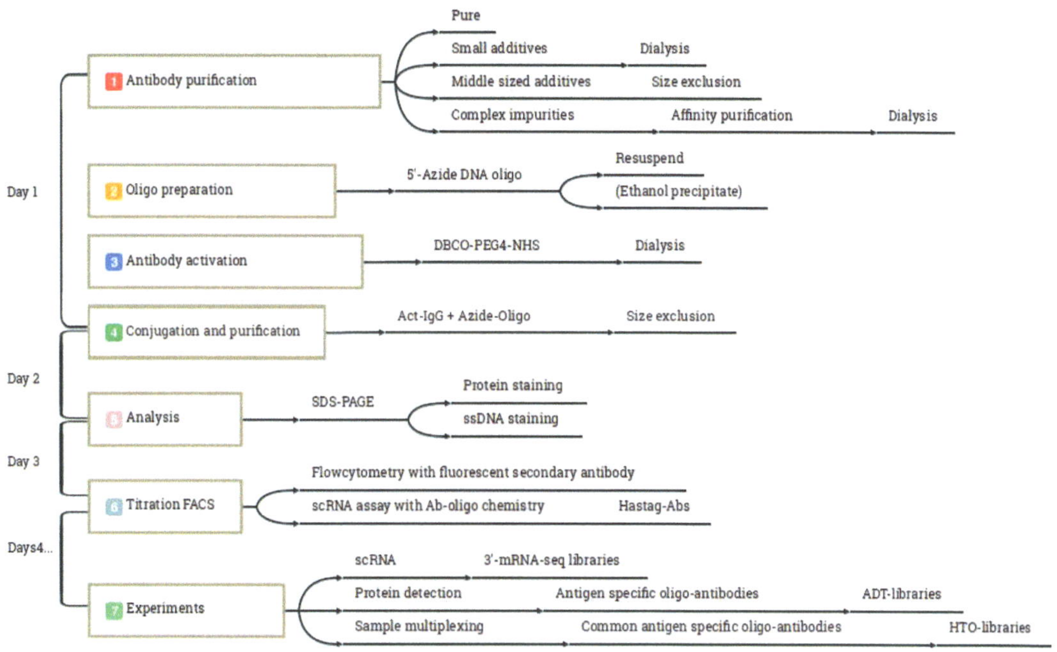

Fig. 1 Overview of the steps of the antibody-oligo conjugation with optimal case timing. The timing is tentative and heavily depends on the steps needed to purify and/or concentrate the target antibody from additives, and on steps needed to purify the final conjugate off from the remaining free oligo

7. Size-exclusion spin columns: Amicon Ultra 0.5 mL 35,000 and 100,000 MWCO or equivalent.
8. Bovine serum albumin (BSA).
9. Microcentrifuge tubes, preferably low protein retention (e.g., Eppendorf Protein LoBind tubes).
10. Microcentrifuge tubes, preferably low DNA retention (e.g., Eppendorf DNA LoBind tubes).
11. Optional: Desalting columns for DNA oligo purification.
12. Stain-free precast gradient gel 4–15% (e.g., Bio-Rad TGX) or equivalent.
13. Silver staining kit (e.g., Bio-Rad Silver Stain Plus) or equivalent.
14. Magnetic stirrer for mixing the dialysis solution.
15. Floating microtube rack to hold the dialysis cups.
16. Microcentrifuge.
17. NanoDrop or spectrophotometer for antibody solution analysis.
18. Agilent Bioanalyzer, Tapestation, Bio-Rad LabChip, or equivalent equipment for nucleic acid analysis.
19. SDS-PAGE equipment.

20. Hemocytometer or an automated cell counter.
21. Droplet or microwell-based single-cell processing platform (e.g., 10× Genomics Chromium Controller).
22. PCR cycler.
23. Access to next-generation sequencing services.

3 Methods

3.1 General Guidelines

1. Keep antibodies cooled and use an ice bath when possible.
2. Store antibodies according to the manufacturer's instructions.
3. Avoid excessive agitation of antibodies. Mix by gentle pipetting or tapping.
4. A fraction of the antibody is lost in every step due to binding to tubes and tips, and in the steps required for monitoring/quality control of the reactions. A good amount of an IgG antibody to start with is 0.1–0.2 mg. Starting with less than 0.05 mg of the antibody for labeling is not recommended.
5. Use autoclaved or sterile filtered solutions in every step to avoid microbial contamination.
6. When processing cells, work as gently and quickly as feasible to prevent stress-induced transcriptomic changes. Avoid excessive centrifugation speeds (>400 × *g*). Pipette slowly and use wide orifice tips whenever possible to prevent damage to the cells from shearing forces. However, regular-bore tips are usually needed to resuspend cell pellets after centrifugation. Keep the cells on ice and start cell capture as soon as the sample is ready, preferably within 30 min.

3.2 Azide-Modified Oligo in Conjugation

1. 5′-Azide-modified oligos are readily available from manufacturers. We have successfully used HPLC-purified 5′-azide-oligos without additional desalting or buffer exchange steps.
2. Rehydrate the lyophilized oligo with PBS to 200 μM concentration.
3. The azide-oligos are stable in PBS for several days in a fridge at 4 °C. For longer term storage, keep them frozen at −20 °C.
4. Phosphorothioate nucleotides at the 3′-end will increase nuclease resistance of the conjugated oligos.
5. The exact sequence design of the conjugated oligo depends on the scRNA-seq platform and chemistry and is out of the scope to go through them in this protocol; *see* [11].
6. The uniformity, size, and integrity of the oligo can be checked with the Agilent Bioanalyzer or Tapestation, Biorad LabChip,

or polyacrylamide gel electrophoresis with appropriate staining (e.g., Sybr Green II, silver staining). Follow the manufacturer's or lab's protocols for using the aforementioned methods.

7. If suspecting problems to arise from the additives in the oligo prep, ethanol precipitation or a desalting column purification can be used for clearing the oligo prep.

3.3 Preparation of the Antibody for Activation

The first step in conjugation is to make sure that the target antibody is free of conjugation-interfering additives, is in a suitable buffer, and is in the right concentration for the activation (*see* **Note 1**). The removal of molecules close in size to the antibody, including gelatin and BSA, require affinity purification, which is beyond the scope of this chapter. Small molecules such as azides or tris can be removed with a simple size-exclusion purification, for which our preferred method is dialysis in a dialysis cup. This requires very little hands-on time, is very affordable and easy, allows simultaneous buffer exchange, and ensures that not much antibody is lost in dialysis. Desalting columns or size-exclusion spin filters can also be used.

1. If the antibody is free of additives (purchased as pure or conjugation ready), adjust concentration to 1 mg/mL with PBS and proceed directly to the activation step. Othervise, start by recording the absorbance profile of the target antibody and a control pure antibody with the nanodrop or other spectrophotometer. A280 absorbance provides an estimate of the antibody concentration, and deviations in the shape of the absorbance curve compared to the pure antibody may reveal presence of additives or impurities in the antibody solution. Azide and many other additives absorb light at wavelengths that will show in the absorbance curve.
2. Prepare the dialysis vessel with pre-cooled PBS and a magnetic stirrer in a cold room or in a fridge.
3. Transfer the antibody solution to a 3.5 kDa MWCO dialysis cup and attach it to a floating rack. Place the rack into the dialysis tank with the dialysis cup membrane in the PBS and start stirring.
4. Dialyze for a minimum of 1 h on 1 L of PBS (overnight dialysis works as well).
5. Change the PBS to fresh.
6. Dialyze for a minimum of 1 h on 1 L of PBS.
7. Depending on the volume of the antibody (>1 mL) and the concentration of the removed molecule (>100 mM), a third round of dialysis may be required, as the removal is dependent on the volume ratios of the antibody solution and the dialysis solution (*see* **Note 2**).

8. Measure the concentration and purity of the antibody with the nanodrop or another available spectrophotometer. If the concentration is close to 1 mg/mL and no obvious impurities are present, proceed to the activation step.
9. If the concentration is lower than 0.75 mg/mL, concentrate the antibody with the 35,000 MWCO spin column (*see* **Note 3**).
10. Alternatively, concentrate the antibody by placing the dialysis cup on a 10 kDa or higher MW dry PEG-bed dialysis membrane, touching the PEG. The PEG will absorb water from the antibody reservoir and concentrate the antibody without antibody loss (a slow process, not recommended) (*see* **Note 4**).
11. Measure the concentration with a nanodrop or other available spectrophotometers. If the concentration is close to 1 mg/mL (6.67 μM), proceed to the activation step or store the antibody in fridge. If the antibody is too concentrated, dilute with sterile PBS.

3.4 Antibody Activation with DBCO-PEG$_4$-NHS

To activate the antibody for azide-oligo conjugation, DBCO-PEG$_4$-NHS molecules are covalently attached to the antibody. The NHS group reacts with the primary amines in lysine residues and the amino-termini of the polypeptide chains. The NHS-amine reaction is very efficient in the right buffer, pH, and concentration of the components (*see* **Note 5** for NHS stability and storage). The strategy in this protocol is to regulate the overall oligo conjugation level by limiting the number of DBCO groups incorporated per antibody. The standard 30 times molecular ratio of DBCO-PEG$_4$-NHS to the antibody recommended in some protocols produces antibodies with a very high number of active DBCO groups. And, if not limited by the availability of the oligo in the conjugation step, will result in antibodies with very high oligo labeling. While this could increase sensitivity, and generally be desired for weak antibody-epitope pairs, we have opted to regulate the level of activation by using the DBCO-PEG$_4$-NHS-to-antibody ratio in the range of 1 to 10–20.

1. Aliquot 100 μg or more of the clean target antibody to a low-protein-binding microcentrifuge tube.
2. Add DBCO-PEG$_4$-NHS to the reaction, mix, and incubate for 30 min at room temperature. The amount of reactive DBCO groups per antibody depends on the molecular ratio of antibody to DBCO-PEG$_4$-NHS in the reaction.
   (a) For a low number of DBCO handles, use 5–10 times the molar excess. For 100 μg of IgG add 0.33–0.66 μL of 10 mM DBCO-PEG$_4$-NHS.
   (b) For a high number of DBCO handles per antibody, use 20 times the molar excess of DBCO-PEG$_4$-NHS to

antibody. For 100 μg of IgG add 1.32 μL of 10 mM DBCO-PEG$_4$-NHS.

3. To quench the NHS present in the free DBCO-PEG$_4$-NHS still in the reaction, add 1/10th of the reaction volume of 1 M tris pH 8.0.
4. Transfer the activated antibody to a 3.5 kDa MWCO dialysis cup; place the cup into a floating rack in the dialysis tank with fresh, cold PBS; and start stirring.
5. Dialyze for 1 h (at minimum) in 1 L of PBS, or overnight for convenience.
6. Transfer the dialysis cup into 1 L of fresh PBS and dialyze for another 1 h.
7. Proceed to the measurement of the activation.

3.5 Verifying DBCO Incorporation with Absorbance Measurement

The DBCO absorbance curve at 235–400 nm differs from that of proteins allowing absorbance based measurement of the level of the DBCO-PEG$_4$-NHS incorporation. DBCO has an absorbance peak at the 309 nm wavelength (A309) at which antibodies do not absorb. This can be used to measure the amount of DBCO in the protein-DBCO solution. To resolve the concentration of the antibody in the mix, the A309 absorbance value with a correction factor (CF) of 1.089 is used to calculate DBCO absorbance at 280 nm (A280), which is then subtracted from the combined A280 to get the A280 value for the antibody in the solution (*see* **Note 6** and Fig. 2 for the DBCO and antibody concentration calculation).

1. Measure the sample with a nanodrop or other available spectrophotometers. If using a nanodrop, record the A280 and A309 and/or save the antibody measurement results as a native nanodrop data file with the .ndv ending. The .ndv file contains the absorbances of the whole measurement range and can be opened in a text editor for the recovery of the A280 and A309 values to be used in calculations or to plot absorbance curves like in Fig. 2.
2. At the lowest suggested DBCO antibody activation ratio (1:5), the amount of attached DBCO is close to the reliable absorbance-based detection limit and may not show up clearly in the A309.
3. The high DBCO derivatization attained with the 1:30 ratio may be above the linear DBCO correction factor range, and thus the calculated IgG concentration and IgG/DBCO ratio values are not accurate in high DBCO incorporation level IgGs. The highest labeling shows up as a size shift in the 50 kDa antibody band in the SDS-PAGE gel (Fig. 3).

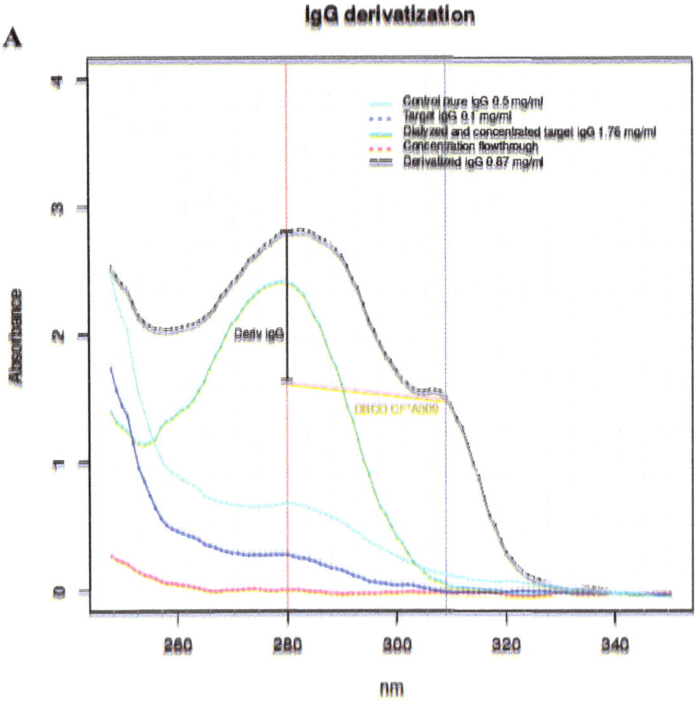

B

$$\frac{DBCO}{IgG} = \frac{204000}{12000} \times \frac{A309}{A280 - A309 \times 1.089}$$

Fig. 2 Example absorbance curves and DBCO ratio calculation. (**a**) Starting, purified, and derivatized IgG and column flow-through show differing absorbance curves. DBCO's unique absorbance peak at A309 is used to calculate the labeling efficiency. (**b**) Equation to calculate DBCO-to-IgG ratio after derivatization

4. If the desired DBCO derivatization is detected, proceed to the oligo conjugation step. A sufficient DBCO-to-IgG ratio is 10–30 molecules of DBCO per IgG.
5. The DBCO in an activated antibody should not react with other biological molecules and is stable for short-term storage at 4 °C. However, we proceed to the oligo conjugation step immediately.

3.6 Antibody Conjugation

The presented strategy limits the oligo labeling at the derivatization step. Hence, it is not absolutely critical to limit the conjugation reaction by limiting the amount of oligo. However, a high concentration of free oligo is difficult to remove from the final conjugate, and therefore, the amount of oligo should be kept reasonable during the conjugation step. We recommend use of 2 to 5 times the molar excess of the oligo to DBCO as calculated from the antibody/DBCO absorbance measurement. The optimal amount of oligo depends on the number of incorporated DBCO groups in each antibody and the desired number of conjugated oligos per antibody. The detection of target antigens with low levels of expression may benefit from a higher number of oligos in the final antibody conjugate. A good starting point is to have 5–10 times more oligo than antibody in the conjugation reaction. The 1 μg/μL IgG antibody concentration corresponds to 6.67 μM molar concentration as calculated with IgG MW of 150,000 g/mol. Hence, for each 100 μg of derivatized IgG, add 16.7–33.3 μL of 200 μM 5′ azide-oligo.

1. Transfer the desired amount of the activated antibody to a protein low-binding microcentrifuge tube. Save a small sample of the activated antibody for SDS-PAGE gel analysis later.
2. Add 200 μM of the azide-oligo solution into the antibody solution, and mix by pipetting. Use 1.67–3.33 μL of the 200 μM azide-oligo solution per 10 μg of activated IgG.
3. Incubate for 3–4 h at room temperature, or for convenience overnight at 4 °C.
4. Proceed to the purification step.

3.7 Removal of Free Oligos from the Antibody-Oligo Conjugate

To avoid a high background signal, the free oligo should be removed from the antibody-oligo conjugate. Also, to improve long term storage, the reaction buffer should be changed to an antibody/oligo storage buffer. In the presented 100 μg scale this may sometimes be problematic, due to losses in purification/exchange. Furthermore, as the chemical properties of the antibody have been modified by oligo addition, some techniques may not work as expected (*see* **Note 7**). Size-exclusion filtering with 100 kDa cutoff spin filters is our recommended option. To avoid lossess due to the antibody binding to the filter, first block the filter with BSA. This should not interfere with the subsequent application and could also help in stabilization of antibodies in the storage. Antibody storage preservatives, such as trehalose and/or antimicrobial agents like sodium azide, can also be added.

1. Wash and block the 100 kDa MWCO spin filter with 1% BSA-PBS.
2. To wash, fill spin filter with 500 μl PBS and spin at 12,000 × *g* for 2 min.
3. To block, add 500 μL of 1% BSA-PBS solution and spin for 5 min.
4. Remove excess blocking solution by spinning the filter shortly at 1000 × *g* upside down in a microtube or by flicking the spin filter empty with hand.
5. Load the conjugated antibody into the spin filter and adjust the volume to 500 μL with sterile PBS.
6. Spin at 12,000 × *g* to concentrate the antibody-oligo conjugate. The time depends on the amount of the antibody, but 3–5 min is usually enough.

7. Discard the flow-through and add 500 μL of fresh PBS into the filter/spin column and spin at 12,000 × *g* until the surface reaches the collection mark. Discard the flow-through.
8. Repeat **step 7** three times.
9. Collect the purified oligo-antibody to the microcentrifuge tube by spinning the filter/spin column at 1000 × *g* for 1 min upside down in a clean low-protein-binding tube.
10. Proceed to the analysis of the oligo-conjugated antibody.

3.8 Analysis of the Oligo-Conjugated Antibody by SDS-PAGE

The produced oligo-conjugated antibody should be analyzed for the level of labeling and if the labeling affects epitope recognition. The conjugation efficiency and the number of oligos per antibody can be estimated by separating the antibody and appropriate controls on a SDS-PAGE gel and staining the gel with protein and single-strand DNA-specific dyes. A convenient way is to have a fluorescent protein dye in the gel (for example stain-free gels with trihalo compound or SYPRO Orange), so that the proteins can be detected immediately after the gel is run and developed with a UV-light. Fluorescent ssDNA-labeling dyes like silver staining or SYBR Green II can be used to detect the oligos in the gel. We use silver staining, which is sensitive and will stain both polypeptides and single-stranded oligos, and can be used directly after protein detection (Fig. 3).

1. Separate the purified antibody-oligo conjugates and appropriate controls in SDS-PAGE gel in standard running and loading

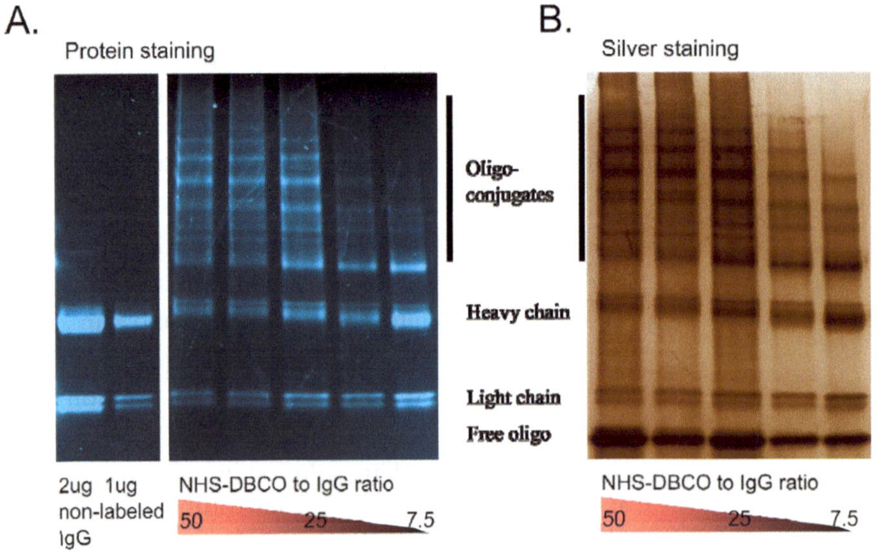

Fig. 3 Oligo-IgG in SDS-PAGE. (**a**) Conjugated anti-CD38 IgG in Bio-Rad stain-free gel in pseudo-color. Left panel: pure-antibody controls; right panel: conjugated IgGs with constant oligo-to-IgG ratio and varying NHS-PEG$_4$-DBCO-to-IgG ratio as indicated with numbers below the gel image. (**b**) The same gel after silver staining. The free oligo travels below the light chain and is clearly visible in pre-purified samples in the gel

buffers and protocol. Preferably use 4–15% gradient gel (e.g., Bio-Rad TGX stain-free precast gel), which separates the target molecules efficiently. Choose the appropriate samples to run from the following:

(a) A 1 μg and/or 2 μg of unconjugated antibody for conjugate quantification:

   (i) Protein bands: antibody light 25 kDa and heavy 50 kDa chains.

   (ii) Use these to estimate the concentration of the purified antibody conjugate.

(b) Conjugated antibody 1–2 μg for estimation of conjugation efficiency, concentration, and clearance of free oligo:

   (i) Protein bands: Should form higher molecular weight ladder above 50 kDa heavy chain.

   (ii) Still contains some non-labeled IgG light 25 kDa and heavy 50 kDa chain bands.

   (iii) ssDNA bands: The non-purified antibody may still contain strong band of free oligo.

   (iv) If already purified, sample should not have a strong oligo band. Note that ssDNA is only visible in appropriate staining. In our case the oligo runs as a ~20 kDa band.

   (v) If BSA was used in the clearing step or was otherwise added to the storage buffer, it shows up as a 66.5 kDa band.

(c) Optional: Activated nonconjugated antibody to visualize the level of derivatization:

   (i) Protein bands: Contains IgG light 25 kDa and heavy 50 kDa chain bands.

   (ii) A detectable size shift in the heavy chain compared to a nonactivated heavy chain indicates successful strong activation.

   (iii) Use free oligo for positive staining control and as a size marker for the oligo. The oligo band is only visible after silver staining (or fluorescent ssDNA staining); in our case it runs as a ~20 kDa band.

2. After running the gel, rinse it briefly with deionized water and document with appropriate fluorecent imaging or scanning apparatus. Alternatively, digital imaging allows for a computer-based quantification of the bands:

(a) Develop the gel with fluorescent protein dye with UV-light and document it on UV-table with digital imager. Different Fluorescent dyes may require specific development. Follow the manufacturer's intructions.

(b) Document the silver stained gel with incandescent light imager or a scanner.

3. Quantify the bands from the gel image with a dedicated gel analysis software, or plug-ins in ImageJ (freely available from the NIH).
4. The ladder-pattern in higher than the 50 kDa molecular weight protein bands indicates strong labeling. The presence of less than 50% of the nonconjugated bands (25 and 50 kDa) in quantification already indicates very strong labeling.

3.9 Testing Antibody Binding on Cells and Initial Titration

Before used in scRNA-seq experiments, the labeled antibody should be tested for antigen recognition specificity, and the right concentration for cell labeling should be determined. If the antibody's antigen recognition site contains lysine residues, it is possible that the antibody activation or oligo conjugation blocks the binding of the antibody to its epitope. This can be tested in flow cytometer assay by labeling the antigen-containing cells with the oligo-conjugated antibody, and then detecting cell bound oligo-IgGs with fluorescently labeled secondary antibody. This method can also be used to find a good antibody concentration for cell staining for scRNA-seq experiments. However, based on our experience, the flow cytometric signal is usually stroger than the standard oligo-antibody staining signal in single-cell sequencing applications. If using flow cytometry based titering, one should calibrate the staining concentration with the used single-cell oligo-sequencing application. Methodologically the oligo-antibody cell surface staining is very similar compared to normal flow cytometry staining protocol. The critical difference is that for scRNA-seq applications, the buffers should not contain additives that would affect subsequent enzymatic steps, or that would compromise intracellular mRNA. These include divalent cations (that affect enzymatic reactions), agents that may compromise membrane integrity (mRNA leakage), and fixatives (mRNA availability). Use a staining buffer composed of Mg^{2+}/Ca^{2+}-free PBS supplemented with 2% BSA and 0.02% Tween is recommended. Otherwise, for scRNA-seq one can use standard flow cytometry staining protocols such as the one described below:

1. Prepare the staining oligo-antibody master mix in staining buffer, with 50 μL per sample. Add each antibody to the predetermined staining concentration and top up to a total volume of 50 μL with the staining buffer. If titering, make three- or fourfold dilution series of the tested antibody.
2. Prepare a single-cell suspension with your cell-specific protocol. We use peripheral blood mononuclear cells purified with a

gradient centrifugation and then washed twice: first with PBS and then with the staining buffer.

3. Pellet the desired amount (>200,000 and <2,000,000) of cells to a microcentrifuge tube.
4. Resuspend the cells into staining buffer and incubate on ice for 30 min.
5. To wash, add 1 mL of staining buffer and spin cells down with $200 \times g$.
6. Repeat wash twice with 1 mL of staining buffer.
7. If doing a fluorescent detection, continue by staining with the appropriate fluorescent secondary antibody and flow cytometry protocol. Calculate the staining index to determine the optimal staining concentration: (MFI of positive population − MFI of negative population)/(2 × SD of negative population) [12].
8. If doing single-cell mRNA sequencing with protein detection, resuspend the cells into scRNA-seq buffer (Mg^{2+}/Ca^{2+}-free PBS supplemented with 0.04% BSA) and continue with the scRNA-seq protocol.

3.10 Preparation of Cells for scRNA-seq

Single-cell transcriptomic analysis can be performed with any platform compatible with the oligonucleotide sequences. For example, the 10× Genomics Chromium Controller with the 3′-single-cell solution can be used, and we recommend adhering to the manufacturer's instructions. A detailed protocol can be found in an earlier volume of this series [13].

1. Stain the cells using titrated concentrations of the antibodies as described above. Keep the cells on ice at all times.
2. Resuspend the cells to 1× PBS (calcium and magnesium free) containing 0.04% weight/volume BSA (400 μg/mL), or another buffer compatible with the Chromium single-cell application (refer to 10× Genomics demonstrated protocol CG000149). The optimal concentration range is from 700 to 1200 cells per μL.
3. We recommend passing the cells through a filter, e.g., a Flowmi Tip Strainer to remove possible cell aggregates. However, note that this step can lead to the loss of some cells.
4. Count the cells using a hemocytometer or an automated cell counter. Measure the viability, e.g., using the trypan blue exclusion method. For improved accuracy, perform four counts from two independent draws for each sample and then calculate the average values. Make sure to include dead cells in the total cell count.
5. Proceed to the single-cell capture according to the manufacturer's instructions.

4 Notes

1. Target antibody: The simplest option is to buy the antibody in a conjugation-ready format. However, in many cases, the antibody clone of interest may not be available as a pure antibody, but rather is stored with stabilizing agents such as additional proteins, primary amine buffers, or azides, or a mixture of them. These additives either prevent the antibody activation step (NHS-primary amine reaction) or react with the DBCO and inhibit azide-oligo conjugation.
2. Dialysis: In each round of dialysis, the molecule to dilute out is diluted by a factor of diluent to the starting solution. If the dialysis buffer volume is 1 L and the starting material is 100 μL, then the theoretical dilution is 1:10000 in this step. The sample volume usually goes up during the dialysis, causing the dilution ratio to change for the subsequent step. For example, if the volume rises to 250 μL, the dilution against 1 L would be 1:4000. For these two steps, the theoretical dilution of the small molecule or the buffer exchange would hence be 1:10,000 $*$ 4000 = 1:40,000,000. Typically, antibodies are stored in a 3.08–7.69 mM sodium azide (0.02–0.05% (w/v)). This is ~1000 more azide than IgG in 1 mg/mL. To ensure the diluting of the sodium azide, we would suggest to dialyze 100 μL of antibody solution at least two times against 1 L of PBS.
3. Spin filter: The use of spin filters is not recommended, as a significant amount of antibody is lost in each concentration step with the spin filter, due to let-through and IgG binding to the membrane.
4. PEG concentrator: Change the PEG to dry if it gets fully wet during concentrating step. Be careful not to concentrate too much, or the antibody may precipitate permanently.
5. NHS storage: NHS is highly reactive and hydrolyzes in aqueous solutions spontaneously in a relatively short time. DMSO is very hygroscopic and condenses water molecules from the air. To prevent the spontaneous degradation of NHS, store DBCO-PEG-NHS dissolved in DMSO in −20 to −70 °C in a sealed container with silica gel beads, or equivalent, as a desiccative. Warm up the DBCO-PEG-NHS tube to the ambient temperature before opening to prevent water condensation on the tube.
6. DBCO measurement: The derivatization efficiency is calculated as the molar ratio of DBCO/IgG. The extinction coefficient for DBCO is 12,000 1/(M*cm), and for IgG it is 204,000 1/(M*cm) [14]; the DBCO A309 to A280 correction factor is 1.089. The simplified equation for the calculation

of DBCO molecules per IgG is 17*A309/(A280 − 1.089*A309).

7. Free oligo removal: For example, a simple ammonium sulfate precipitation does not efficiently remove free oligos from the conjugate. We have had mixed success with dialysis with 50 kDa cutoff filters, which should let the 21 kDa free oligos through.

Acknowledgments

The Academy of Finland has supported EK (decision 308913) and TL (decision 311081). This work was supported by the Foundation for Pediatric Research in Finland and the Väre Foundation. Eliisa Kekäläinen, and Tapio Lönnberg contributed equally to this work.

References

1. Chen H, Ye F, Guo G (2019) Revolutionizing immunology with single-cell RNA sequencing. Cell Mol Immunol 16:242–249
2. Papalexi E, Satija R (2018) Single-cell RNA sequencing to explore immune cell heterogeneity. Nat Rev Immunol 18:35–45
3. Vegh P, Haniffa M (2018) The impact of single-cell RNA sequencing on understanding the functional organization of the immune system. Brief Funct Genomics 17:265–272
4. Azizi E, Carr AJ, Plitas G, Cornish AE, Konopacki C, Prabhakaran S et al (2018) Single-cell map of diverse immune phenotypes in the breast tumor microenvironment. Cell 174:1293–1308.e36
5. Savas P, Virassamy B, Ye C, Salim A, Mintoff CP, Caramia F, Salgado R, Byrne DJ, Teo ZL, Dushyanthen S, Byrne A, Wein L, Luen SJ, Poliness C, Nightingale SS, Skandarajah AS, Gyorki DE, Thornton CM, Beavis PA, Fox SB, Kathleen Cuningham Foundation Consortium for Research into Familial Breast Cancer (kConFab), Darcy PK, Speed TP, Mackay LK, Neeson PJ, Loi S (2018) Single-cell profiling of breast cancer T cells reveals a tissue-resident memory subset associated with improved prognosis. Nat Med 24:986–993
6. Villani A C, Satija R, Reynolds G, Sarkizova S, Shekhar K, Fletcher J et al (2017) Single-cell RNA-seq reveals new types of human blood dendritic cells, monocytes, and progenitors. Science 356(6335):eaah4573
7. Song Q, Hawkins GA, Wudel L, Chou P-C, Forbes E, Pullikuth AK et al (2019) Dissecting intratumoral myeloid cell plasticity by single cell RNA-seq. Cancer Med 8:3072–3085
8. Peterson VM, Zhang KX, Kumar N, Wong J, Li L, Wilson DC, Moore R et al (2017) Multiplexed quantification of proteins and transcripts in single cells. Nat Biotechnol 35:936–939
9. Vogel C, Marcotte EM (2012) Insights into the regulation of protein abundance from proteomic and transcriptomic analyses. Nat Rev Genet 13:227–232
10. Liu Y, Beyer A, Aebersold R (2016) On the dependency of cellular protein levels on mRNA abundance. Cell 165:535–550
11. Stoeckius M, Hafemeister C, Stephenson W, Houck-Loomis B, Chattopadhyay PK, Swerdlow H et al (2017) Simultaneous epitope and transcriptome measurement in single cells. Nat Methods 14:865–868
12. Cossarizza A, Chang HD, Radbruch A et al (2017) Guidelines for the use of flow cytometry and cell sorting in immunological studies. Eur J Immunol 47:1584–1797
13. De Simone M, Rossetti G, Pagani M (2019) Chromium 10× single-cell 3′ mRNA sequencing of tumor-infiltrating lymphocytes. Methods Mol Biol 1979:87–110
14. Kotagiri N, Li Z, Xu X, Mondal S, Nehorai A, Achilefu S (2014) Antibody quantum dot conjugates developed via copper-free click chemistry for rapid analysis of biological samples using a microfluidic microsphere array system. Bioconjug Chem 25:1272–1281

Chapter 4

Cellular Fatty Acid Analysis in Macrophage Using Stable Isotope Labeling

Kevin J. Williams and Steven J. Bensinger

Abstract

Fatty acids (FAs) are essential for building complex lipids, posttranslational modifications, and energetics. FAs can be imported from extracellular sources or synthesized by cells. The analysis of fatty acid methyl esters (FAMEs) by gas chromatography–mass spectrometry (GC–MS) allows for the quantitative analysis of long-chain and very-long-chain fatty acid content of cells. When coupled with isotopic labeling, this approach can elucidate the synthetic pathways being engaged by the cells, and the relative contribution of synthesis and import to maintain lipid content. Here, we describe a method for total cellular fatty acid analysis in macrophages.

Key words Fatty acids, FAME analysis, GC–MS, Stable isotope labeling, Macrophages

1 Introduction

The analysis of fatty acid methyl esters (FAMEs) by gas chromatography–mass spectrometry (GC–MS) was first employed on human tissue samples in 1958 [1]. Since that time, a number of methods have been developed to assess total fatty acid content through the conversion of complex lipids into their constitutive fatty acids coupled with the derivatization of liberated fatty acids into FAMEs for GC-MS analysis. Three methodologies have emerged: (1) saponification followed by methylation [2], (2) acid-catalyzed transesterification, and (3) base-catalyzed transesterification [3]. Acid-catalyzed and base-catalyzed transesterification (methods 2 and 3) have the advantage of being a single-step procedure, and as such have become the most commonly used approaches.

The stable isotope labeling of metabolites is a technique employed to provide information about the movement of a metabolite through a cell's enzymatic machinery. Using this technique, one can label a precursor pool, for example glucose, and monitor the incorporation of the isotope into downstream metabolites

Suresh Mishra (ed.), *Immunometabolism: Methods and Protocols*, Methods in Molecular Biology, vol. 2184,
https://doi.org/10.1007/978-1-0716-0802-9_4,

using mass spectrometry. Isotopic labeling data can be used to determine the relative fluxes through a biochemical pathway and the contribution of specific metabolites to the total metabolite pool. While not absolute information, these approaches do provide additional insights that static measurements of pools cannot provide. The employment of isotopic labeling techniques and the modeling of isotopic labeling data from fatty acids have increased the value of the conventional fatty acid analysis. The incorporation of stable isotope labels (e.g., 13C or deuterium) into specific fatty acids allows for the discrimination of newly synthesized fatty acids from those that are in preexisting pools, or fatty acids that have been imported from an extracellular source.

Macrophages are immune cells that regulate a variety of inflammatory, host defense, and wound repair processes. As such, they have been studied extensively and are an excellent model system for studying cell differentiation, gene regulation, and signal transduction. Well-established procedures exist to isolate, culture, and activate mouse and human macrophages. In this chapter, we describe a procedure for stable isotope labeling of cultured macrophages. We also provide methods on the collection and derivatization of fatty acids from macrophages, and how to measure these labeled FAMEs by GC–MS.

2 Materials

This method requires the use of three major instruments: a high-content fluorescence imager, an automated liquid handler, and a GC–MS. The configuration and programming of these instruments are described in as much detail as possible. The method assumes an operational understanding of these hardware systems and their software controls. The modeling of stable isotope labeling data is also beyond the scope of this chapter. Once the data is acquired at the conclusion of this protocol, it can be applied to isotopomer spectral analysis (ISA) or mass isotopomer distribution analysis (MIDA) modeling [4, 5].

All reagents, especially glucose, should be tissue culture grade and pyrogen free if possible. As 13C-glucose is expensive, labeling medium should be made fresh and in appropriate quantities for the experiment at hand. In most cases, a 50:50 mix of U-13C-glucose and unlabeled glucose is added back to the appropriate glucose-free "dropout" medium. All reagents used in the derivatization and extraction process should be of the highest purity available (HPLC grade or better). The amount of cellular material required for the FAME analysis is largely a function of the background within the system. Fatty acids are common contaminants within organic solvents and can be leeched from plastics requiring more sample material to achieve signal above background.

2.1 Macrophage Cell Culture, Labeling, and Collection

1. 24-Well culture plates: For imaging-based cell counting, select an optically pure plate (e.g., Cellvis #P24-0-N).
2. Labeling medium: Weigh out appropriate amounts of U-13C-glucose (Cambridge Isotope Lab) and D-glucose. Add to DMEM with no glucose to bring the concentration up to 4.5 g/L. As the glucose products are not sterile, this labeling media should be filter sterilized through a 0.22 μm filter. Then add 5% FBS, 1% penicillin-streptomycin (10,000 U/mL), 1% L-glutamine (200 mM), 0.5% sodium pyruvate (100 mM), and recombinant M-CSF to DMEM medium (*see* **Note 1**).

2.2 Imaging and Sample Collection

1. Calcein AM dye stock: Prepare as a 5 mM stock with DMSO. Prepare aliquots to minimize freeze/thaws and light exposure. For each well in a 24-well dish, 15 μL of calcein-AM should be combined with 30 μL of media and mix before being added to the well; this is the calcein-AM working mix.
2. Fluorescence imager: Use the ImageXpress Micro XL imager (Molecular Devices) with MetaXpress software for image analysis or any other equivalent equipment. Several other high-content fluorescence imagers are on the market. We have created the plate profile for the Cellvis 24-well culture plate (above). The acquisition program for the imager requires the 10× Plan Fluor objective lens and the illumination setting for FITC.
3. Guanidine solutions: Prepare a 6 M guanidine hydrochloride and a 3 M methanolic guanidine hydrochloride solution with molecular biology-grade guanidine hydrochloride. The 6 M solution should be made with LC-grade water and the 3 M solution with LC-grade methanol.
4. Sample collection rack: A 96-well sample collection rack is assembled from an aluminum base (Chromtech).
5. 2.5 mL flat bottom vials, 9 × 75 mm
6. A molded PTFE-lined silicone mat.
7. An aluminum cover plate secured with screws (Chromtech).

2.3 Fatty Acid Derivatization

1. Standard curve mix: Prepare a 5 mg/mL FAME standard mix stock for long-chain fatty acids (14:0, 16:0, 16:1, 18:0, 18:1, 18:2, 18:3) from a FAME mix with a ratio of 2:30:3:14:41:7:3 (Nu-Chek Prep) in hexane. The exact amount of FAME standard in the standard curve mix will be adjusted to the type of cells and conditions in your experiment. A good starting mix will contain 12 μL of FAME standard mix (5 mg/mL) brought up to 500 μL with hexane.
2. Acid methanolysis mix: Prepare 1 mg/mL internal standards from methyl nonadecanoate (Nu-Chek Prep). 720 μL of acid

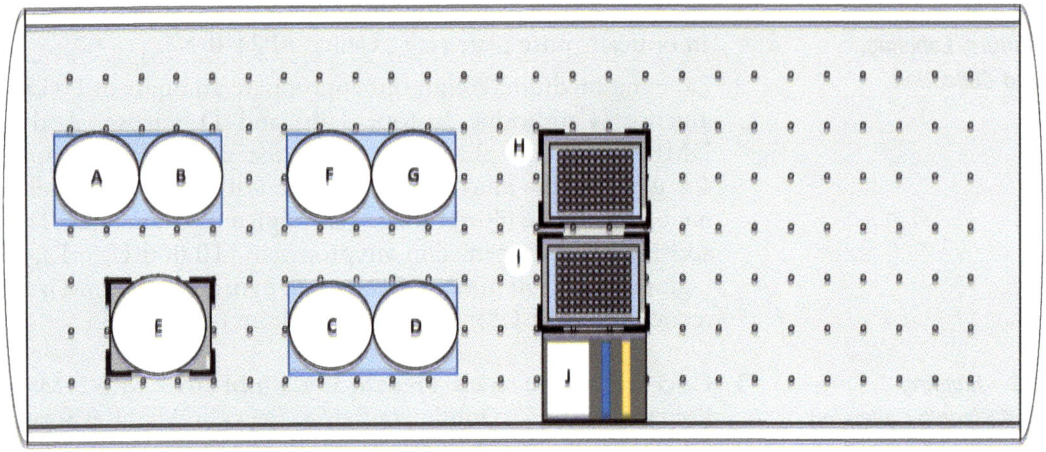

Fig. 1 Example stage layout for liquid handler that is compatible with all three liquid handler programs. (**a**–**c**) Hexane wash bottles, (**d**) optional fourth hexane wash bottle, (**e**) waste beaker, (**f**, **g**) reservoir bottles, (**h**) sample collection rack, (**i**) running rack, (**j**) wash station

methanolysis mix is required for each sample. Prepare this by combining 647 μL methanol, 72 μL toluene, and 1.02 μL concentrated hydrochloric acid. A master mix of acid methanolysis solution should be prepared in a bottle compatible with your liquid handler (*see* **Note 2**). Internal standard (0.15–0.3 μg) is added to each sample; the average mixture will require 20–80 μL of 5 mg/mL internal standard mix. The standard needs to be within an order of magnitude of the unknowns.

3. Automated liquid handler: To measure out organic solvents accurately and without the contamination of solvents with plasticizers, an automated liquid handler is used. Several other similar instruments from Perkin Elmer, Tecan, and Hamilton are currently on the market. Use microplate adaptors to fix the sample rack and running rack, which are used to set the reservoir, wash, and wash bottles in fixed locations. For the reservoir and wash bottles, use 16 oz. straight-sided jars. Use a standard 1 L beaker as a waste container (*see* Fig. 1 for sample configuration). These items (bottles, waste beaker, and 96-well sample rack) can be programmed into the Multiprobe control software (WinPREP) as custom hardware (*see* **Note 3**).
4. Incubator: A forced-air incubator or oven that can maintain 45 °C.

2.4 Extraction/Mass Spec Prep

1. 40 mM NaCl solution: Prepare 500 mL with LC-grade water.
2. Tabletop centrifuge: Beckman Coulter Allegra X-15R with SX4750 microplate carriers. Many such carriers are available on the market. Note that the microplate carrier must have a

clearance of at least 8.5 cm in order to hold the sample collection rack.

3. Running rack: Assemble the aluminum base (Chromtech) with a 1.5 mL flat bottom vials (9 × 44 mm) and pierceable mat (Chromtech) for use in a GC–MS autosampler deep-well plate holder.

2.5 GC–MS Analysis

1. Gas chromatography–mass spectrometry: Agilent 7890B/5977A GC–MS with CTC sampler with 120 cm rail. While a number of GC–MS instruments can be used for this application, this instrument is optimally configured for this application. Five key features streamline the application: (1) the CTC sampler with a refrigerated drawer is capable of holding the running rack, (2) a multimode inlet that enables large-volume injections with solvent venting, (3) a purged union in the GC that allows a midpoint method on the column that significantly reduces the time of the GC–MS run, (4) the "fast" oven option that increases the speed of oven heating, and (5) an EI extractor source that improves sensitivity for lower cell count samples.
2. GC column: Agilent DB-WAX, Ultra Inert, 30 m, 0.25 mm diameter, 0.25 μm film. Cut this 30 m column into a 10 and 20 m column and install with the 10 m column between the inlet and the purged union and the 20 m column between the purged union and the MS.
3. Splitless inlet liner: For Agilent GC/MS use an ultra-inert splitless, single-taper inlet liner.
4. Long/very-long-chain FAME standard mix: Prepare the GLC-96 standard mix (Nu-Chek Prep) in a 5 mg/mL stock and dilute into a standard curve mix identically to the GLC-20a mix with identical amounts of internal standard for use in backflush development.

3 Methods

This method is optimized for macrophage cultures. Macrophages should be plated in an optically pure 24-well culture dish (*see* **Note 4**). A confluent or near-confluent well, at the time of collection, is ideal for fatty acid measurement (*see* **Note 5**).

3.1 Macrophage Cell Culture and Labeling

1. Change regular culture medium with labeling medium and begin cell treatment (e.g., Toll-like receptor agonists, cytokines, drugs) concurrently. Divide the warm labeling media into appropriate volumes (500 μL per well) and combine with desired treatments. Completely aspirate the old medium and add new medium to the well.

3.2 Culture Imaging and Sample Collection

1. 24–48 h after labeling and treatment, add 30 μL of calcein AM mix per well to the 24-well dish. Return dish to incubator and incubate for 10–15 min.
2. Perform imaging using the ImageXpress Micro XL imager. Set acquisition program to acquire 21 sites per well in a 7 × 6 checkboard, providing 21.83% coverage of the well (a total of 504 images per plate). The focus is set to "laser and image focus." The imager focuses on the plate and well bottom with a post-laser offset of 23.22 μm. Image-based autofocus is based on the range of ±50 μm with a maximum of seven Z-movements. Analyze acquired images by the object counting feature of the ImageXpress analysis software for cell count. The sum total of counts in the 21 images is multiplied out to represent the entire well (4.58× multiplier).
3. Prior to running the acquisition program, the imager should be centered on a test well to perform a test focus. Following the successful test focus, the exposure should be auto-adjusted to optimal. Then plate can be imaged and removed from imager to harvest cells.
4. Aspirate media and add 50 μL of 6 M guanidine-HCl solution to each well. Place the sample on a rocker or rock plate periodically for 5 min. Allow cells to disassociate.
5. One at a time for each well, add 100 μL of 3 M methanolic guanidine-HCl, immediately pipette up and down six times, and transfer the entire well to 2.5 mL vial in an aluminum base. Prepare a 96-well plate map to track sample location through running and data analysis (*see* **Note 6**).
6. If not derivatizing immediately, once all samples have been collected, apply mat to tubes and screw on cover. Place sealed tube rack in −80 °C freezer (*see* **Note** 7).

3.3 Derivatization of Sample

1. Just prior to derivatization, add the standard curve to the first column on the plate. Create a six-point standard curve using 1, 5, 10, 20, 40, and 80 μL of standard curve mix with a Hamilton syringe. Use the remaining two wells in the first column as blanks (*see* **Note 8**).
2. The stage of the Multiprobe II liquid handler is fitted with a set of three hexane wash bottles, a waste container, the acid methanolysis mix reservoir bottle, and the sample rack containing samples and standards (*see* Fig. 1 for sample-stage configuration) (*see* **Note 9**).
3. There are three basic steps to the derivatization program.
   (a) First, wash the fixed tips/tubing by aspirating 1 mL of hexane from each of the three wash bottles in fixed order and dispensing to waste.

   (b) Second, transfer 720 μL of acid methanolysis mix from the reservoir bottle to each sample in the 96-well sample rack. Dump the first aliquot within each tip to waste in order to improve consistency (*see* **Note 10**).
   (c) Third, wash the fixed tips/tubing by aspirating 1 mL of hexane from each of the three wash bottles in fixed order and dispensing to waste.
4. Recover the sample rack with the PTFE mat and secure the cover plate with screws. Place the plate in a 45 °C incubator or oven overnight (10–14 h).

3.4 Extraction/Mass Spectrometry Preparation

1. Remove the sample rack from the 45 °C incubator and allow to cool for 15–20 min before removing cover plate.
2. Return the sample rack to the Multiprobe II liquid handler for the addition of solvents needed for the hexane extraction.
3. Prepare the stage of the Multiprobe II liquid handler with a set of three hexane wash bottles, a waste container, a reservoir bottle of hexane, and a reservoir bottle of 40 mM NaCl.
4. There are four basic steps to the solvent addition program:
   (a) First, wash the fixed tips/tubing by aspirating 1 mL of hexane from each of the three wash bottles in fixed order and dispensing to waste.
   (b) Second, transfer 0.88 mL of hexane from the reservoir bottle to each tube in the rack.
   (c) Third, transfer 0.88 mL of 40 mM NaCl from the reservoir bottle to each tube in the rack.
   (d) Fourth, wash the fixed tips/tubing by aspirating 1 mL of hexane from each of the three wash bottles in fixed order and dispensing to waste.
5. Following the solvent addition, the mat can be replaced on the rack and the cover secured with screws once again. One should observe an interface forming in the wells between an upper phase of hexane and a lower phase of methanol/NaCl solution.
6. Carefully invert the rack 5–6 times and spin it in the centrifuge at 500 × *g* for 5 min in a tabletop centrifuge with microplate adaptor. For a single rack, a centrifuge balance will have to be prepared with a second rack containing an identical number of tubes and an equivalent mass of solvent.
7. Following centrifugation, the interface should be well defined and the upper phase should be clear. Carefully remove the cover and mat and return the rack to the Multiprobe II liquid handler for extraction.
8. The stage of the Multiprobe II liquid handler should be prepared with a set of three hexane wash bottles, a waste

container, and the sample running rack with 1.5 mL tubes (*see* **Note 11**).

9. There are two steps in the extraction program:
   (a) First, wash pins by transferring 1 mL of hexane between the three successive hexane wash bottles and the waste, followed by a transfer of 0.7 mL of upper phase from the sample rack (2.5 mL tubes) to the running rack (1.5 mL tubes). With an eight-pin liquid handler, this cycle allows the instrument extract one column, wash all tips, and proceed to extract the second column.
   (b) Second, wash the fixed tips/tubing by aspirating 1 mL of hexane from each of the three wash bottles in fixed order and dispensing to waste.
10. The running rack can then be covered by a pierceable mat and placed in the refrigerated drawer on the GC–MS.

3.5 Mass Spec Run

1. Configure the PAL autosampler with a 100 µL syringe for large-volume injections and two large-volume wash bottles with LC-grade hexane.
2. The settings for the GC/MS program are as follows:
   (a) PAL autosampler: Injection volume is 30 µL. Three pre-injection washes of 50 µL from wash bottle 2. The filling speed is 20 µL/s, and there is one filling stroke of 10 µL. For solvent venting, the injection speed is 3.5 µL/s, with a 400 ms pre- and postinjection delay (*see* **Note 12**).
   (b) Multimode inlet: Set the inlet to "Solvent Vent" mode. Initially, set the inlet temperature to 90 °C for 0.4 min. Then, ramp it at 600 °C/min to 325 °C and hold throughout the run. Keep the vent pressure at 5 psi until 0.4 min has passed (during solvent vent). Set the vent flow to 100 mL/min and the septum purge flow to 3 mL/min. Set the purge flow to split vent at 80 mL/min at 1.2 min. Turn the gas saver on after 3 min and set to 15 mL/min (*see* **Note 13**).
   (c) Oven: Set the oven to an initial temperature of 55 °C with a hold time of 1.4 min (this is the solvent focusing period). Then ramp it at 90 °C/min up to 120 °C. A second ramp of 15 °C/min brings the temperature to 205 °C with a 2-min hold. A third ramp of 55 °C/min brings the temperature to 250 °C with a 1.5-min hold. A final ramp of 55 °C/min to 260 °C with a 0.3-min hold clears the column at the end of the run. The entire run time should be approximately 13.5 min (*see* **Note 14**).
   (d) Column: Set the 10 m DB-WAX column to a "Ramped Flow" with an initial flow rate of 1.4 mL/min for 9.8 min.

Then set 2.4 mL/min for the remainder of the run (this is the backflush period). Then, set 20 m DB-WAX to a "Constant Flow" run of 1.8 mL/min (*see* **Note 15**).

(e) Transfer line and source: Set the transfer line to 250 °C, the MS Quad to 150 °C, and the MS source to 230 °C. Tune the MS for use of the extractor source by the MassHunter automated ATUNE and ETUNE programs prior to each major run.

(f) Mass spec acquisition: Set the MS to SIM mode. Ion is acquired with a 20 ms dwell time and ranges are acquired at one-AMU intervals. There are five scheduled acquisitions as follows (*see* **Note 16**):

(i) 14-carbon (m/z): 242.3–257.3

(ii) 15-carbon (m/z): 256.3–272.3

(iii) 16-carbon (m/z): 268.3–283.3

(iv) 17-carbon (m/z): 282.3–302.3

(v) 18, 19-carbon (m/z): 294.3–317.3

The timing of these acquisition windows should be determined empirically with test runs. The beginning of data acquisition should be no earlier than 3–4 min after the start of the run (this is the solvent delay) and should be moved forward to 30–60 s prior to the retention time of the first FAME in order to save filament life. Once all the other settings and the backflush timing have been established for the run, a test sample should be performed, acquiring the four most abundant FAME ions (57.1, 68.1, 74.1, and 79.1) without scheduling. This will establish the retention time of each FAME and allow the user to schedule ion collection between these features (*see* Table 1). For example, the first transition to collecting the 15-carbon ions should occur at a time halfway between the retention time of 14:0 and 15:0. After establishing the scheduling, the program can be tested on a labeled experimental sample. The ion ranges listed above accommodate the highest numbers of labels possible on each FAME. The number of ions should be adjusted down in order to avoid collecting unnecessary ions; this modification is key to maximizing the number of points acquired over each feature.

3. The above program has an optimal sample-to-sample injection time of ~15 min. This allows an entire 96-well rack of samples to be run in approximately 24 h. Chromatograms should be spot-checked throughout the run to ensure consistent, well-resolved peaks (*see* Fig. 2 for an example of a chromatogram).

Table 1
List of common chromatographic features within the long-chain fatty acid GC–MS run for macrophages

Notation	Retention time (min)	Systematic name	Unlabeled ion	Expected labeled ions
14:0	7.35	Methyl tetradecanoate	242.3	242.3–257.3
15:0	8.25	Methyl pentadecanoate	256.3	256.3–272.3
16:0	9.3	Methyl hexadecanoate	270.3	270.3–283.3
16:1	9.45	Methyl 9-hexadeeenoate	268.3	268.3–281.3
17:0	10.22	Methyl heptadecanoate	284.3	284.3–302.3
18:0	11.25	Methyl octadecanoate	298.3	298.3–317.3
18:1	11.45	Methyl 9-octadecanoate	296.3	296.3–315.3
	11.48	Methyl 11-octadecenoate	296.3	296.3–315.3
18:2	11.9	Methyl 9,12-octadecadienoate	294.3	294.3
19:0	12.1	Methyl nonadecanoate	312.3	312.3

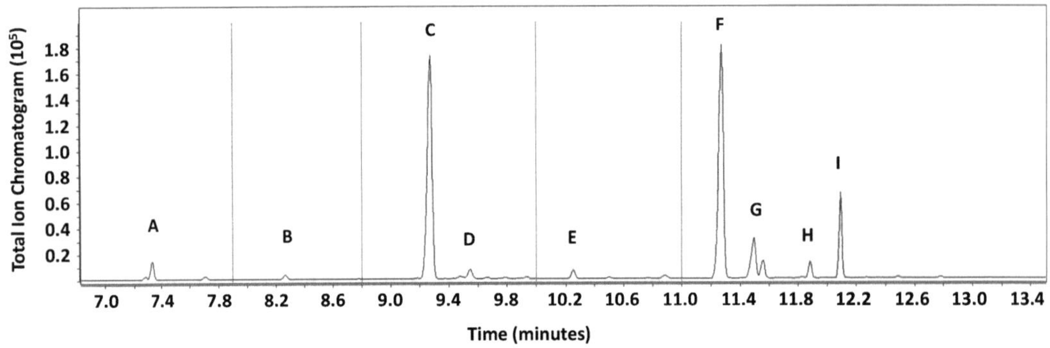

Fig. 2 Example chromatogram from a sample of 300,000 bone marrow-derived macrophages. Measured fatty acids include (**a**) methyl tetradecanoate (14:0), (**b**) methyl pentadecanoate (15:0), (**c**) methyl hexadecanoate (16:0), (**d**) methyl 9-hexadecenoate (16:1), (**e**) methyl heptadecanoate (17:0), (**f**) methyl octadecenoate (18:0), (**g**) methyl 9-octadecanoate, methyl 11-octadecenoate (18:1), (**h**) methyl 9,12-octadecadienoate, (**i**) methyl nonadecanoate (19:0)

3.6 Data Analysis

1. MassHunter Qualitative Analysis is ideal for the first steps in analyzing data. All chromatograms can be overlaid to establish the average retention time of each FAME.
2. These retention times can be applied to an integration program within the MassHunter Quantitative Analysis (*see* Table 1 for the listing of chromatographic features).
3. The "Agile" integrator within the MassHunter efficiently integrates sufficiently large peaks that are not near adjacent peaks. The "General" integrator allows the user to set the parameters

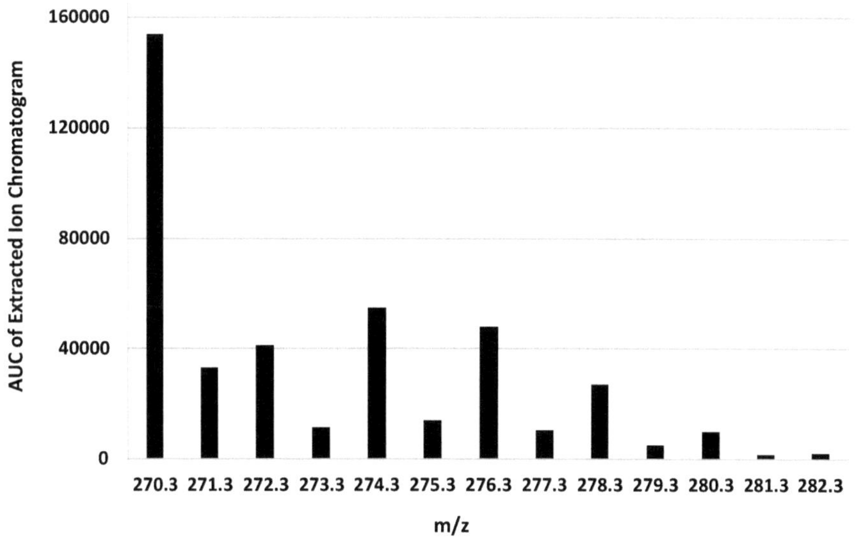

Fig. 3 Isotopic labeling pattern of palmitate (16:0) represented by the area under the curve (AUC) values for each ion in the labeled distribution

that can allow more selective integration of peaks (*see* **Note 16**).

4. Integrations should be reviewed in the "compound-at-a-glance" feature. Manual integrations of poorly integrated peaks can be performed during this review.
5. The area under the curve (AUC) for each ion peak can be exported from MassHunter into Excel for analysis. See examples of AUC values for palmitate (16:0) (*see* Fig. 3).
6. The labeled distribution can be analyzed by modeling with ISA or MIDA to measure the amount of synthesis in each fatty acid.
7. For quantification purposes, all ions of a given FAME should be summed and normalized against the internal standard. This should be carried out both for the unknown and standard curve values.
8. These normalized values can then be quantified against the normalized standard curve and the resulting lipid amounts can be divided by cell numbers for reporting purposes.

4 Notes

1. This assay directly measures the amount of synthesized fatty acid synthesis in cells. Thus, it is important to optimize the amount of FBS in cultures to optimize the influence of the treatments or genetics on synthesis in your model. In general,

increasing the amount of FBS (thus increasing the amount of lipids) in cultures decreases the amount of fatty acid synthesis the cells will perform. Conversely, decreasing the FBS amount in the medium increases fatty acid synthesis.

2. Generally, this mixture is scaled to 100–300 reactions depending on the actual number of samples in the experiment and the volume in the reservoir bottle.

3. For each piece of custom hardware, a profile needs to be developed that defines the number of tubes or well in a piece of hardware, the footprint of the hardware (such that the liquid handler arm can travel safely around it), and how deep the pins should enter the well to aspirate or dispense liquid into them. In general, the dispensing height for all hardware should be just below the mouth of the tube or bottle (to avoid dipping tips into samples or waste). The aspiration height for all reservoir or wash bottles should be close to the bottom of the bottle (to avoid large dead volumes). If the hardware is rearranged on the stage, it is important to reverify the appropriate locations before running a program.

4. Care must be taken in the preparation of macrophage media since macrophages are sensitive to pyrogens, and their activation will have significant effects on lipid metabolism.

5. Practice platings should be performed prior to the experiment to ensure confluency and maximize consistency among the wells. Although the assay is normalized to the image-based cell count, we have observed that significant variation in the ratio between cells and media will influence the lipid metabolic state of the cells, and can confound results.

6. In general, the first column should be kept empty for standard curve and blanks. Since the liquid handler works vertically across the columns, samples should generally be added one column at a time, rather than across a row.

7. Cells collected in the guanidine solution are stable for several weeks at −80 °C. Sample racks should be placed into a Ziploc bag prior to freezing to avoid excess frost and moisture on the mat, tubes, and aluminum base.

8. For greatest consistency in standard curve, the excess hexane in the standard curve tubes should be evaporated off over a nitrogen or argon flow.

9. Each liquid transfer step is governed by a performance profile that sets the operating parameters of the syringe pumps (speed and magnitude of movement). These settings need to be optimized for each solvent transferred. The accuracy and consistency of liquid transfer should be tested and adjusted by gravimetric analysis. The profile for the acid methanolysis mix

is most important because all quantitative measurements are being normalized by the internal standard containing within this solution.

10. Prior to performing the extraction step, the aspiration height of the tips should be checked in an empty tube to verify that the height of the tip is safely above the interface. As an added precaution, the aspiration speed for this step can be decreased to reduce the chance of aspirating the interface layer.
11. The accuracy/consistency of injections within this assay is not very important because the internal standard corrects for variations in the entire sample prep process, including the sample injection. If greater accuracy and consistency are required, an additional sample wash and additional sample strokes can be performed.
12. Following each sample run, the instrument needs to cool down before the next sample can be injected. While the oven is force air heated/cooled, the conventional split/splitless inlet is cooled by diffusion (thus it is slowest to cool). The multimode inlet has the option to receive cooling from a liquid nitrogen tank vapor valve, or a zero-grade air tank/gas generator. Force cooling of the inlet will reduce injection-to-injection times by a minute or two per sample and is recommended for very large experiments.
13. The ability of a GC oven to ramp the temperature will vary over the temperature range. In general, ovens can ramp faster within lower temperature ranges than higher ranges. It is important that the programmed ramps be within the capability of the heater. If the ramp rate is beyond the ramp capacity of the instrument, the user will observe inconsistencies in chromatography from injection to injection depending on how fast the ramp rate actually was in that particular sample injection.
14. The most recent version of the MassHunter instrument control software (Agilent) includes a "Backflush Wizard" that can be used to optimize the ideal backflush time. You should begin the "Backflush Wizard" with the above GC–MS program without the backflush and with a 250 °C hold time of 20 min (rather than 1.5 min). The Wizard should be used to analyze the GLC-96 mix. After the first run, the wizard will give the user the opportunity to select the last peak of interest (the 19:0 peak). The Wizard will then set the backflush time and run the same sample with the new program. When the user has verified that there are no more peaks detected after the last peak of interest, the program can be shortened accordingly. This is critical because biological samples will contain very-long-chain fatty acids, as well as cholesterol, which have extremely long retention times and will foul the column over successive injections.

15. The acquisition of odd-chain FAMEs is optional. These may be taken up from the FBS or the product of synthesis within the cell. Some subset of these odd-chain species may be branched-chain fatty acids. They are generally low in abundance and their isotopic labeling (in cases in which they are labeled) cannot be modeled. However, changes in their relative abundance may be informative.
16. The monounsaturated FAMEs (16:1 and 18:1) form doublets in most biological samples due to different double-bond locations. Depending on the condition of the column and the oven program, these doublets will have varying degrees of resolution. The peaks can be integrated individually or together (as generic 16:1 and 18:1). If they are integrated individually, the ideal quantification would use a supplement (the alternative monounsaturated FAMEs) into the standard mix. For example, the GLC20a mix contains methyl oleate (18:1 (n-9)) and would need methyl vaccenate (18:1 (n-7)) supplemented into the standard.

References

1. Bottcher CJ, Keppler JG, Ter Haarromey-Wachter CC, Boelsma-Van Houte E, Van Gent CM (1958) Analysis of lipids of the arterial wall. Lancet 2:1207–1209. https://doi.org/10.1016/s0140-6736(58)92358-4
2. Doss M, Oette K (1965) Rapid method for the preparation of fatty acid methyl esters for gas chromatographic analysis. Z Klin Chem Klin Biochem 3:125–129
3. Eder K (1995) Gas chromatographic analysis of fatty acid methyl esters. J Chromatogr B Biomed Appl 671:113–131. https://doi.org/10.1016/0378-4347(95)00142-6
4. Kharroubi AT, Masterson TM, Aldaghlas TA, Kennedy KA, Kelleher JK (1992) Isotopomer spectral analysis of triglyceride fatty acid synthesis in 3T3-L1 cells. Am J Phys 263:E667–E675. https://doi.org/10.1152/ajpendo.1992.263.4.E667
5. Hellerstein MK, Neese RA (1992) Mass isotopomer distribution analysis: a technique for measuring biosynthesis and turnover of polymers. Am J Phys 263:E988–E1001. https://doi.org/10.1152/ajpendo.1992.263.5.E988

Chapter 5

Proteomics Network Analysis of Polarized Macrophages

Jayanta K. Chakrabarty, Abu Hena Mostafa Kamal, A. D. A. Shahinuzzaman, and Saiful M. Chowdhury

Abstract

Macrophages play a critical role in innate immunity through Toll-like receptor (TLR) signaling. Lipopolysaccharides (LPS) are a ligand of microbial origin that can trigger cell signaling in macrophages through TLRs and production of pro-inflammatory cytokines. Statin, a hypercholesterolemia drug, on the contrary, can reduce inflammatory cytokine production, and inflammation at large. Discovery-based quantitative proteomics is a useful method for unraveling complex protein networks and inter-protein interactions. Here, we describe protocols for studying the inflammatory proteomics network in RAW 264.7 cells (a model murine macrophage cell line) with the singular or sequential treatment of LPS and statin. We provide detailed protocols, including a quantitative proteomic analysis by mass spectrometry data, a protein network analysis by bioinformatics, and a validation of target through biochemical methods (e.g., immunocytochemistry, immunoblotting, gene silencing, and real-time PCR).

Key words Macrophages, Toll-like receptors, Quantitative proteomics, Inflammatory proteomic network, Protein-protein interactions, Immunocytochemistry, Immunoblotting, Statin, Lipopolysaccharide, Mass spectrometry

1 Introduction

Macrophages are immune cells, responsible for detecting, engulfing, and destroying pathogens [1, 2]. They play vital roles in innate immunity by triggering host defense mechanisms. The Toll-like receptors (TLRs) are pattern recognition receptor proteins (PRRs), usually expressed in macrophages. After being stimulated by microbial ligands, they play a critical role in inducing several kinds of pro-inflammatory cytokines [3]. In this protocol, lipopolysaccharide (LPS) is used as a bacterial ligand to activate RAW 264.7 cells [4]. Statins, as known as 3-hydroxy-3-methylglutaryl-CoA (HMG-CoA) reductase inhibitors, are a commonly prescribed medicine for hyperlipidemic patients [5]. Recently, it has been reported that statins reduce the risk associated with sepsis by downregulating TLR-4 and NF-κB phosphorylation [6–12], and

Suresh Mishra (ed.), *Immunometabolism: Methods and Protocols*, Methods in Molecular Biology, vol. 2184,
https://doi.org/10.1007/978-1-0716-0802-9_5, © Springer Science+Business Media, LLC, part of Springer Nature 2020

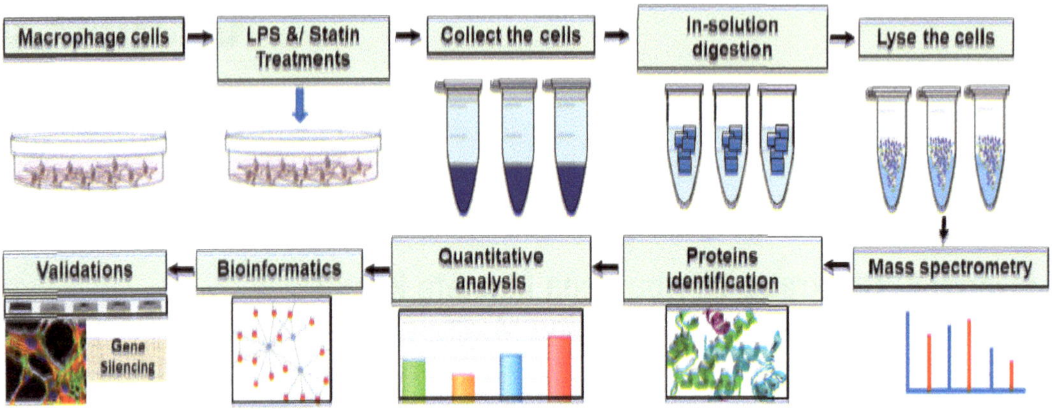

Fig. 1 A schematic diagram showing the workflow for quantitative proteomics analysis of polarized macrophages

consequently causing the reduction of transcription and lymphocyte cytokine production and overall reduced inflammation [12, 13].

Discovery-based quantitative proteomics, along with liquid chromatography and tandem mass spectrometry (LC-MS/MS), is a powerful method to investigate cell signaling pathways and their downstream consequences [11]. As an example, here we describe protocols for studying the inflammatory proteomics network in RAW 264.7 cells in response to pro-inflammatory agents (LPS) and anti-inflammatory agents (statin) [11, 12]. A flowchart is shown in Fig. 1.

We will describe the preparation of the protein sample, the mass spectrometric analysis, the bioinformatic analysis of proteomic data, and a candidate protein validation using molecular biology and proteomics techniques. In addition, we will describe the network analysis conducted by the Ingenuity Pathway Analysis (IPA) software tool.

2 Materials

Prepare all solutions with 18 Milli-Q ultrapure water at room temperature. Use reagents that are analytical grade.

2.1 Cell Culture and Treatments

1. RAW 264.7 cells.
2. Cell culture media: Dulbecco's modified Eagle medium (DMEM). Add 10% fetal bovine serum (FBS) and 1% penicillin-streptomycin (PS) to the DMEM medium to maintain the growth of the cultured cells, and to prevent the microbial/fungal growth.

3. Lipopolysaccharide: Prepare the stock solution (5 mg/mL) using endotoxin-free water.
4. Simvastatin: Prepare 10 μM simvastatin solution using dimethyl sulfoxide (DMSO).
5. Phosphate buffer saline (PBS): For making 1 L of 10× PBS, take 800 mL of distilled water. Add 80 g of NaCl, 2 g of KCl,1.44 g of Na_2HPO_4, and 0.24 g of KH_2PO_4 and stir properly to dissolve completely. Then adjust the pH to 7.4 with HCl. Finally, add distilled water up to a total volume of 1 L.

2.2 Protein Extraction, Purification, and Digestion

1. RIPA buffer cocktail: Radioimmunoprecipitation assay buffer (RIPA buffer) from Thermo Fisher Scientific (IL, USA) mixed with protease inhibitor at 10:1 ratio as a cell lysis buffer. Take one tablet of protease inhibitor and dissolve in 1 mL ultrapure water. For the cocktail preparation, mix RIPA buffer with a protease inhibitor in 10:1 ratio. For 1 mL of cocktail preparation, mix 900 μL of RIPA buffer with 100 μL of protease inhibitor.
2. BCA assay: BCA Protein Assay Kit.
3. Proteolytic enzyme: Sequencing-grade modified trypsin.
4. C18 Tips, 100 μL for desalting the peptides.
5. Analytical grade methanol.
6. Chloroform.
7. Iodoacetamide (IAA).
8. Reducing agent dithiothreitol (DTT).

2.3 Sample Analysis: Nano-LC-MS/MS

1. Instrument: LTQ Velos pro mass spectrometry coupled with a UHPLC (Ultimate 3000, Dionex, USA). High-resolution instrument is preferred if available.
2. A nano viper analytical C18 column: Acclaim PepMap, 75 μm, 150 mm, 2 μm, 100 Å.
3. Solvent A: 0.1% Formic acid in H_2O.
4. Solvent B: 95% Acetonitrile, 5% water, and 0.1% formic acid.

2.4 SDS Polyacrylamide Gel

1. Resolving gel buffer: 1.5 M Tris–HCl, pH 8.8. Weigh 181.7 g Tris–HCl and transfer to the 1 L graduated cylinder. Add 500 mL of water first to dissolve and make up to 1 L with water. Adjust the pH, filter properly, and store at 4 °C.
2. Stacking gel buffer: 0.5 M Tris–HCl, pH 6.8. Weigh 60.6 g Tris–HCl and prepare a 1 L solution as in the previous step.
3. 30% Acrylamide/Bis solution (29.2:0.8).
4. 10% Ammonium persulfate (APS).

5. Sodium dodecyl sulfate: Make 10% solution in ultrapure water.
6. TEMED: N,N,N,N′-tetramethyl-ethylenediamine.
7. Protein ladder: Blue pre-stained protein standards (10–250 kD).

2.5 Immunoblotting

1. Filter paper: 3 M Filter papers.
2. Nitrocellulose membrane (pore size 0.45 μm).
3. Western blot transfer buffer: 10× Transfer buffer (Tris base: 30 g glycine: 144 g ddH_2O: 1 L).
4. Tris-buffered saline (TBS; 10×): 1.5 M NaCl, 0.1 M Tris–HCl, pH 7.4.
5. TBST buffer: 0.05% Tween-20 in TBS buffer.
6. Blocking solution: 5% Skim milk or 5% bovine serum albumin (BSA) in TBST buffer. Store at 4 °C.
7. Primary antibody: Specific to protein of interest.
8. Secondary antibody: Horseradish peroxidase or fluor probe-conjugated secondary antibody.
9. Developing solution: Enhanced chemiluminescent substrate.

2.6 Immunocytochemistry

1. Coverslips.
2. 1 M HCl.
3. Ethanol.
4. Methanol.
5. PBS buffer (pH 7.4).
6. PBST buffer: PBS buffer containing 0.05% Tween-20.
7. Blocking buffer: 1% BSA, 22.52 mg/mL glycine in PBST (0.1% Tween-20).
8. Mounting medium.
9. Primary antibody: Anti-plectin.
10. Secondary antibody: Goat anti-rabbit Alexa Flour 488.
11. Filamentous actin strainer: Phalloidin-iFlour 488 Cytopainter.
12. Confocal microscope.

2.7 Transfection with Plectin siRNA

1. Plectin siRNA.
2. Scramble siRNA.
3. Transfection reagent.

2.8 RNA Extraction and Real-Time PCR (qRT-PCR)

1. Reagents for qRT-PCR: Trizol reagent.
2. Chloroform (without amylene).
3. Isopropanol, ethanol.

4. RNase-free water.
5. Instruments: PCR thermal cycler and real-time detection system.

3 Methods

3.1 Cell Culture and Treatments

3.1.1 RAW264.7 Macrophages (ATCC® TIB-71™)

1. Culture RAW264.7 cells as per the ATCC protocols (https://www.atcc.org/products/all/TIB-71.aspx). Warm growth medium at 37 °C for 30 min.
2. Warm cell vials (containing frozen cells) in a 37 °C water bath for 1 min. Transfer thawed cells to the warm growth medium into a 15 mL falcon tube and mix gently. Centrifuge at 400 × *g* for 3 min, remove the medium, and suspend the cell pellet into 1 mL of growth medium.
3. Disperse 10 mL of the growth medium into the petri dish, then suspend cells, and mix into the growth medium. Mix the cells properly in the growth medium, and then incubate it at 37 °C with a 5% CO_2 in the incubator.
4. After washing with PBS, prepare a subculture of the cells by scraping. After centrifugation at 400 × *g* for 3 min, remove the medium, and suspend the cell pellets into 1 mL of growth medium.
5. Disperse 10 mL of growth medium into each petri dish, and then mix 200 μL of suspended cells into the growth medium. Mix the cells properly within the growth medium, and then incubate it at 37 °C with a 5% CO_2.

3.1.2 Bone Marrow-Derived Macrophage (BMDM) Culture

1. Culture the BMDMs collected from mouse femurs and tibias in L-cell-conditioned IMDM (Iscove's modified Dulbecco's medium: containing 10% FBS, 1% nonessential amino acid, and 1% penicillin-streptomycin) for 6 days to differentiate into macrophages [14, 15].
2. Subculture the BMDMs into 6-well plate to prepare for immunocytochemistry as well as immunoblotting.

3.1.3 Simvastatin and Lipopolysaccharide (LPS) Treatments

1. Treat the cells with simvastatin at 10 μM of final concentration for 24 h and stimulate with LPS at 1 μg/mL for 1 h.
2. After the treatments, rinse cells with phosphate buffer saline (PBS) three times, and then harvest the cells by scraping.
3. Centrifuge at 400 × *g* for 3 min, and remove the medium. Wash the cell pellet in PBS and process it for protein extraction.

Table 1
Relative ratio/amount of materials used for methanol-chloroform precipitation method with different sample volumes

Sample	150 μL	100 μL	75 μL
MeOH	600	400	300
Chloroform	150	100	75
Water	450	300	150
MeOH	450	300	150

3.2 Protein Extraction, Purification, and Digestion

3.2.1 Protein Extraction

1. Add 350 μL of RIPA buffer containing a protease inhibitor to the cell pellets and then disperse properly.
2. Vortex the cells for 30 s and incubate in ice for 30 min.
3. After 15 min, vortex again and sonicate for 30 min.
4. Repeat vortexing and incubate in ice for another 30 min.
5. Finally, centrifuge the cells at 12,000 × *g* at 4 °C for 30 min and carefully transfer the supernatant into a new Eppendorf tube.
6. Quantify the isolated proteins using BCA protein assay.

3.2.2 Methanol-Chloroform Purification

Methanol-chloroform purification is a widely used method for precipitation of proteins. The steps are given below:

1. Take 75–150 μg proteins and proceed with methanol:chloroform purification. The concentration of the sample should be maintained at 1 μg/μL. Refer to Table 1 for details.
2. Add 600 μL of methanol to the sample and then vortex for 30 s.
3. Then, add 150 μL of chloroform and vortex for 30 s.
4. Add 450 μL of water and vortex for 30 s.
5. Centrifuge at 12,000 × *g* for 10 min at room temperature to separate the two phases.
6. Discard the upper phase, keeping the white precipitate disk.
7. Add 450 μL of methanol slowly (carefully not to break the disk) and mix gently (*see* **Note 1**).
8. Centrifuge the sample at 12,000 × *g* for 10 min at room temperature.
9. Discard the supernatant and dry the pellet for 5 min in speed vacuum.
10. Reconstitute the dried pellet in 50 mM ammonium bicarbonate for in-solution digestion.

3.2.3 Protein Digestion and Desalting

1. Add 10 μL of 10 mM dithiothreitol (DTT) and incubate at 56 °C for 45 min in the dark.
2. Add 10 μL of 55 mM iodoacetamide (IAA) and incubate at 37 °C for 30 min in the dark.
3. Add 50 mM NH_4HCO_3 to increase volume up to 1 mL.
4. Add 10 μL of trypsin solution (trypsin:protein = 1:50 w/w ratio) and incubate at 37 °C for 12–16 h (*see* **Note 2**).
5. Add 10 μL of 0.1% formic acid to quench the activity of trypsin and dry the sample afterwards in a speed vacuum.
6. Finally, reconstitute the samples with 100 μL 0.1% formic acid for desalting.
7. Desalt the samples by C18 column (zip tip) according to the manufacturer's instructions.
8. After drying the peptides, reconstitute the peptides with 0.1% formic acid or trifluoroacetic acid.

3.3 Sample Analysis (Nano-LC-MS/MS)

1. Analyze the peptides using the ESI-LTQ velos Pro mass spectrometer combined with a Dionex Ultimate 3000 UHPLC (Thermo Fisher Scientific) or any similar or better platform.
2. Perform a reverse-phase chromatographic separation using the nano-viper Acclaim PepMap C18 column with the following dimensions: 15 cm length, an inner diameter 75 μm, particle size 2 μm, and pore size 100 Å.
3. Perform a multistep binary gradient system with a run time of 90 min for the separation of peptides.
4. Inject samples in a partial mode with a set flow rate of 300 nL/min.
5. Use Nano-ESI as an ionization source, and operate the instrument in the positive ion mode with capillary spray voltage set at 2.50 kV and source temperature at 275 °C.
6. Operate the mass spectrometer in normal scan mode, and perform full MS scans with 1 micro scan at a resolution of 3000 and a mass range of m/z 350–1500.
7. MS and MS/MS acquisition should be regulated by the data-dependent acquisition (DDA) mode. The three precursor ions with the most intense signal in a full MS scan need to be consecutively isolated and fragmented to acquire their corresponding MS2 scans.
8. Consecutive ions should be separated in every 30 ms (dynamic exclusion time), and mass spectrum data acquisition should be set for 90 min. MS/MS data processing should be performed by Xcalibur™ software or any other similar software.

9. In CID fragmentation mode, the normalized collision energy should be set to 35%, along with an isolation width of 1.5 Da, activation Q of 0.25, and activation time of 10 ms. Fragment ion spectra produced via high-energy collision-induced dissociation (CID) should be acquired in the linear ion trap with a resolution of 0.05 FWHM (full width half maximum) with an Ultra Zoom-Scan between *m/z* 50 and 2000.

3.4 Data Analysis and Quantification

1. Proteins can be identified through the Proteome Discoverer software using the UniProt mouse (*Mus musculus*) protein sequence database (*see* **Note 3**).
2. SEQUEST search parameters for all normal peptides should be performed selecting the carbamidomethylation of cysteine as the static modification, and the oxidation of methionine as the dynamic modification.
3. Enzyme trypsin should be indicated as the protease, with at least two missed cleavages. Peptide and fragment mass tolerance were set at ±1.6 and 0.6 Da, respectively, and precursor mass range of 350–5000 Da should be used, with peptide charges set excluding +1.
4. SEQUEST results should be filtered with the target PSM validator to improve the sensitivity and accuracy of the peptide identification.
5. Decoy searches must be performed concurrently to account for a false discovery. Target false discovery rates (FDR) for peptide identification of all searches should be set at 1% with at least two peptides per proteins, and the peptides should be strictly filtered by ΔCn (<0.01), *X*corr (≥1.5), and peptide spectral matches (PSMs) with high confidence (with a *q*-value of ≤0.05).
6. For label-free quantitative proteomics, the total spectrum count of identified proteins should be used. However, additional criteria should be applied to increase confidence that the PSMs must be present in all three biological replicate samples.
7. The normalization of identified PSMs among LC-MS/MS runs should be done dividing individual PSMs of proteins with total PSMs. An average of the % PSM counts should be utilized for calculating fold changes for different treatments.
8. For the contrasting relative intensities of proteins among the control, LPS, statin-LPS, and statin groups, the samples should be evaluated using cumulative confident normalized PSM values.

3.5 Data Visualization and Bioinformatics

1. Functionally categorize protein-coding genes based on gene ontology classification such as biological processes, molecular activity, and cellular components by the PANTHER gene classification system [16].

2. Protein abundance can be visualized as a heat map. Apply hierarchical clustering with complete linkage based on Pearson distance metrics. Generate the heat map by MeV software (ver. 4.9; http://www.tm4.org/) [17]. Other similar software could be utilized.
3. The proteomic data set, which included UniProt identifiers and fold changes of total identified proteins, should be submitted into Ingenuity Pathway Analysis (IPA) for core analysis [18].
4. The core analysis in IPA provides the relationships among the proteins based on their biological mechanisms, functions, and pathways.
5. An analysis of upstream regulators might help to understand the regulation of mechanism with expression changes.
6. Integrating results from the IPA analysis enables the identification of the potential regulators, as well as their mechanistic interactions.
7. The matched proteins' encoding genes from the Ingenuity Knowledge Base will generate molecular networks according to biological and molecular functions. These include canonical pathways, upstream regulatory analyses, and disease-based functional networks, which will help to discover the list of biomarkers.
8. The IPA core analysis should be carried out with the settings of indirect and direct relationships between focused molecules based on experimentally observed data, and the mouse databases in the Ingenuity Knowledge Base should be considered as the data sources in those analyses [18].
9. Perform the right-tailed Fisher's exact test to determine the probability that the biological function and/or disease in question was related to the proteins.

3.6 Immunoblotting

3.6.1 SDS-PAGE

1. Prepare the resolving buffer through the mixing of 2.5 mL of a 1.5 M Tris buffer, 4 mL of an acrylamide mixture, and 3.4 mL of water in a 50 mL falcon tube. Add 100 μL of the 10% SDS, 50 μL of 10% ammonium persulfate, and 10 μL of TEMED, and pour the mixture into a 7.25 cm × 10 cm × 1.5 mm gel cassette to cast the resolving gel.
2. Prepare the stacking gel by mixing 1.25 mL of stacking buffer of a 0.5 M Tris buffer, 1 mL of the acrylamide mixture, and 2.65 mL water in a 50 mL falcon tube. 100 μL of SDS, 25 μL of ammonium persulfate, and 5 μL of TEMED should also be added. Pour this mixture on the top of the resolving gel and then place a 10-well gel comb on top carefully, without introducing air bubbles.

3. Carefully remove the binding clips and comb from the gel. Place the sandwich of the glass plates into the electrophoresis core. Then place the core assembly into the running tank. Add running buffer (1×) to the tank.
4. Mix the protein sample with the sample buffer (Laemmli buffer) and heat for 5–10 min at 95 °C. Load samples into the wells using a 200 μL pipette tip. Usually, 20 μL of the sample is loaded into each well.
5. Once electrodes are attached, proteins move toward the anode. Run gel at a 180 V voltage until the dye reaches the bottom of the gel.
6. Turn off the power, disconnect electrodes, and remove the gel sandwich from the tank. Wash the gel in 200 mL of ultrapure water with gentle shaking for 15 min.

3.6.2 Protein Transfer

1. Prepare enough transfer buffer to fill the transfer tank, plus an additional 200 mL to equilibrate the gel and membrane and filter paper.
2. Immerse the gel in transfer buffer for 10–30 min.
3. Soak 3 M filter paper and the membrane, equilibrated in the transfer buffer for 30 s.
4. Place a foam (fiber) pad on one side of the cassette.
5. Place a sponge, 3 M filter paper, gel, nitrocellulose membrane, 3 M filter paper, and sponge as a stack (*see* **Note 4**).
6. Place the assembled unit in an ice bucket in the tank and fill up with transfer buffer (*see* **Note 5**).
7. Connect the anode lead and cathode lead to their corresponding power outputs. Turn on the power supply and transfer protein for 90 min at 100 V.
8. After the transfer is complete, remove the cassette holder from the tank and carefully disassemble the transfer membrane.

3.6.3 Antibody Incubations

1. Wash the membrane with TBST buffer.
2. Block the membrane with skim milk in TBST buffer for 2 h at room temperature or overnight at 4 °C.
3. Wash the membrane with the TBST buffer and incubate with the primary antibody overnight at 4 °C.
4. Wash the membrane with the TBST buffer three individual times, 10 min each.
5. Incubate the membrane in HRP-conjugated secondary antibody for 2 h at room temperature. Wash with TBST three times, 10 min each.

3.6.4 Detection

1. Prepare the substrate working solution (ECL) by mixing equal parts of the peroxide solution and the luminol enhancer solution (1:1). Mix just enough substrate (1 mL) to cover the membrane.
2. Incubate the membrane in ECL solution for 30–60 s at room temperature and allow the ECL solution to flow over the membrane.
3. Place the membrane in the cassette and remove bubbles using an absorbent towel.
4. Working in a darkroom with a safelight, place the covered membrane in a film cassette with the protein side facing up. Place an X-ray film on top of the membrane and leave it exposed for 10–60 s. Adjust exposure time to achieve optimal results.
5. Scan the film with the X-ray machine, and analyze the image using ImageJ. Alternatively, the image could be acquired using a gel-imaging system.

3.7 Immunocytochemistry

1. Dip the coverslips in 1 M HCl overnight at room temperature.
2. Wash with water.
3. Wash with 100% ethanol.
4. Dry and sterilize in a fume hood under UV for 3–4 h.
5. Incubate the cells in 100% methanol (chilled at −20 °C) at RT for 5 min.
6. Wash cells three times with ice-cold PBS.
7. Incubate the samples for 10 min with PBST.
8. Wash the cells three times with PBST for 5 min each.
9. Incubate the cells with 1% BSA and 22.52 mg/mL glycine in PBST for 30 min.
10. Incubate the cells overnight at 4 °C with the diluted primary antibody which is prepared with 1% BSA in PBST.
11. Decant the solutions and then wash cells three times with PBST, 5 min each.
12. Incubate the cells with a secondary antibody solution and wash three times in PBST, 5 min each in the dark.
13. Mount coverslips with a drop of mounting medium.
14. Seal coverslip with nail polish to prevent drying and movement under a microscope.
15. Store in the dark at 4 °C.

3.8 Transfection with siRNA

3.8.1 siRNA Preparation

Prepare a stock solution of 20 μM from the lyophilized siRNA. Dilute this to 5 μM of working solution. Perform the transfection in 6-well plates.

1. Plate cells the day before the transfection, and perform the transfection only when the cell confluency is around 50–70%.
2. Change the media, and keep in the incubator for 30 min to equilibrate with the environment.
3. Add 120 μL of the medium in 5 μL of 5 μM siRNA and 20 μL of siTRAN, and incubate at RT for 15 min.
4. Add the mixture dropwise onto the media and mix gently.
5. Keep it in a CO_2 incubator for 5 h, and change the fresh media after 48 h.
6. The transfected cells are now ready for RNA extraction, and immunocytochemistry.

3.8.2 Actin Filamentous Staining

1. Dip coverslips in 1 M HCl overnight at room temperature.
2. Wash with water.
3. Wash with 100% ethanol.
4. Dry and sterilize in a fume hood under UV light.
5. Place sterilized coverslips in the 6-well clear-bottom plates. Grow the RAW 264.7 macrophage cells in a 6-well clear-bottom plate until they reach confluence (70–80%).
6. Aspirate the cell culture medium with care to avoid dislodging cells.
7. Wash twice in the PBS buffer (pH 7.4).
8. Incubate the cells with a 4% formaldehyde solution at room temperature for 10 min.
9. Wash the cells three times with PBS buffer.
10. Add Phalloidin Alexa 488 (1:1000 ratio) in 1% bovine serum albumin (BSA) in PBS (pH 7.4)-conjugated working solution. Incubate at room temperature for 20–30 min.
11. Rinse cells 2–3 times with PBS, 5 min per wash.
12. Add 500 μL of propidium iodide for 5 min at room temperature.
13. Rinse cells 2–3 times with PBS, 5 min per wash.
14. Mount coverslip with a drop of mounting medium.
15. Seal coverslip with nail polish to prevent drying and movement under a confocal microscope.
16. Store in the dark at 4 °C.
17. Observe the cells at excitation/emission at 493/517 nm, respectively.

3.9 Quantitative Real-Time PCR (qRT-PCR)

3.9.1 RNA Extraction

1. After washing with PBS (pH 7.4), place 1 mL of TRIzol reagent directly into the petri dish. Incubate at room temperature for 1 min. After pipetting the cells up and down several times (10–15 times), place the sample into 1.5 mL Eppendorf tube at 4 °C.
2. Incubate the cells at room temperature for 5 min to permit the complete dissociation of nucleoprotein complexes.
3. Add 200 μL chloroform (without amylene) and vortex well.
4. Centrifuge at 12,000 × *g* for 2 min at 4 °C. As a consequence, the mixture will separate into a lower colorless, white interphase, and a colorless upper aqueous phase.
5. Transfer the upper phase into a fresh tube without disturbing the interphase.
6. Add 200 μL isopropanol and vortex for 30 s.
7. Incubate the solution at room temperature for 10 min.
8. Centrifuge the solution at 12,000 × *g* for 15 min at 4 °C.
9. Remove supernatant completely.
10. Wash pellet with 70% ethanol with RNase-free water.
11. Centrifuge at 12,000 × *g* for 1 min at 4 °C.
12. Dry pellet and dissolve in 20–30 μL RNase-free water.
13. Measure the RNA concentration and purity using NanoDrop.

3.9.2 cDNA Synthesis

1. Perform first-strand cDNA synthesis using a cDNA synthesis kit according to the manufacturer's protocols.
2. Thaw all components on ice (except the enzyme), mix thoroughly, and centrifuge before use.
3. Prepare 20 μL of the reaction mixture with the RNA sample (1 μg/μL RNA, 14 μL RNase-free water, 4 μL 5× cDNA synthesis mix, and 1 μL reverse transcriptase).
4. Gently vortex the mixture and place in the thermocycler as follows: 1 cycle: 22 °C, 5 min; 1 cycle: 42 °C, 30 min; 1 cycle: 85 °C, 5 min; 4 °C hold.
5. After the completion of the cDNA synthesis, use 1/10th of the first-strand reaction for qRT-PCR amplification.

3.9.3 qRT-PCR Analysis

1. Thaw the SYBR green supermix and other frozen reaction components according to the manufacturer's protocol.
2. Prepare 10 μL of the reaction mixture along with the cDNA sample as follows: 1× SYBR green supermix, 250–500 nM each of forward and reverse primers, 100 ng cDNA, and variable RNase-free water.

3. Run a real-time PCR system as follows: polymerase activation and DNA maturation (30 s at 95 °C), denaturation (5 s at 95 °C), annealing (30 s at 60 °C), 40 cycles, and melt curve analysis (65–95 °C increment s/step).

4 Notes

1. Add methanol very gently at 100 μL, four times, and mix gently by hand to end this step.
2. Mix 20 μg trypsin with 200 μL of the trypsin buffer (100 ng/μL).
3. The reviewed protein sequences of the mouse can be downloaded from the UniProt protein database (www.uniprot.org).
4. When the gel is placed, the size marker should be placed on the left side. The cassette holder should be closed and placed in the transfer tank so that the gel side of the cassette holder will be facing the cathode (−) and the membrane side will be facing the anode (+).
5. The black cathode lead (−) should be inserted into the cathode jack and the red anode lead (+) into the anode jack on the transfer unit.

Acknowledgments

The authors would like to acknowledge funding support from grant UA5GM113216-01, NIGMS, NIH.

References

1. Arango Duque G, Descoteaux A (2014) Macrophage cytokines: involvement in immunity and infectious diseases. Front Immunol 5:491. https://doi.org/10.3389/fimmu.2014.00491
2. Wynn TA, Chawla A, Pollard JW (2013) Macrophage biology in development, homeostasis and disease. Nature 496:445–455. https://doi.org/10.1038/nature12034
3. Anderson KV (2000) Toll signaling pathways in the innate immune response. Curr Opin Immunol 12:13–19. https://doi.org/10.1016/S0952-7915(99)00045-X
4. Bonta IL, Ben-Efraim S (1993) Involvement of inflammatory mediators in macrophage antitumor activity. J Leukoc Biol 54:613–626
5. LaRosa JC, He J, Vupputuri S (1999) Effect of statins on risk of coronary disease: a meta-analysis of randomized controlled trials. JAMA 282:2340–2346
6. Bu D, Griffin G, Lichtman AH (2011) Mechanisms for the anti-inflammatory effects of statins. Curr Opin Lipidol 22:165–170. https://doi.org/10.1097/MOL.0b013e3283453e41
7. Jain MK, Ridker PM (2005) Anti-inflammatory effects of statins: clinical evidence and basic mechanisms. Nat Rev Drug Discov 4:977–987. https://doi.org/10.1038/nrd1901
8. Liao JK, Laufs U (2005) Pleiotropic effects of statins. Annu Rev Pharmacol Toxicol 45:89–118
9. Yaniv A, Alexander S, Victor N et al (2004) Prior statin therapy is associated with a decreased rate of severe sepsis. Circulation 110:880–885
10. Hackam DG, Mamdani M, Li P, Redelmeier DA (2006) Statins and sepsis in patients with cardiovascular disease: a population-based

cohort analysis. Lancet (London, England) 367:413–418. https://doi.org/10.1016/S0140-6736(06)68041-0
11. Kamal AHM, Chakrabarty JK, Udden SMN et al (2018) Inflammatory proteomic network analysis of statin-treated and lipopolysaccharide-activated macrophages. Sci Rep 8:164. https://doi.org/10.1038/s41598-017-18533-1
12. Kamal AHM, Fessler MB, Chowdhury SM (2018) Comparative and network-based proteomic analysis of low dose ethanol- and lipopolysaccharide-induced macrophages. PLoS One 13:e0193104. https://doi.org/10.1371/journal.pone.0193104
13. Kim Y, Lee EJ, Jang HK et al (2016) Statin pretreatment inhibits the lipopolysaccharide-induced epithelial-mesenchymal transition via the downregulation of toll-like receptor 4 and nuclear factor-kappaB in human biliary epithelial cells. J Gastroenterol Hepatol 31:1220–1228. https://doi.org/10.1111/jgh.13230
14. Weischenfeldt J, Porse B (2008) Bone marrow-derived macrophages (BMM): isolation and applications. CSH Protoc 2008:pdb.prot5080. https://doi.org/10.1101/pdb.prot5080
15. Udden SMN, Peng L, Gan J-L et al (2017) NOD2 suppresses colorectal tumorigenesis via downregulation of the TLR pathways. Cell Rep 19:2756–2770. https://doi.org/10.1016/j.celrep.2017.05.084
16. Mi H, Muruganujan A, Casagrande JT, Thomas PD (2013) Large-scale gene function analysis with the PANTHER classification system. Nat Protoc 8:1551–1566. https://doi.org/10.1038/nprot.2013.092
17. Saeed AI, Sharov V, White J et al (2003) TM4: a free, open-source system for microarray data management and analysis. BioTechniques 34:374–378. https://doi.org/10.2144/03342mt01
18. Ellis CN, LaRocque RC, Uddin T et al (2015) Comparative proteomic analysis reveals activation of mucosal innate immune signaling pathways during cholera. Infect Immun 83:1089–1103. https://doi.org/10.1128/IAI.02765-14

Chapter 6

The Purification and Characterization of Exosomes from Macrophages

Eran Fridman, Lana Ginini, Ziv Gil, and Neta Milman

Abstract

Macrophages play an essential role in diverse biological processes, from the immune response to inflammatory and neurodegenerative disorders, to various cancers. A macrophage subpopulation, known as tumor-associated macrophages (TAMs), has been shown to promote tumorigenesis, metastasis, and immune escape of cancer cells. Some of the pro-tumorigenic effects of TAMs are mediated via the secretion of nano-vesicles (exosomes) from macrophages to neighboring cells. In this chapter, we describe peritoneal macrophage isolation methods, polarization of TAMs, and purification and characterization of macrophage-derived exosomes.

Key words Mouse peritoneal macrophages, Exosomes, Differential centrifugation, Exosome characterization, Tumor-associated macrophages (TAMs)

1 Introduction

Macrophages are prominent cells of the innate immune system. As such, they play a role in many biological processes, including development, infection, atherosclerosis, inflammatory diseases, neurodegenerative disorders, metabolism, and cancers [1]. In response to factors in their environment (cytokines and ligands activating toll-like receptors), macrophages change their activity to a pro-inflammatory state or to an inflammation-resolving state. In vitro, macrophages can be polarized to a pro-inflammatory phenotype by exposure to γ-interferon, lipopolysaccharide, or granulocyte macropahge colony-stimulating factor, or to an anti-inflammatory state by exposure to IL4/IL13 or macrophage colony-stimulating factor. Analogous to type 1 and type 2 T helper (Th1/Th2) cell polarization, in vitro-induced pro-inflammatory macrophages and anti-inflammatory macrophages are referred to as M1 and M2 macrophages, respectively.

Tumors are highly abundant in macrophages [2]. Tumor-associated macrophages (TAMs) were shown to play a role in

Suresh Mishra (ed.), *Immunometabolism: Methods and Protocols*, Methods in Molecular Biology, vol. 2184,
https://doi.org/10.1007/978-1-0716-0802-9_6, © Springer Science+Business Media, LLC, part of Springer Nature 2020

tumorigenesis, angiogenesis, promotion of tumor cell invasion, migration, suppression of antitumoral immune response, and metastasis [3]. Our group has shown that TAMs augment chemotherapy resistance via transfer of microRNAs (miRNA) from TAMs to pancreatic cancer cells through exosomes [4]. Moreover, macrophage-derived exosomes exert metabolic changes in pancreatic cancer cells, which increase resistance to chemotherapy. Exosomes are extracellular nano-vesicles ranging in size from 40 to 150 nm that are released from donor cells and internalized in recipient cells. They serve as cell-to-cell vehicles for the transfer of genetic components, proteins, metabolites, and more. Their role in cancer has been thoroughly investigated and is well recognized [5].

In this chapter, we describe methods used to study exosome-based communication between macrophages and pancreatic cancer cells. These include the isolation and polarization of mouse peritoneal macrophages, purification of exosomes from the conditioned media of the polarized macrophages, and characterization of the resulting exosomes. This protocol may be used to study exosome-based communication in other experimental settings as well.

2 Materials

2.1 The Isolation of Peritoneal Macrophages

2.1.1 The Induction of Chemical Peritonitis in Donor Mice

1. Sterile 3% Brewer thioglycolate medium.
2. 10 mL Syringes, 18-gauge needle for aspiration, and 27-gauge needles for injection.
3. 8–12-Week-old, specific pathogen-free C57BL/6 mice.

2.1.2 The Isolation of Peritoneal Macrophages

1. Isoflurane.
2. A chamber for CO_2 euthanasia.
3. Cold PBS (Dulbecco's phosphate-buffered saline without calcium and magnesium).
4. 10 mL Syringes.
5. 23-gauge needles.
6. 70% Ethanol.
7. Surgical scissors and forceps.
8. 50 mL Centrifuge conical tube.
9. Complete culture medium: RPMI supplemented with 10% fetal bovine serum, 2.5 g/L glucose, 4 mM glutamine, and 0.1% pen/strep.

2.2 Culture and Polarization of Peritoneal Macrophages

1. Recombinant murine IL-4.
2. Recombinant murine IFN-γ.
3. Lipopolysaccharides (LPS) from *Escherichia coli.*
4. Exosome culture media: RPMI supplemented with 2.5 g/L glucose, 4 mM glutamine, and 0.1% pen/strep.
5. RNA extraction kit.
6. cDNA synthesis kit.
7. Real-time PCR primers.
8. SYBR Green reaction mix.

2.3 The Purification of Exosomes

1. Conditioned medium (*see* Subheading 3.2 below).
2. Cold PBS (Dulbecco's phosphate-buffered saline without calcium and magnesium).
3. 50 mL Conical sterile polypropylene centrifuge tubes.
4. 50 mL Polypropylene tubes suitable for high-speed centrifugation with sealing caps.
5. 38.5 mL Ultra-Clear™, thin-wall centrifuge tubes.
6. Benchtop centrifuge.
7. High-speed centrifuge.
8. Ultracentrifuge.
9. Syringe filter with pore size of 0.2 μm.

2.4 Characterization of the Exosomes

2.4.1 Western Blot Analysis for Exosomal Surface Markers

1. Protein lysis buffer containing 150 mM NaCl, 1 mM EDTA, 25 mM Tris (pH 7.5), and 1% Triton.
2. Micro BCA protein assay kit.
3. Microplate spectrophotometer.
4. Protein assay kit.
5. Loading buffer: 4× Laemmli sample buffer with 2-mercaptoethanol at a 1:9 ratio.
6. Protein size marker.
7. SDS-PAGE casting and electrophoresis system.
8. SDS-PAGE gel transfer system.
9. Hydrophobic polyvinylidene fluoride (PVDF) 0.45 μm.
10. Tris-buffered saline, 0.1% Tween 20 (TBST).
11. Blocking solution: 5% Bovine serum albumin in TBST.
12. Anti-CD63 antibody.
13. Anti-ALIX antibody.
14. Anti-LAMP2 antibody.
15. A secondary antibody against the host of the primary antibody, conjugated to horseradish peroxidase.
16. Chemiluminescence reagent.

2.4.2 Nanoparticle Tracking Analysis

1. PBS.
2. 0.02 μm Syringe filters.

2.4.3 Exosome Internalization Assay

1. Carboxyfluorescein diacetate succinimidyl ester (CFDA-SE).
2. PBS.
3. Complete culture media: RPMI with 10% exosome-depleted FBS.
4. CFDA-SE-stained exosomes.
5. Exosome target cells.
6. 12-Well plate.
7. 10 mm diameter glass coverslips.
8. PBS.
9. 4% Paraformaldehyde.
10. Triton-X100.
11. 1% BSA in PBS.
12. Alexa Fluor™ 647 Phalloidin.
13. DAPI 1 μmol/L 4′,6-diamidino-2-phenylindole.
14. 0.25% Trypsin and 0.02% EDTA solution.
15. Fixable viability dye.

3 Methods

3.1 Isolation of Peritoneal Macrophages

To isolate peritoneal macrophages, we follow a protocol published by Zhang et al. [6], which we have slightly adapted. First, inject thioglycolate in the peritoneum cavity of a mouse to elicit inflammation. Inflammation-eliciting drugs such as thioglycolate, proteose peptone, and others induce monocyte migration into the peritoneal cavity, thus increasing the yield per mouse. Isolate the peritoneal exudate and culture surface adherent macrophages as described below.

3.1.1 Induction of Peritonitis: 4 Days Prior to the Harvest of Peritoneal Monocytes

1. Inject 0.75–1.25 mL/10 g of sterile thioglycolate into the peritoneal cavity of each of the donor mice. Use a 10 mL syringe with an 18-gauge needle for aspiration of thioglycolate and then switch to a 27-gauge for injection (*see* **Note 1**).
2. Tilt the mouse in a head-down position so the head is slightly lower than the base of tail (*see* **Note 2**).
3. After injection, keep the mice in specific pathogen-free conditions for 4 days.

3.1.2 The Isolation of Peritoneal Macrophages

Four days after the induction of peritonitis, euthanize the mice (*see* **Note 3**). This can be accomplished via cervical dislocation or with CO_2 (*see* **Note 4**). Prepare a workspace using a Styrofoam surface for easy fixation (as explained below), which is covered with a disposable absorbent pad to absorb excess fluids.

1. Spray the abdomen of the mouse with 70% ethanol. Then use scissors to make a small incision along the midline of the abdomen, just bellow the xiphoid. Care should be taken not to incise the peritoneum.
2. Using a blunt retractor, retract the skin flaps to the caudal and cephalic margins until the entire abdomen is exposed. This will assure minimal injury to the peritoneal cavity.
3. Affix the mouse to the work surface. Use four syringe needles to attach the limbs to the Styrofoam. This will assure immobilization during the peritoneal lavage (Fig. 1a).
4. Fill a 10 mL syringe with sterile, cold PBS and attach a 23-gauge needle to it.
5. Spray the exposed peritoneal wall with 70% ethanol.
6. Inject the PBS into the peritoneal cavity. Inject the PBS into the peritoneal cavity by inserting the needle in the direction of the lower abdomen. Be sure that the needle bevel is pointing up to

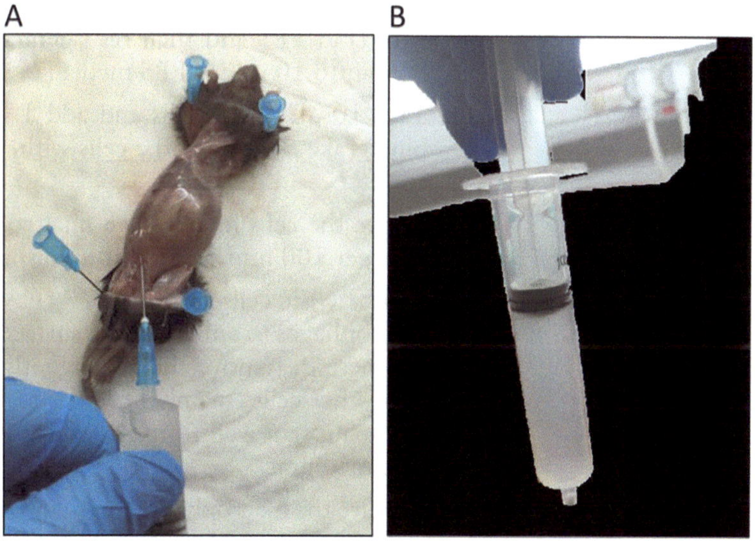

Fig. 1 Isolation of mouse peritoneal macrophages: (**a**) After the sacrifice, a small incision was made in the midline of the mouse skin over the abdomen, a 3–5 mm bellow the xiphoid, and the abdomen was exposed. Four needles were used to attach the mouse limbs into a Styrofoam surface. 10 mL of cold PBS were injected to the mouse using a syringe and a 23G needle. The needle and syringe were left in place and the plunger was retracted slowly to recover the peritoneal lavage. (**b**) Peritoneal lavage recovered from the peritoneal cavity. The fluid appears turbid due to the presence of peritoneal cells

avoid the accidental puncture of any organ. Upon the insertion of the needle, begin injecting the PBS immediately, as the PBS stream will push the organs of the peritoneal cavity (mainly the bowel). This is a delicate procedure and care should be taken not to puncture any organs, as this would increase the chances of infection and necessitate discarding both the fluid and the mouse (*see* **Note 5**).

7. Leaving the needle and the syringe in place, slowly retract the plunger of the syringe while paying close attention to the needle bevel to avoid drawing an organ into the needle. If an organ is entrapped in the needle (usually the small intestine or the pancreas), gently push the plunger to release the organ. Keep aspiring the fluid. Typical recovery is 5–7 mL for the first aspirate (Fig. 1a).
8. Gently disconnect the syringe from the mouse, leaving the needle in place. Then dispense the fluid into a 50 mL conical centrifuge tube and store the sample on ice. The fluid recovered from the mouse should look turbid due to the presence of peritoneal cells (Fig. 1b).
9. Connect a new syringe containing 10 mL sterile cold PBS to the needle and repeat the PBS injection process until the fluid being recovered from the mouse is clear (*see* **Note 6**).
10. Pellet the peritoneal cells by centrifugation (5 min at 400 × *g*, 4 °C). Discard the supernatant. Wash the pelleted cells twice with cold PBS and then resuspend them by gently pipetting them into 10 mL of PBS media (*see* **Note 7**).
11. Take 10 μL of the cells and add 1:1 solution of 0.4% trypan blue. Take a count of the cells with a hemocytometer or a cell counter.
12. Dilute the cells in PBS to a final concentration of 15 million cells per 100 mm tissue culture dish.
13. Incubate the cells for 2 h at 37 °C with 5% CO_2 to allow the macrophages to adhere to the surface of the dish. At the end of this period, remove the PBS, gently wash the cells to ensure that other cells (mainly fibroblasts) which may loosely adhere are being removed from the plate, and add the complete culture medium.
14. Culture the macrophages for 1 day before adding the cytokines for polarization. Macrophages can be maintained in culture for approximately 3 weeks (*see* **Note 8**).

3.2 Culture and Polarization of Peritoneal Macrophages

For the purpose of exosome isolation, mouse peritoneal macrophages (and other macrophage cell lines) are cultured in exosome culture media without FBS. This is because FBS has been found to contain significant amounts of exosomes [7]. Other macrophage cell lines, which require FBS for growth, can be grown with exosome-depleted FBS (*see* **Note 9**). In our experience, macrophages can be maintained for 3 weeks, during which the time-conditioned media is collected twice a week.

1. To polarize macrophages toward the M1 phenotype, add to the culture media LPS and interferon γ to a final concentration of 100 ng/mL and 20 ng/mL, respectively [8].
2. To polarize macrophages toward the M2 phenotype, add IL-4 to a final concentration of 20 ng/mL.
3. Culture macrophages in the presence of the indicated cytokines at 37 °C and 5% CO_2 for 72 h. Then collect the conditioned media and directly proceed to the exosome purification step.
4. Conduct a real-time PCR analysis, verifying the polarization results by assessing the relative quantities of M1 markers (iNOS and CD86) and M2 markers (arginase-1 and CD206) by real-time PCR analysis [4].
5. For RNA extraction from the M1, M2, and non-treated macrophages, use a spin column based commerical kit for isolation of total RNAtas per the manufacturer's instructions.
6. For the synthesis of cDNA: Use a first strant cDNA synthesis commercial kit
7. Use the SYBR Green method Fast SYBR™ Green Master Mix, as per the manufacturer's instructions.
8. Assess fold change in gene expression by calculating the ΔCt of M1 and M2 macrophages compared to their levels in untreated macrophages [9]. Use β-actin as a housekeeping gene. A representative analysis is presented in Fig. 2.

3.3 Purification of Exosomes

Our exosome purification process is a modification of a protocol published by Thery et al. [10]. In our protocol, three centrifugations, executed at differential and increasing speeds, result in the elimination of specific cell populations from the media. Dead cells and large-sized debris are eliminated from the media during the first centrifugation. Large microvesicles are eliminated from the media in the second centrifugation. In the third centrifugation, which involves high-speed ultracentrifugation, exosomes are sedimented and the pellet is collected. Figure 3 shows a schematic description of the method.

All steps must be performed in aseptic conditions and under a biological hood. Centrifugations should be carried out at 4 °C. All samples must be stored on ice throughout the procedure.

1. Collect the conditioned media from the cultured macrophages in a 50 mL conical centrifuge tube. Store the conditioned media on ice.

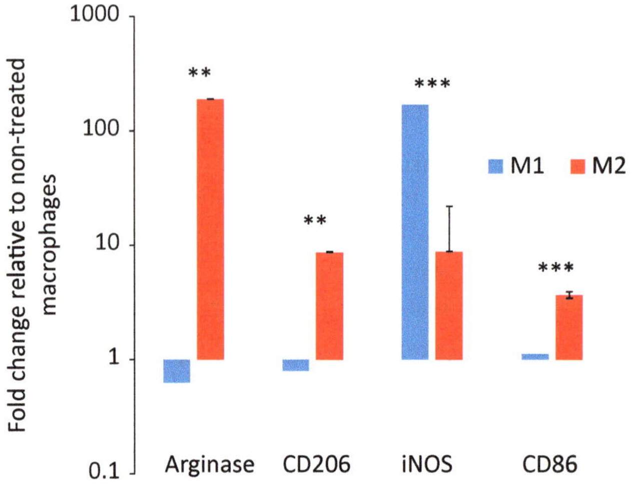

Fig. 2 Verification of macrophage polarization. Fold change in the transcript levels of indicated genes, after treatment of peritoneal macrophages with 20 ng/mL γ-interferon and 100 ng/mL LPS (M1 polarization) or with 20 ng/mL IL-4 to (M2 phenotype). The transcript levels were assessed by qRT-PCR. CD86 and iNOS were used as M1 markers; arginase and CD206 were used as M2 markers. The graph depicts fold change of indicated genes over non-treated macrophages. Student's *T*-test was performed to determine significance ($*p < 0.01$, $**p < 0.001$, $***p < 0.0001$)

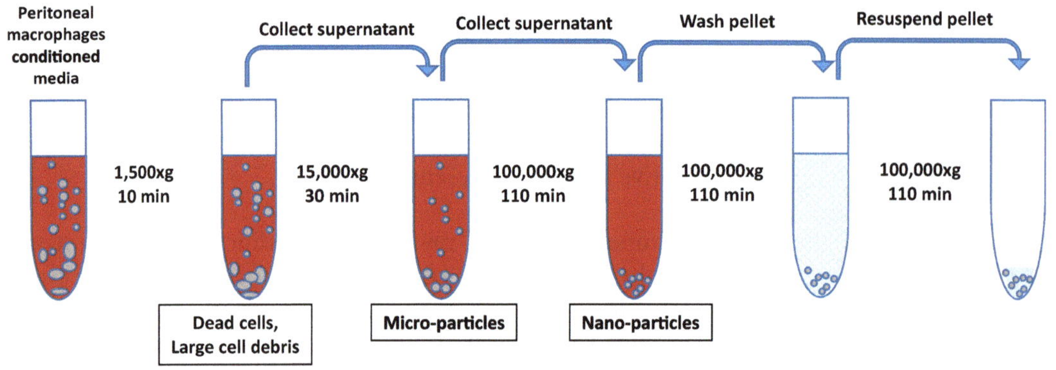

Fig. 3 A schematic description of the purification of exosomes by the differential centrifugation method. Peritoneal macrophages conditioned media is processed by the indicated centrifugation steps. Dead cells, large-cell debris, microparticles, and nanoparticles are sedimented in each step accordingly

2. Cool a benchtop centrifuge to 4 °C. Centrifuge the conditioned media for 10 min at 1500 × *g*.
3. Collect the supernatant, transfer it to polypropylene tubes suitable for high-speed centrifugation, and discard the pellet. Centrifuge the media at 15,000 × *g* for 30 min in a high-speed centrifuge.

4. Carefully collect the supernatant and transfer it to ultracentrifugation tubes. Discard the pellet. Ultracentrifuge the supernatant at 100,000 × *g* for 110 min.
5. After centrifugation has been completed, carefully remove the supernatant, perform a PBS wash, resuspend the pellet (which is not visible) with cold PBS, and then repeat the high-speed ultracentrifugation step (*see* **Note 10**).
6. After the final centrifugation, remove the supernatant and resuspend the 100,000 × *g* pellet in cold PBS (*see* **Note 11**).
7. Pass the resulting suspension through a 0.2 μm syringe filter, aliquot, and store at −80 °C for future use (*see* **Note 12**).

3.4 Characterization of Exosomes

The 100,000 × *g* pellet of macrophage-conditioned media can be charaterized by several experiemntal approaches. Of particular importance is the use of cryotransmission electron microscopy (cryo-TEM) to characterize the structure and size of the pellet, and to verify the presence of membrane-coated nano-vesicles [4] (Fig. 4a). Despite being a low-throughput and labor-intensive technique, this method is considered to be a gold standard for the identification of exosomes. It is also worth noting that during the sample's preparation (dehydration and embedding), exosome loss can occur.

To evaluate the number of particles and their size distribution, use a nanoparticle tracking analysis (NTA) (Fig. 4b). In this method, the number and size of particles in a solution are determined through an analysis of Brownian motion via light scattering. This method only detects particles larger than 50 nm.

To further characterize the pellet, prepare protein lysates and determine the amount of total protein per dose by micro BCA protein assay. Perform a Western blot analysis to examine the expression of exosomal surface markers [11]. This method does not permit to assess the number of nano-vesicles or their size (Fig. 4c). Finally, to evaluate the integrity of the exosomes, perform an assay to determine the internalization efficiency of fluorescent exosomes in a pancreatic cancer cell line (Fig. 4d, e). We recommend carrying out this assessment in tandem with NTA and Western blot analyses on these preparations. In the next section, the experimental protocols for these three procedures are briefly described.

3.4.1 Western Blot Analysis for the Expression of Exosome Surface Markers

1. Pellet the desired quantity of exosomes via centrifugation at 100,000 × *g* for 110 min. Remove the supernatant and place the pellet on ice.
2. Add 60 μL of cold protein lysis buffer to each dose of exosomes. Incubate with gentle agitation at 4 °C for 30 min and

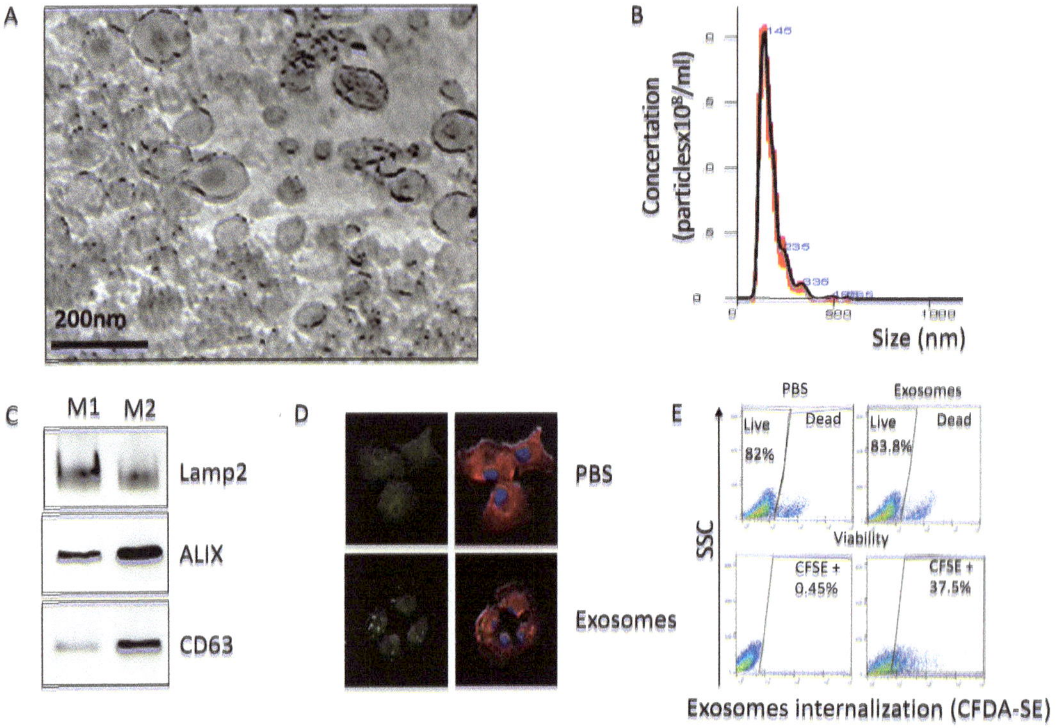

Fig. 4 Characterization of mouse peritoneal macrophage conditioned media 100,000 × *g* conditioned media: (**a**) Cryo-TEM image of exosomes in macrophages media (bar 200 nm). (**b**) Nanoparticle tracking analysis of exosomes derived from M2 polarized macropahages exosomes. Nanoparticle tracking analysis of exosome samples diluted 1:100. The graph depicts the concentration of particles as a function of size in nanometers. Particles exhibit a peak with a mean diameter of 145 nm. (**c**) Western blot of analysis of exosomal surface markers from M1 and M2 polarized macrophages. (**d**) A representative image from an exosome internalization assay. Murine KPC cells adhered to glass coverslips were incubated with PBS (upper panel) or macrophage-derived exosomes stained with CFDA-SE (lower panel, green) for 3 h. The cells were fixed and permeabilized and then stained with phalloidin (red) to visualize the actin cytoskeleton and with DAPI to visualize nuclei (blue). Images were acquired with scanning confocal microscopy. (**e**) Evaluation of exosome internalization by pancreatic cancer cells by flow cytometry. Macrophage-derived exosomes stained with CFDA-SE were incubated with KPC murine cell line. Cells were then lifted with trypsin, stained with viability dye, and subsequently analyzed with a flow cytometer. The live-cell fraction was isolated (upper panel) and the percentage of CFDA-SE-positive cells was evaluated (lower panel)

then centrifuge for additional 30 min in a chilled benchtop centrifuge at 13,000 × *g*.

3. Collect the supernatant. Evaluate the protein concentration using a protein assay kit (*see* **Note 13**).
4. Cast a 12% SDS-PAGE.
5. Add 4× Laemmli buffer with 2-mercaptoethanol at a ratio of 1:9 to 2 μg of protein lysate, heat in a heat block set to 95 °C for 5 min, and load the protein lysates on the gel.

6. Electrophorese the lysate proteins alongside a protein size marker and then transfer them to a PVDF membrane.
7. Block the membrane for 1 h with a blocking buffer.
8. Add the primary antibody, diluted in a blocking buffer, and incubate overnight at 4 °C (*see* **Note 14**).
9. Wash the membrane with TBST (3 × 5-min washes) and then incubate with a secondary antibody against the host of the primary antibody, conjugated to horseradish peroxidase for 1 h.
10. Add a chemiluminescence reagent to visualize the immune-reactive bands, as per the manufacturer's recommendations.

3.4.2 Nanoparticle Tracking Analysis (NTA)

1. Thaw the desired amount of exosome doses. Keep the exosomes on ice throughout the procedure.
2. Centrifuge at 100,000 × *g* for 110 min in a chilled ultracentrifuge.
3. Filter PBS solution with a 0.02 μM syringe filter to obtain particle-free PBS.
4. Aspirate the supernatant and dilute the exosome pellet with 100 μL particle-free PBS.
5. Dilute the exosomes in filtered PBS to obtain an optimal number of exosomes per field of view (*see* **Note 15**).
6. Inject the exosomes and acquire the data (*see* **Note 16**) [4].

3.4.3 Exosome Internalization Assays

Perform this assay to assess the internalization efficiency of fluorescently stained exosomes by their target cells. We have found that macrophage-derived exosomes are internalized efficiently by pancreatic cancer cells. The exosomes are labeled with carboxyfluorescein diacetate succinimidyl ester (CFDA-SE), a membrane-permeable dye that covalently binds to proteins in the lumen of the exosome, thereby undergoing esterification, which causes them to fluoresce [12]. Next, we analyze the percentage of cells that internalized CFDA-SE-positive particles by fluorescent microscopy or by flow cytometry.

1. Resuspend the 100,000 × *g* exosomal pellet in 100 μL pre-warmed PBS.
2. Prepare the CFDA-SE as per the manufacturer's instructions. Add it to your desired final concentration (*see* **Note 17**).
3. Incubate for 10 min at room temperature in the dark.
4. Stop the labeling reaction through the addition of 0.5 cold complete media containing 10% exosome-depleted serum. Incubate on ice for five additional minutes.
5. Wash three times with cold complete media followed by centrifugation at 100,000 × *g* for 110 min at 4 °C. Resuspend the pellet of the last wash in PBS (*see* **Note 18**).

6. For fluorescent microscopy-based exosome internalization assays, insert glass coverslips into a 12-well plate.
7. Plate the target cells in the wells (*see* **Note 19**) and allow them to settle overnight.
8. The following morning, incubate the cells in the presence/absence of CFDA-SE-labeled exosomes at 37 °C under 5% CO_2 for the required time (*see* **Note 20**).
9. Remove unbound exosomes through extensive washing with PBS.
10. Fix cells by incubation with 4% paraformaldehyde for 15 min at room temperature.
11. Permeabilize the cells by incubation in 0.1% Triton-X100 for 3 min.
12. Wash the cells with PBS.
13. Block the cells for nonspecific binding with 1% BSA in PBS for 20 min.
14. Add Alexa Fluor™ 647 Phalloidin, diluted in PBS containing 0.1% BSA, to the cells for 20 min at room temperature.
15. Wash in PBS.
16. Stain the nuclei with DAPI for 10 min in the dark.
17. Mount the coverslips with mounting media.
18. Acquire images with a fluorescent microscope or a confocal laser scanning microscope.
19. Using image analysis software, analyze the number of cells that internalized exosomes.
20. For flow cytometry-based exosome internalization assays, apply the CFDA-SE-labeled exosomes to pancreatic cancer cells in culture for 3 h.
21. Wash unbound exosomes with PBS.
22. Add a 0.25% trypsin/0.02% EDTA solution to remove cells from the plate. Incubate for 5 min at 37 °C under 5% CO_2. Verify that the cells have detached with a microscope.
23. Wash the cells twice in PBS.
24. To discriminate live from dead cells, stain the cells with a Fixable Viability Dye.
25. Analyze the cells with a flow cytometer (or microscope as applicable). Detectable fluorescence indicates the internalization of exosomes rather than just surface binding.
26. Using flow cytometry software, analyze the percentage of live cells that internalized CFDA-SE-stained exosomes.

4 Notes

1. Use the one-hand restraint method using your nondominant hand, leaving your dominant hand free for injection.
2. Holding the mouse in this way permits the abdominal viscera to shift toward the head. This minimizes the chances of unintentionally puncturing organs within the peritoneal sac.
3. Start with isolating macrophages from 10 to 15 mice processing 5 at a time.
4. Euthanize the mice with CO_2. Transfer them into a sealed chamber where they are anesthetized by way of an isoflurane-soaked cotton ball that is placed inside the chamber. Next, flow CO_2 into the chamber for 5 min.
5. Typically inject 10 mL depending on the size of the mouse. The injection should be administered at a slow and steady rate to avoid rupturing the peritoneal sac as the peritoneum sac has a limited compliance. When rupturing occurs, it typically evidences itself in the form of a small breach with fluid leakage.
6. We usually repeat this step 2–3 times per mouse.
7. Occasionally a red layer of red blood cells is observed after the first wash. This is typically due to minor bleeding that occurs during the procedure. In such instances, the addition of a red blood cells lysis step is recommended. Use the ACK Lysing Buffer, as per the manufacturer's instructions.
8. Optional: After 3 days of culture, remove adherent cells from the plate and perform phenotypic staining using the CD11b leukocyte surface marker and the F4/80 mouse macrophage marker.
9. Exosome-depleted FBS is available commercially. Alternatively, it can be prepared through the ultracentrifugation of FBS at $100{,}000 \times g$ for 110 min.
10. Typically, resuspend the pellet in 2 mL of cold PBS and use micro-ultracentrifugation for this washing step.
11. Add 100 μL PBS to the $100{,}000 \times g$ pellet that originated from the 30 mL conditioned culture media. However, depending on the quantity of protein/number of exosomes/physiological activity, it can be resuspended in any volume.
12. We define one dose as a pellet originating from the 30 mL conditioned culture media.
13. A typical dose of exosomes contains approximately 5 μg.
14. Use antibodies for the following exosomal surface markers: CD63, LAMP-2, and ALIX.

15. In general, dilute the exosomes 1:100 in PBS to obtain ~100 exosomes in the field of view.
16. We use the NanoSight NS300 nanoparticle tracking analysis (NTA) system (or any other equivalent NTA) with an electronic injector. Utilizing NTA 2.2 analytical software, we record three videos of 30–60 s for each sample [4].
17. We use a CFDA-SE concentration of 7.5 μM.
18. Generally, stained exosomes are stable in the dark at 4 °C for at least 24 h.
19. We use a Pdx1- Cre, lox-stop-lox-KrasG12D/+, lox-stop-lox-tp53R172H/+ (KPC) mouse-derived cell line and plate 50,000–75,000 cells per well in a 12-well culture dish.
20. In our system, kinetic analyses showed exosome internalization of approximately 30% of the cells after 3 h of incubation.

References

1. Wynn TA, Chawla A, Pollard JW (2013) Macrophage biology in development, homeostasis and disease. Nature 496:445–455
2. Sica A, Mantovani A (2012) Macrophage plasticity and polarization: in vivo veritas. J Clin Invest 122:787–795
3. Qian B-Z, Pollard JW (2010) Macrophage diversity enhances tumor progression and metastasis. Cell 141:39–51
4. Binenbaum Y et al (2018) Transfer of miRNA in macrophage-derived exosomes induces drug resistance in pancreatic adenocarcinoma. Cancer Res 78:5287–5299
5. Milman N, Ginini L, Gil Z (2019) Exosomes and their role in tumorigenesis and anticancer drug resistance. Drug Resist Updat 45:1–12
6. Zhang X, Goncalves R, Mosser DM (2008) The isolation and characterization of murine macrophages. Curr Protoc Immunol Chapter 14:Unit 14.1.doi: https://doi.org/10.1002/0471142735.im1401s83
7. Shelke GV, Lasser C, Gho YS, Lotvall J (2014) Importance of exosome depletion protocols to eliminate functional and RNA-containing extracellular vesicles from fetal bovine serum. J Extracell Vesicles 3. https://doi.org/10.3402/jev.v3.24783
8. Murray PJ et al (2014) Macrophage activation and polarization: nomenclature and experimental guidelines. Immunity 41:14–20
9. Nolan T, Hands RE, Bustin SA (2006) Quantification of mRNA using real-time RT-PCR. Nat Protoc 1:1559–1582
10. Thery C, Amigorena S, Raposo G, Clayton A (2006) Isolation and characterization of exosomes from cell culture supernatants and biological fluids. Curr Protoc Cell Biol Chapter 3:Unit 3.22
11. Li J, Liu K, Liu Y (2013) Exosomes mediate the cell-to-cell transmission of IFN-α-induced antiviral activity. Nat Immunol. 14 (8):793–803. https://doi.org/10.1038/ni.2647
12. Lannigan J, Erdbruegger U (2017) Imaging flow cytometry for the characterization of extracellular vesicles. Methods. 112:55–67. https://doi.org/10.1016/j.ymeth.2016.09.018

Chapter 7

Isolation and Characterization of T Lymphocyte-Exosomes Using Mass Spectrometry

Inbar Azoulay-Alfaguter and Adam Mor

Abstract

Exosomes are cell-derived vesicles that have been implicated in the pathogenesis of many inflammatory diseases. In the immune system, it has been shown that T lymphocyte-derived exosomes are able to induce diverse cellular responses. There are several methods to isolate and to characterize exosomes, each with their own advantages and disadvantages. Here, we describe a centrifugation approach, combined with mass spectrometry characterization, as a means to study exosomes derived from primary human T lymphocytes. This method is sensitive and therefore can be applied when a limited amount of sample is available.

Key words Centrifugation, Exosomes, Mass spectrometry, Signaling, T cell

1 Introduction

Similar to microvesicles and apoptotic bodies, exosomes are a type of extracellular vesicle. The nomenclature of the different vesicle types depends on their size, morphology, function, and cell of origin. Exosomes show standard round morphology in cryo-electron microscopy, ranging in size between 30 and 100 nm [1]. They are secreted by nearly all types of cells and are widely distributed in various body fluids [2]. Unlike the larger microvesicles (size above 100 nm), exosomes are either released from the cells when multivesicular bodies fuse with the plasma membrane or discharged directly from the plasma membrane [3]. Since their discovery over 50 years ago, exosomes have been found to play a vital role in many biological processes, including intercellular communication, viral replication, development and differentiation of stem cells, cell signaling, and tissue regeneration [4]. It is well established that exosomes released by immune cells can modulate responses through antigen presentation and transfer of pro- or anti-inflammatory mediators. Exosomes also play a key role in the pathogenesis of asthma and rheumatoid arthritis, and have a potential role in cancer immuno-editing [5].

Suresh Mishra (ed.), *Immunometabolism: Methods and Protocols*, Methods in Molecular Biology, vol. 2184,
https://doi.org/10.1007/978-1-0716-0802-9_7,

Exosomes contain a variety of bioactive molecules derived from their parent cells. These include nucleic acids involving DNA and RNA, proteins such as growth factors and cytokines, and lipid mediators. The functional impact of exosomes is conveyed by the molecular components that they carry [6]. The cargo of exosomes is not a result of a random process, but rather involves a complex sorting mechanism that favors specific biomolecules over others [7, 8]. Although structural proteins are commonly found in large numbers in these vesicles regardless of their cellular origin [9], additional knowledge about the specific cargo of T cell-derived exosomes is crucial to better understand the function of the acquired immune responses. Thus, we have taken advantage of a proteomic approach to more accurately characterize the unique set of proteins that are found in T cell-derived exosomes. Our findings support a mechanistic role for exosomes in immune cell activation, and also imply that exosomes should be considered as a biomarker for the activity of inflammatory diseases [10].

There are various methods to isolate exosomes from biological fluids, including centrifugation, immunological separation, polymer-based precipitation, and filtration [11]. These methods have made the isolation process faster and easier; however, each method has its own strengths and weaknesses [11]. Differential centrifugation utilizes several centrifugation cycles of different centrifugal forces and durations to isolate exosomes based on their density and size differences from other components in a sample of cells, large vesicles, and debris. This method requires little technical expertise, is affordable over time (i.e., one ultracentrifuge machine for long-term use), and can be used to isolate exosomes from biological fluids and large volumes of media faster than other methods. For these reasons, ultracentrifugation-based techniques have become a rather popular option among researchers in exosome research. The main weakness of the method is its low efficiency when isolating exosomes from viscous fluids such as serums. Density gradient centrifugation combines ultracentrifugation with a sucrose density gradient to allow for the separation of low-density exosomes based on their size, mass, and density [12]. Immunoaffinity separation incorporates magnetic beads bound to specific antibodies that recognize exosome-specific surface markers [13]. This allows for the isolation of selective subsets of exosomes, but is limited by a small sample volume and high cost of reagents. Polymer-based precipitation involves mixing the samples with polymer-containing precipitation solutions, incubation, and centrifugation at low speeds [14–16]. The advantage of this method is the minimal harming effect of the process on the isolated exosomes. Difficulties include the precipitation of non-vesicular contaminants and a limited ability to perform downstream functional analysis. Finally, a filtration membrane is used to separate exosomes from other macromolecules while concentrating the exosomes into a small volume [17]. Limitations vary from the stickiness of the exosomes to the filtration membranes and subsequent loss of material. In addition, the use of force may result in the deformation and breakup of large vesicles, which may potentially skew the results of the downstream analysis [16].

The various technologies applied for exosome characterization include biophysical, molecular, and microfluidic methods. Biophysical methods use size range to characterize exosomes. These include optical particle tracking [18], photon correlation spectroscopy [19], resistive pulse sensing [20], atomic force microscopy [21], transmission electron method [22], and cryo-electron microscopy [23]. Molecular approaches, such as flow cytometry, are used for the characterization of exosome surface proteins. Moreover, flow cytometry allows for the measurement of both the size and the structure of the exosomes [24]. The microfluidic-based methodologies are used for the characterization of exosomes by binding exosomes to specific antibodies on microfluidic channels followed by the elution of bound vesicles [25]. Several methods have been developed to analyze exosome RNA content, including microarray analysis, next-generation sequencing, and digital droplet PCR [26]. Exosome protein analysis can be done by western blotting and fluorescence-based cell sorting [27, 28].

In this chapter, we describe an alternative method of collecting and preparing exosomes from primary human T cells for proteomic analysis using high-resolution mass spectrometry. This approach can be easily modified to study vesicles derived from other type of cells.

2 Materials

Prepare all solutions using distilled water unless mentioned otherwise. Filter and sterilize all solutions through a 0.45 μm filter. Prepare all reagents in a sterile hood. Store all reagents at 4 °C unless mentioned otherwise. Keep all solutions and reagents sterile and diligently follow waste disposal regulation.

2.1 Human T-Cell Isolation

1. A whole-blood sample (e.g., Leukopacks).
2. Sterile phosphate-buffered saline (PBS).
3. Human CD3 T-cell enrichment cocktail.
4. Multiple 50 mL polypropylene tubes.
5. Lymphoprep.
6. Roswell Park Memorial Institute (RPMI) enriched cell culture medium 1640, containing 10% fetal bovine serum (FBS), 1% L-glutamine, 1% nonessential amino acids, 1% sodium pyruvate, and 1% HEPES with 100 IU IL-2.
7. T75 tissue culture flasks.

2.2 Preparation of Magnetic Beads for Cell Stimulation

1. Dynabeads M-270 Epoxy antibody coupling kit.
2. DynaMag-2 magnet.
3. Analytic weight.
4. Mixer allowing rotation or tilting of tubes.
5. Anti-CD3 antibody.
6. Anti-CD28 antibody.
7. Anti-human IgG isotype antibody.

2.3 T-Cell Activation

1. Bovine serum albumin (BSA).
2. PBS with 0.1% BSA.
3. Exosome-free RPMI enriched media 1640, containing 10% exosome-free fetal bovine serum, 1% L-glutamine, 1% nonessential amino acids, 1% sodium pyruvate, and 1% HEPES with 100 IU IL-2.

2.4 Isolation of Exosomes

1. Refrigerated centrifuge running at 10,000 × *g*.
2. Ultracentrifuge.
3. Beckman Ti84 rotor.
4. Beckman centrifuge tubes.
5. Sterile PBS.
6. Mass spectrometry collection buffer (8 M urea in 100 mM Tris–HCl at pH 8.5).
7. Mass spectrometry sample vials with snap lid.

2.5 Mass Spectrometry In-Solution Digestion

1. 1 mM Dithiothreitol (DTT) in 100 mM ammonium bicarbonate.
2. 50 mM Iodoacetamide (IAA) in 100 mM ammonium bicarbonate (pH 8).
3. 50 mM Ammonium bicarbonate.
4. Trypsin/Lys-C mix (MS grade).
5. Sequencing-grade modified trypsin.
6. 0.1% Trifluoroacetic acid (TFA) (v/v) in water (LC–MS grade).
7. 0.1% Formic acid (v/v) in water (LC–MS grade).
8. Buffer A (0.1% formic acid).
9. Buffer B (80% acetonitrile, LC–MS grade/0.1% formic acid).
10. C18-StageTip column (3 M).
11. 75 μm i.d. × 50 cm long EASY-spray PepMap columns.
12. Speed-vac vacuum concentrators.
13. Acetonitrile solution (2% acetonitrile/0.1% TFA).

2.6 Liquid Chromatography (LC)–Mass Spectrometry (MS) Analysis

1. EASY-nLC1000 HPLC coupled to the Q-Exactive (QE).
2. 75 μm i.d. × 50 cm long EASY-spray PepMap columns.

3 Methods

Carry out all procedures at room temperature unless otherwise specified. All procedures should be done under sterile conditions unless otherwise specified.

3.1 Isolation of Human Peripheral Blood T Lymphocytes

1. Obtain healthy donor human blood (Leukopacks). Allow the blood to reach room temperature before proceeding to the next step.
2. Split the blood into four sterile tubes (50 mL tubes), 25 mL of blood in each tube.
3. Add RosetteSep human CD3 T-cell enrichment cocktail to the tubes. To each tube of 25 mL of blood, add 1250 μL of RosetteSep cocktail (50 μL of RosetteSep cocktail per 1 mL of blood).
4. Mix the blood using a 25 mL sterile pipette and let it rest for 20 min at room temperature allowing the antibodies in the cocktail to bind to the cells.
5. Dilute the blood samples with an equal volume of PBS at room temperature. Add 25 mL of sterile PBS to each of the tubes containing 25 mL blood.
6. Prepare eight 50 mL tubes and add 20 mL of a density gradient medium (Lymphoprep) to each tube. Procedure is done at room temperature in a sterile hood. The volume of the density gradient medium can be changed based on the specific requirements. Every 20 mL of density gradient medium can be layered by up to 25 mL of diluted blood. Calculate the number of tubes required based on the amount of blood needed for the T-cell isolation.
7. Layer the diluted blood samples on the density gradient medium, being careful to minimize their mixing. For each density gradient medium tube, layer not more than 25 mL diluted blood.
8. Centrifuge the tubes at a speed of 1200 × *g*, at room temperature, for 20 min with the "brake off."
9. Harvest the enriched cell layer with a plastic pipette and transfer it to a clean sterile 50 mL collection tube. Add all the cells harvested into the same 50 mL collection tube.
10. Wash the enriched cells. Top up the tube with sterile PBS.

11. Centrifuge at 400 × *g*, room temperature, for 10 min, "brake off" low.
12. Discard the supernatant by gentle aspiration. Be careful not to disrupt the cell pellet.
13. Repeat the wash one more time (**steps 10–12**).
14. Resuspend cells in 10 mL of sterile PBS.
15. Count cells using a hemocytometer. Add 15 μL of cell suspension to the hemocytometer chamber and cover the glass. Count the number of cells in all four outer squares and divide this number by four (the mean number of cells/square). The number of cells per square $\times 10^4 =$ the number of cells in 1 mL of suspension (*see* **Note 1**).
16. Primary T cells can be maintained at a concentration of 1×10^6 cells per 1 mL of RPMI enriched media at 5% CO_2 at 37 °C with 100 IU of IL-2 for up to a period of 1 week.

3.2 Preparation of Magnetic Beads

Coat the magnetic beads with anti-CD3 and anti-CD28 antibodies. Use 2 μg of the anti-CD3 (0.5 mg/mL) and 2 μg of the anti-CD28 (0.5 mg/mL) antibodies to couple 1 mg of magnetic beads. As a negative control, coat another set of beads (1 mg) with 4 μg of an anti-human IgG isotype antibody (10 mg/mL) (*see* **Notes 2** and **3**).

1. Disinfect the magnets to prevent sample contamination.
2. Weigh out 3 mg of magnetic beads. Weight can be adjusted based on the total amount of desired reagents.
3. Wash the beads with 1 mL of C1 buffer (provided with the kit) and mix by pipetting up and down.
4. Place the tube on a magnet for 1 min and allow the beads to collect at the tube wall. Gently remove the supernatant.
5. Add the appropriate volume of antibody to the C1 buffer to reach a volume of 150 μL.

   For example, to couple 3 mg of beads, the required quantity of antibodies is 12 μL of anti-CD3 and 12 μL of anti-CD28 to be added to 126 μL of C1 buffer. For the control beads, the anti-human IgG isotype should be diluted 1:10 (final concetntration 1.0 mg/mL) before coupling 3 mg of beads with 12 μL of diluted anti-human IgG isotype and 138 μL of C1 buffer.
6. Add 150 μL of C2 buffer (provided with the kit) and mix by gently pipetting up and down.
7. Incubate on a roller at 37 ° C overnight. Adjust the speed of the roller to ensure thorough mixing and to avoid the beads from settling.
8. Place the tubes on the magnet for 1 min and allow the beads to collect on the tube wall. Remove the supernatant.

9. Add 800 μL of HB wash buffer (provided with the kit) and mix by pipetting up and down.
10. Place the tubes on a magnet for 1 min and allow the beads to collect on the tube wall. Remove the supernatant.
11. Add the 800 μL of LB wash buffer (provided with the kit) and mix by pipetting up and down.
12. Place the tubes on the magnet for 1 min and allow the beads to collect on the tube wall. Remove the supernatant.
13. Add the 800 μL of SB wash buffer (provided with the kit) and mix by pipetting up and down.
14. Place the tube on a magnet for 1 min and allow the beads to collect on the tube wall. Remove the supernatant.
15. Repeat the wash described in **step 13** once more.
16. Add the 800 μL of SB wash buffer and mix by pipetting up and down.
17. Incubate on a roller at room temperature for 15 min.
18. Place the tube on a magnet for 1 min and allow the beads to collect on the tube wall. Remove the supernatant.
19. Resuspend antibody-coupled beads in 300 μL SB buffer and store at 4 °C until use. The final bead concentration is 10 mg/mL.

3.3 The Activation of Human Peripheral Blood T Lymphocytes (See Note 4)

1. Wash coated beads for 5 min in PBS with 0.1% BSA.
2. Prepare an exosome-free enriched RPMI medium (*see* **Note 5**).
3. Count primary human T cells using a hemocytometer, as described above. Use 80×10^6 T cells per condition.
4. Cells should be pelleted by centrifugation at $400 \times g$, at room temperature for 5 min. Replace the media with 20 mL of fresh exosome-free enriched RPMI medium.
5. Add anti-CD3/CD28 or IgG control-coated magnetic beads to the cells in a 1:2 ratio (number of beads to number of cells). Mix the cells with the beads by pipetting up and down.
6. Spin down the cells with the antibody-coupled beads, at a speed of $400 \times g$, at room temperature for 1 min.
7. Transfer the cells with the beads using 10 mL sterile pipette to a T75 flask.
8. Incubate the cells with the beads for 18 h at 5% CO_2 at 37 °C.

3.4 Isolation of Exosomes (See Note 6)

1. Collect media from 80×10^6 resting, stimulated (with beads coated with anti-CD3/CD28 antibodies), and controlled (beads coated with IgG isotype control antibody) primary human T cells (*see* **Note 7**).
2. Centrifuge supernatants at $800 \times g$ at 4 °C for 5 min.

3. Collect supernatants and recentrifuge at 3200 × *g* at 4 °C for 10 min to discard dead cells.
4. Collect supernatants and recentrifuge at 10,000 × *g* at 4 °C for 30 min to discard cell debris.
5. Supernatants should be ultracentrifuged at 100,000 × *g* at 4 °C for 60 min (*see* **Notes 8** and **9**).
6. Discard supernatant by using 1 mL pipette tips (*see* **Notes 8** and **9** below) and wash with 1 mL sterile PBS (*see* **Notes 10** and **11**).
7. Ultracentrifuge at 100,000 × *g* at 4 °C for 60 min.
8. Discard supernatant by pouring the supernatant or use 1 mL pipette tips (*see* **Note 12**).
9. Resuspend and collect exosomes using 50 μL MS collection buffer (*see* **Note 13**). Transfer the samples to MS sample vials (*see* **Note 14**). Let samples rest at room temperature for 30 min.

3.5 In-Solution Digestion

More details about this section are found at doi.org/10.1074/mcp.RA119.001362 [29].

1. Reduce the samples with 1 mM dithiothreitol (DTT) at room temperature for 30 min.
2. Alkylate samples with 5 mM iodoacetamide (IAA) for 30 min in the dark (*see* **Note 15**).
3. Dilute the samples fourfold with 50 mM ammonium bicarbonate.
4. Digest lysates with trypsin/Lys-C mix (1:100 for the enzyme-to-protein ratio) and sequencing-grade modified trypsin (1:50 enzyme-to-protein ratio) at 37 °C overnight.
5. Acidified resulting peptides with trifluoroacetic acid (TFA) and force the acidified peptide sample through the C18-StageTip column (3 M Empore). Separate peptides on 75 μm i. d. × 50 cm long EASY-spray PepMap columns packed with 2 μm C18 material with 100 Å pore size.
6. Wash the column with buffer A. Elute the peptides from the C18 material using 20–30 μL buffer with a gradient of 7–28% buffer B, at a flow rate of 300 nL/min, over a gradient of 210 min. Elute directly into a microfuge tube or autosampler plate.
7. Carefully dry samples in the speed-vac without heating, until all acetonitrile has evaporated (2–3 μL of final volume) (*see* **Note 16**).
8. Resuspend the dried peptides in 2% acetonitrile/0.1% TFA.
9. Samples can be stored for a short term at 4 °C in a refrigerator, or for a long term at −80 °C in a freezer.

3.6 Liquid Chromatography (LC), Mass Spectrometry (MS), and Mass Spectrometry (MS) Analysis (See Note 17)

1. Analyze peptides by LC using the EASY-nLC1000 HPLC coupled to the Q-Exactive (QE) Plus or Q-Exactive HF mass spectrometers.
2. Perform MS acquisition in a data-dependent manner, with a positive-ion mode and selection of the top ten peptides from each MS spectrum for fragmentation and MS/MS analysis.
3. Acquire the full MS spectra at a resolution of 70,000 (QE-Plus) or 60,000 (QE-HF), m/z range of 300–1800 Th, with AGC target of 3E+06 ions and maximal injection time of 20 ms (QE-Plus) or 100 ms (QE-HF).
4. Peptides were isolated for fragmentation with an isolation window of 1.6 m/z. Perform higher energy collisional dissociation (HCD) fragmentation with normalized collisional energy of (NCE) 25 (QE-Plus) or 27 (QE-HF).
5. Acquire the MS/MS spectra at a resolution of 17,500 (QE-Plus) or 30,000 (QE-HF), with AGC target of 1E+05 (QE-Plus) or 5E+04 (QE-HF) and a maximal injection time of 60 ms (QE-Plus) or 50 ms (QE-HF). Set the dynamic exclusion to 30 s.

3.7 Database Searching and Protein Identification

1. Analyze the raw MS files in the MaxQuant software (version 1.5.2.18) and the Andromeda search engine.
2. MS/MS spectra are searched against the UniProt database (UniProt, http://www.uniprot.org) [30], a decoy, reverse database of the same size, and a list of common contaminants.
3. Searching parameters include fixed modification: carbamidomethyl-cysteine, and variable modifications: N-terminal acetylation and methionine oxidation.
4. MaxQuant search parameters for the initial mass recalibration of the precursors were 20 ppm, and in the main search, the mass tolerance for precursor and fragment ions was 4.5 ppm and 20 ppm, respectively.

4 Notes

1. For an accurate determination of the total number of cells, the cell count should be between 20 and 50 cells per one single square of the hemocytometer. If the cell density is over 100 cells per square, the cell suspension should be diluted.
2. The choice of antibody or ligand is the most important factor for successful target capture. Not all antibodies are suitable for functional assays. The coupling of antibodies or other proteins stabilized in glycerol is not recommended. The presence of antibody aggregates in the antibody stock used for coupling

can result in antibody leakage during the downstream assay. To help reduce this, we recommend removing the antibody aggregates from the antibody stock by centrifugation at 16,000 × *g* for 10 min at 4 °C. The moisture on the unused beads deactivates the reactive groups necessary for covalent antibody coupling. To avoid condensation on unused beads, ensure that the beads are at room temperature prior to opening the bottle.

3. Dynabeads-coupled antibodies can be stored for up to 2 weeks at 4 °C or may be made fresh a day before usage.
4. It is preferred to perform T-cell activation as soon as possible after the isolations, and no more than a day after T-cell isolation.
5. The enriched RPMI medium should be warmed up to 37 °C before use.
6. All centrifugations for exosome isolation and purification should be performed at 4 °C. Ensure that none of the pellet is collected and contaminates the supernatant. Use a pipette (rather than pouring off the supernatant), and leave behind half a centimeter of liquid above the pellet.
7. Overall, the contamination of isolated exosomes with other particles should be avoided. Exosomes are to be isolated from cultured media; an important consideration is to use either serum-free media or exosome-free fetal bovine serum.
8. Tubes should be weighed with samples to ensure that weight distribution is equal during centrifugation. It is crucial to properly balance ultracentrifuge tubes.
9. For ultracentrifugation, mark each ultracentrifuge tube with a waterproof marker and orient the tube in the rotor with the mark facing up. The mark is a reference for the location of a pellet following centrifugation.
10. After the first ultracentrifugation the exosome pellet will not be visible. For swinging-bucket rotors, the pellet is at the bottom of the tube. For fixed-angle rotors, the pellet is on the side of the tube near the base. When removing the supernatant with the pipette, hold the tube at an angle, so that the pellet is always covered with supernatant, and leave 2 mm of supernatant above the pellet.
11. Exosome sterility is not necessary in cases of biochemical analyses. The use of filtered and sterile PBS is recommended to ensure that no contaminants are included during preparation and solubilization.
12. For a complete supernatant removal following an ultracentrifugation wash, for fixed-angle rotors, pour the supernatant rather than using a pipette. For swinging-bucket rotors, remove the supernatant with the pipette.

13. To characterize extracellular vesicles as exosomes, it is important to demonstrate the expression of common exosome proteins using immunoblotting. The commonly used markers are CD63 (tetraspanin CD63), CD81 (tetraspanin CD81), TSG101 (tumor susceptibility gene 101), Alix (PDCD6IP, programmed cell death 6 interacting protein), and HSP70 (heat-shock protein 70).
14. MS sample vials should be low binding.
15. IAA solution pH should be exactly 8.0.
16. When drying samples in the speed-vac to evaporate acetonitrile, do not completely dehydrate/overdry the peptide sample.
17. The experiment should be carried out with at least three biological replicates of each treatment. Biological replicates are essential and can be easily obtained by different blood donors.

Acknowledgments

This work was supported by the National Institute of Health (R01 AI125640) and Cancer Research Institute grants to A.M.

References

1. Raposo G, Stoorvogel W (2013) Extracellular vesicles: exosomes, microvesicles, and friends. J Cell Biol 200(4):373–383
2. Zhang Y, Liu Y, Liu H et al (2019) Exosomes: biogenesis, biologic function and clinical potential. Cell Biosci 9:19
3. Hessvik NP, Llorente A (2018) Current knowledge on exosome biogenesis and release. Cell Mol Life Sci 75(2):193–208
4. Rashed HM, Bayraktar EK, Helal G et al (2017) Exosomes: from garbage bins to promising therapeutic targets. Int J Mol Sci 18:538
5. Sastre B, Canas JA, Rodrigo-Muñoz JM et al (2017) Novel modulators of asthma and allergy: exosomes and microRNAs. Front Immunol 8:826
6. Properzi F, Logozzi M, Fais S (2013) Exosomes: the future of biomarkers in medicine. Biomark Med 7:769–778
7. Abels ER, Breakefield XO (2016) Introduction to extracellular vesicles: biogenesis, RNA cargo selection, content, release, and uptake. Cell Mol Neurobiol 3:301–312
8. Stevanato L, Thanabalasundaram L, Vysokov N et al (2016) Investigation of content, stoichiometry and transfer of miRNA from human neural stem cell line derived exosomes. PLoS One 11(1):e0146353
9. Ventimiglia LN, Alonso MA (2016) Biogenesis and function of T cell-derived exosomes. Front Cell Dev Biol 4:84
10. Azoulay-Alfaguter I, Mor A (2018) Proteomic analysis of human T cell-derived exosomes reveals differential RAS/MAPK signaling. EJI 48(11):1815–1917
11. Li P, Kaslan M, Lee SH et al (2017) Progress in exosome isolation techniques. Theranostics 7 (3):789–804
12. Witwer KW (2013) Standardization of sample collection, isolation and analysis methods in extracellular vesicle research. J Extracell Vesicles 2. https://doi.org/10.3402/jev.v2i0.20360
13. Chen C, Skog J, Hsu CH et al (2010) Microfluidic isolation and transcriptome analysis of serum microvesicles. Lab Chip 10(4):505–511

14. Yamamda T, Inoshima Y, Matsuda T et al (2012) Comparison of methods for isolating exosomes from bovine milk. J Vet Med Sci 74 (11):1523–1525
15. Zeringer E, Barta T, Li M et al (2015) Strategies for isolation of exosomes. Cold Spring Harb Protoc 2015(4):319–323
16. Batrakova EV, Kim MS (2015) Using exosomes, naturally-equipped nanocarriers, for drug delivery. J Control Release 219:396–405
17. Quintana JF, Makepeace BL, Babayan SA et al (2015) Extracellular Onchocerca-derived small RNAs in host nodules and blood. Parasit Vectors 8:58
18. Dragovic RA, Gardiner C, Brooks AS et al (2011) Sizing and phenotyping of cellular vesicles using nanoparticle tracking analysis. Nanomedicine 7(6):780–788
19. Lim J, Yeap SP, Che HX et al (2013) Characterization of magnetic nanoparticle by dynamic light scattering. Nanoscale Res Lett 8(1):381
20. Graham M (2013) The coulter principle: imaginary origins. Cytometry A 83:1057–1061
21. Sharma S, Gillespie BM, Palanisamy V et al (2011) Quantitative nanostructural and single-molecule force spectroscopy biomolecular analysis of human-saliva-derived exosomes. Langmuir 27:14394–14400
22. Aras O, Shet A, Bach RR et al (2004) Induction of microparticle- and cell-associated intravascular tissue factor in human endotoxemia. Blood 103:4545–4553
23. Petersen K, Manangon E, Hood JL et al (2014) A review of exosome separation techniques and characterization of B16-F10 mouse melanoma exosomes with AF4-UV-MALS-DLS-TEM. Anal Bioanal Chem 406:7855–7866
24. Pospichalova V, Svoboda J, Dave Z et al (2015) Simplified protocol for flow cytometry analysis of fluorescently labeled exosomes and microvesicles using dedicated flow cytometer. J Extracell Vesicles. 4:25,530
25. He M, Crow J, Roth M et al (2014) Integrated immunoisolation and protein analysis of circulating exosomes using microfluidic technology. Lab Chip 14:3773–3780
26. Ramirez MI, Amorim MG, Gadelha C (2018) Technical challenges of working with extracellular vesicles. Nanoscale 10(3):881–906
27. Qiu H, Shi S, Wang S et al (2018) Proteomic profiling exosomes from vascular smooth muscle cell. Proteomics Clin Appl 12(5):e1700097
28. Higginbotham JN, Zhang Q, Jeppesen DK et al (2016) Identification and characterization of EGF receptor in individual exosomes by fluorescence-activated vesicle sorting. J Extracell Vesicles 5:29,254
29. Barnabas GD, Bahar-Shany K, Sapoznik S et al (2019) Microvesicle proteomic profiling of uterine liquid biopsy for ovarian cancer early detection. Mol Cell Proteomics 18 (5):865–875
30. Alpi E, Griss J, da Silva AW (2015) Analysis of the tryptic search space in UniProt databases. Proteomics 15(1):48–57

Chapter 8

Analysis of Immune-Tumor Cell Interactions Using a 3D Co-culture Model

Tanya N. Augustine

Abstract

Three-dimensional (3D) cultures are better able to reflect the tumor microenvironment than two-dimensional (2D) monolayer cultures by facilitating cell-cell interactions in the appropriate spatial dimensions. Here I describe the isolation and co-culture of immune cells with tumor cell lines in a three-dimensional system facilitated by a basement membrane extract. This allows for further downstream applications to analyze interactions between these cell types.

Key words 3D cell models, 3D culture, Lymphocyte culture, Co-culture, Tumor microenvironment

1 Introduction

Three-dimensional (3D) culture systems are a suggested intermediary between reductionist two-dimensional (2D) monolayer culture models, which may be unable to recapitulate the in vivo microenvironment [1–3], and animal models in which human immunological processes may not be adequately simulated [4]. Here we detail a partial recreation of the tumor microenvironment, wherein magnetically isolated immune cell subsets from human participants are cultured with tumor cell lines in a biologically relevant scaffold. This allows for the assessment of immune-tumor cell interactions in spatial dimensions that reflect the in vivo tumor microenvironment. This model may be adapted for different cell types, and for determining responses to various therapeutic agents.

2 Materials

Prepare all solutions as per the relevant local health and safety regulations, and collect samples as per the appropriate ethics

Suresh Mishra (ed.), *Immunometabolism: Methods and Protocols*, Methods in Molecular Biology, vol. 2184,
https://doi.org/10.1007/978-1-0716-0802-9_8,

approvals. Use personal protective equipment (such as gloves) and work within a biohazard safety unit with an aseptic technique.

2.1 The Isolation of Peripheral Blood Mononuclear Cells (PBMC)

1. Ethylenediaminetetraacetic acid (EDTA)-vacutainers, 18-gauge vacutainer needles (*see* **Note 1**).
2. Phosphate-buffered saline (PBS), 8.5 g NaCl, 1.07 g Na_2HPO_4, 0.39 g $Na_2H_2PO_4*2H_2O$, 1 L distilled H_2O. pH to 7.4 or 7.2 as needed. Autoclave to sterilize.
3. Ficoll-Hypaque (1.077 g/cm^3) (*see* **Note 2**).
4. 0.4% Trypan blue dye, a hemocytometer, or an automated cell counter (*see* **Note 3**).
5. Sorting buffer, 0.5% bovine serum albumin (BSA), 2 mM EDTA in 1 M PBS pH 7.2 (*see* **Note 4**).

2.2 Magnetic Cell Sorting

1. MACS® separation columns (Miltenyi Biotec) for positive isolation of lymphocytes, magnetic separator and stand, collection tubes. CD4$^+$ Multisort Microbead Kit, MACS Multisort Release Reagent, MACS Multisort Stop Reagent, CD25 Microbeads (*see* **Note 5**).
2. Activation medium A: RMPI 1640 supplemented with 0.1% penicillin-streptomycin, 10% fetal bovine serum (FBS), phytohemagglutinin (PHA) 1 μg/mL, and IL-2 0.8 ng/mL (filter sterilize) (*see* **Note 6**).
3. Activation medium B: RMPI 1640 supplemented with 0.1% penicillin-streptomycin, 10% FBS, and IL-2 0.8 ng/mL (filter sterilize) (*see* **Note 6**).

2.3 3D Immune Cell-Tumor Cell System

1. Growth factor-reduced Matrigel (GFRM) (*see* **Note** 7).
2. RPMI 1640 supplemented with 10% FBS and 0.1% penicillin-streptomycin.

2.4 Harvesting 3D Cultures

1. β-Mercaptoethanol in lysis buffer.
2. FBS.
3. 4% Paraformaldehyde (*see* **Note 8**).
4. Eosin.
5. Thrombin 20 U/mL in PBS.
6. A series of absolute alcohols: 50%, 70%, 95% diluted with distilled water.
7. Chloroform.
8. Paraffin wax.

3 Methods

3.1 The Isolation of Peripheral Blood Mononuclear Cells (PBMC)

1. Collect venous blood (approximately 30 mL) via venipuncture in EDTA-vacutainers.
2. Dilute blood 1:1 with 1 M PBS (pH 7.4) in 50 mL conical tubes. Use an appropriately sized sterile serological pipette and controller to gently layer the solution at a ratio of 3:2 onto Ficoll-Hypaque in 50 mL tubes (*see* **Notes 1** and **2**).
3. Centrifuge at 400 × *g* for 30 min at 20 °C.
4. Transfer the white PBMC layer using a pipette tip (volume of 1000 μL or 5 mL) and resuspend in 1 mL PBS.
5. Centrifuge at 400 × *g* for 5 min at 20 °C to remove excess Ficoll-Hypaque (*see* **Note 2**).
6. Resuspend the PBMC pellet in 1 mL PBS and determine the number of viable cells by the trypan blue exclusion assay using a hemocytometer or automated cell counter (*see* **Note 3**).
7. Wash the pellet by centrifuging the suspension again at 400 × *g* for 5 min at 20 °C.
8. Discard the supernatant and resuspend the pellet in sorting buffer, in preparation for magnetic cell sorting (*see* **Note 4**).

3.2 Magnetic Cell Sorting: Isolation of a $CD4^+CD25^+$ Lymphocyte Population (See Note 9)

1. Place the MACS® column in the magnetic field by attaching it to the separator on the stand. Place a collection tube for effluent under the column. Prime the column by applying 3 mL sorting buffer to the reservoir and allow the solution to filter through it. Discard the effluent and replace with a new 15 mL tube to collect unlabeled cells, should you wish to isolate other cell types (*see* **Notes 4** and **5**).
2. For a positive selection isolation of $CD4^+$ lymphocytes use a $CD4^+$ Multisort Microbead Kit.

   Incubate PBMC with 20 μL CD4 Multisort Microbeads per 1×10^7 cells for 15 min at 8 °C in the dark.
3. Centrifuge cell suspension in 1 mL sorting buffer at 400 × *g* for 10 min.
4. Resuspend in 1 mL sorting buffer and add to the column reservoir. Allow suspension to flow through (*see* **Note 10**).
5. Once the reservoir is empty, add 3 mL of the sorting buffer and allow for flow through. Repeat twice.
6. Retain the unlabeled PBMC for further sorting of populations if required.
7. Remove the column from the separator, and place over a new collection tube. Add 3 mL of sorting buffer to the reservoir and flush out the $CD4^+$ cells using a plunger.

8. For the removal of residual magnetically labeled cells and enrichment of the CD4⁺ population: incubate the CD4⁺ cell suspension with 20 μL of the MACS Multisort Release Reagent per 3 mL of sorting buffer for 10 min at 8 °C in the dark.
9. During the incubation period, add a new MACS® column to the separator. Prime with 3 mL of the sorting buffer and empty the discard collection tube. Place a new 15 mL collection tube under the column.
10. After the incubation period has lapsed, apply the cell suspension to the column reservoir and allow it to flow through. When the reservoir is empty, add 3 mL of sorting buffer to the column. Repeat twice.
11. Discard the column. Retain the cells that have passed through the column into the 15 mL collection tube.
12. Centrifuge the cell suspension at 400 × *g* for 5 min.
13. Resuspend the purified CD4⁺ cell pellet in 1 mL of sorting buffer to assess the number of viable cells using the trypan blue exclusion assay.
14. Centrifuge the cell suspension at 400 × *g* for 5 min.
15. For the removal of Multisort Microbeads from the enriched CD4⁺ cells: resuspend the cell pellet at a final concentration of 1×10^7 cells per 50 μL sorting buffer. Add 30 μL MACS Multisort Stop Reagent per 1×10^7 cells and mix well.
16. For the labeling of (CD4⁺) CD25⁺ cells, add 10 μL CD25⁺ Microbeads to the cell suspension. Add the sorting buffer to adjust to a final volume of 100 μL and incubate for 15 min at 8 °C.
17. During this incubation period, prime a new MACS® column with 3 mL of sorting buffer and retain the discard collection tube.
18. When the incubation period has lapsed, add 1 mL of sorting buffer per 1×10^7 cells to the cell suspension.
19. Centrifuge at 400 × *g* for 10 min and resuspend the cell pellet (up to 1×10^8 cells) in 500 μL of sorting buffer.
20. Add the cell suspension to the column reservoir and allow it to flow through. When the reservoir is empty, add a further 3 mL of sorting buffer, twice.
21. Remove the column from the separator and place over a new 15 mL collection tube. Add 3 mL of sorting buffer to the reservoir and flush out the labeled cells (CD4⁺CD25⁺) with a plunger.
22. Count cells using the trypan blue activation assay. Centrifuge the cell suspension at 400 × *g* for 5 min.

3.3 Activation and Expansion of Lymphocyte Populations

1. Discard the supernatant and resuspend isolated cells at a maximum of 1×10^5 cells/100 μL in activation medium A per well of a 96-well plate for 18 h.
2. To continue expansion, remove the cell suspension from each well into 1.5 mL microcentrifuge tubes and add 1 mL of PBS (pH 7.4).
3. Centrifuge at 400 × *g* for 5 min.
4. Discard the supernatant and resuspend in activation medium B for a further 30 h. Proceed to Subheading 3.4, **step 2** (*see* **Note 11**).

3.4 3D Immune Cell-Tumor Cell Co-culture

1. Thaw the cryopreserved tumor cells (e.g., breast tumor luminal phenotype MCF-7 cell line) at room temperature. Resuspend in 1 mL of PBS pH 7.4 and centrifuge at 400 × *g* for 5 min, twice. Discard the supernatant and resuspend the cells in 1 mL of RPMI 1640 medium supplemented with 10% FBS and 0.1% penicillin-streptomycin. Determine the number of viable cells.
2. Aspirate the expanded lymphocyte suspensions and resuspend in 1 mL PBS. Centrifuge at 400 × *g* for 5 min. Discard the supernatant and resuspend the cell pellet in 1 mL of RPMI 1640 cell culture medium supplemented with 10% FBS and 0.1% penicillin-streptomycin. Determine the number of viable cells.
3. Determine the appropriate number of cells required at specific ratios (e.g., immune cell 1:2 tumor cells). Calculate the required volume of cell suspensions.
4. Combine cell suspensions (immune cells and tumor cells) in appropriate volumes in microcentrifuge tubes. Centrifuge at 400 × *g* for 5 min. Discard the supernatant.
5. Working on a cooling block, resuspend each cell pellet in 66 μL of RPMI 1640 cell culture medium and add 200 μL of GFRM. Mix gently (*see* **Note 12**).
6. Plate in quadruplicate, approximately 65 μL/well, in a 96-well plate on a cooling block.
7. Allow the GFRM to settle at room temperature for 5 min. Incubate at 37 °C for 30 min to allow the GFRM to solidify.
8. Overlay the 3D cultures with 100 μL of the RPMI 1640 cell culture medium supplemented with 10% FBS and 0.1% penicillin-streptomycin.
9. Incubate at 37 °C in 5% CO_2 for 72 h.
10. Check cell culture media levels daily and add cell culture medium if required.

3.5 Harvesting of 3D Cultures for RNA Extraction

1. Remove culture medium.
2. Rinse cultures in 1 M PBS, pH 7.4.
3. Add 150 μL of 0.01% β-mercaptoethanol in lysis buffer to each culture.
4. Transfer the lysate to 1.5 mL microcentrifuge tubes and homogenize using a 21-gauge needle.
5. Snap freeze in liquid nitrogen and store at −80 °C until it is time for the standard RNA extraction procedure.

3.6 Harvesting of 3D Cultures for Immunocyto-chemistry (See Note 13)

1. Remove culture medium.
2. Rinse cultures in 1 M PBS, pH 7.4.
3. Incubate with 30 μL FBS at 37 °C for 1 min.
4. Add 30 μL (20 U/mL) of thrombin, mix gently, and incubate for approximately 1 min at 37 °C.
5. Incubate at room temperature for 5 min to ensure clot formation.
6. Add 200 μL of fresh 4% paraformaldehyde and incubate at room temperature for 1 h.
7. Remove fixative and add two drops of eosin with a transfer pipette (*see* **Note 14**).
8. Remove sample with the transfer pipette and sandwich in 0.45 μm filter paper (*see* **Note 15**).
9. Place into cassette and follow tissue-processing methodology (50% alcohol for 20 min, 70% alcohol for 20 min, three changes of 95% alcohol for 5 min, three changes of absolute alcohol for 15 min, two changes of chloroform for 10 min, and two changes of paraffin wax for 10 min).
10. Embed samples in paraffin wax and store in fridge for subsequent use.

4 Notes

1. When collecting blood, keep vacutainers in a rack in a polystyrene box with ice. Keep cool and process immediately. Note that the gauge of the vacutainer needle employed should be large enough to limit any shearing of lymphocytes and thus reduce viability, and small enough to lessen donor discomfort.
2. To layer diluted blood onto Ficoll-Hypaque, use an appropriately sized sterile serological pipette to contain the volume. Tilt the 50 mL tube slightly and pipette along the side of the tube, gently and quickly to layer diluted blood onto Ficoll-Hypaque. Avoid allowing blood to sink beneath the Ficoll-Hypaque. Following centrifugation, when transferring the PBMC layer,

take care not to aspirate the Ficoll. Wash steps are limited to preserve viability but should excess Ficoll be aspirated, a further wash step may be needed. PBS and FBS may be brought to room temperature, but keep the Ficoll cool.

3. For the trypan blue exclusion assay: dilute trypan blue (filtered) with cell suspension at a 1:1 ratio. 10 μL should suffice.
4. Prepare the sorting buffer the day prior to use. The sorting buffer should be filter sterilized (0.22 μm). Consider using a sonicator if available. Careful pipetting will avoid the development of bubbles during magnetic cell sorting, which can prevent the flow-through of cell suspension in the MACS® sorting column.
5. All reagents used in magnetic cell sorting must be maintained at 8 °C during the experiment.
6. Prepare the activation media fresh when the cell numbers have been determined.
7. Store aliquots of GFRM at −20 °C and defrost at 8 °C overnight, prior to experimentation.
8. Conduct standard preparation in a fume hood. Take all health and safety precautions. Make fresh the day prior to experimentation or freeze aliquots at −20 °C and thaw overnight in fridge.
9. Lymphocyte subgroups can be isolated using either magnetic cell sorting or other techniques, including flow cytometry. Subheading 3.2, **step 1**, is an example; note that an unlabeled fraction may be kept for the subsequent isolation of additional subgroups for multiple groups. This ensures the best use of a small volume of blood. Note that wash steps are kept to a minimum to ensure the retention of viability.
10. Allow the cell suspension to flow through the column unaided. Pipette gently to ensure that the BSA in the sorting buffer does not bubble.
11. It is advised to conduct flow cytometry analysis prior to and post-expansion/activation of lymphocytes to validate phenotype and viability of cells. Note that positive selection may result in the downregulation of markers if the same markers are assessed using flow cytometry.
12. Place all solutions and pipette tips on ice when working with GFRM. When plating cells within the GFRM, use a cooling block (e.g. stainless steel block that had been frozen for 24 h).

    Resuspend cells at a minimum requirement of 1×10^5 cells in a final volume of 50 μL of GFRM, diluted with 16.5 μL of the cell suspension (3:1)—this would account for growth for 72 h, as well as allow a bit of fluidity of the GFRM (it is very viscous) for accurate pipetting while maintaining the ability to solidify. Use larger pipette tips (1000 μL) when working with the GFRM. The volume of GFRM chosen is for cost efficiency for the number of lymphocytes able to be isolated and expanded from the initial volume of blood, and for the number of co-cultures to be generated.

13. The integrity of the sample depends on the activity of the cells, the number of cells, and the duration of the culture. For example, excessive migration can render the GFRM more fluid than gel-like. This morphology is difficult to process for immunocytochemistry and should be optimized further.
14. The eosin only allows for visualization of the sample.
15. Cut off the tip of the transfer pipette to make the aperture wider. This will assist in transferring the sample.

References

1. Augustine TN, Dix-Peek T, Duarte R et al (2015) Establishment of a heterotypic 3D culture systems to evaluate the interaction of TREG lymphocytes and NK cells with breast cancer. J Immunol Methods 426:1–15. https://doi.org/10.1016/j.jim.2015.07.003
2. Gonzalez MG, Cichon I, Scislowska-Czarnecka A et al (2018) Challenges in 3D culturing of neutrophils: assessment of cell viability. J Immunol Methods 457:73–77. https://doi.org/10.1016/J.JIM.2018.02.015
3. Domaica CI, Fuertes MB, Rossie LE et al (2009) Tumour-experienced T cells promote NK cell activity through trogocytosis of NKG2D and NKp46 ligands. EMBO Rep 10 (8):908–915
4. Wager LE, DiFazio RM, Davis MM (2018) Advanced model systems and tools for basic and translational human immunology. Genome Med 10:73. https://doi.org/10.1186/s13073-018-0584-8

Chapter 9

Studying Adipocyte and Immune Cell Cross Talk Using a Co-culture System

Jennifer M. Monk, Danyelle M. Liddle, Amber L. Hutchinson, and Lindsay E. Robinson

Abstract

The co-culture of adipocytes and immune cells, such as macrophages or T cells ($CD4^+$ or $CD8^+$ subsets), is a novel experimental approach used to study paracrine interactions (or the cross talk) between cultured cell types in isolation, in order to understand their role in obese adipose tissue (AT) inflammation and dysfunction. Here we describe the general methodologies required for the co-culture of mature adipocytes (differentiated 3T3-L1 pre-adipocyte cell line) with primary immune cell subsets purified from mouse splenic mononuclear cells using a magnetic MicroBead positive selection, wherein multiple immune cell populations can be purified sequentially from a single mouse spleen, thereby providing diversity in the types of immune cells that can be co-cultured with adipocytes. Additionally, we describe experimental procedures for co-culturing adipocytes and immune cells in two different co-culture systems, including a cell contact-dependent co-culture system, wherein the cells are in direct physical contact, and a cell contact-independent, soluble mediator-driven co-culture system wherein the cells are physically separated by a trans-well semipermeable membrane. Finally, we discuss how these co-culture models can be utilized to recapitulate the AT microenvironment in obesity by utilizing physiologically relevant ratios of adipocytes: immune cells (specifically $CD11b^+$ macrophages, $CD4^+$ T cells, or $CD8^+$ T cells) and lipopolysaccharide stimulation that mimics endotoxin concentrations observed in obesity.

Key words Adipocyte, Macrophage, T cell, Co-culture, Obesity, Inflammation

1 Introduction

The cellular composition of adipose tissue (AT) is diverse and includes both adipocytes and various immune cell types that comprise the stromal vascular fraction. Paracrine interactions (or cross talk) between these cell types promote the secretion of inflammatory mediators resulting in increased local (i.e., within AT) and systemic inflammation, and the development of metabolic dysfunction [1–4]. Furthermore, in obesity, immune cells are recruited to

Jennifer M. Monk and Danyelle M. Liddle contributed equally to this work.

Suresh Mishra (ed.), *Immunometabolism: Methods and Protocols*, Methods in Molecular Biology, vol. 2184, https://doi.org/10.1007/978-1-0716-0802-9_9,

the inflamed AT, which alters the cellular ratio of adipocytes to specific immune cell populations and further promotes the sustained (i.e., chronic) production of inflammatory and immune cell chemotactic signals that are central components of dysfunctional obese AT [1–6]. Therefore, the adipocyte and immune cell cross talk in AT represents a critical intervention point to limit the severity of obesity-related chronic inflammation and metabolic dysfunction. Due to the diversity of immune cell types that contribute to AT dysfunction in obesity [1, 3–5], it is important to understand how the paracrine interactions between adipocytes and specific immune cell populations contribute to overall AT inflammation, recruitment of other immune cell populations into the expanding AT, and subsequent metabolic dysfunction. This prompted our research team to develop adipocyte-primary immune cell co-culture models to recapitulate the obese AT cellular microenvironment, with a specific focus on two immune cell types: macrophages and T cells ($CD8^+$ and $CD4^+$ subsets) [7–12]. The methodologies required to establish these adipocyte-primary immune cell co-culture models are discussed herein; however, these experimental procedures can be utilized with any purified primary immune cell population of interest (isolated from any tissue site of interest; the spleen is described herein) or cell line. For example, we have also utilized an adipocyte-macrophage co-culture model exclusively with cell lines (i.e., 3T3-L1 pre-adipocytes differentiated into mature adipocytes and RAW264.7 macrophages) [13] and primary visceral AT conditioned media co-cultured with RAW264.7 macrophages [14], thereby demonstrating some possible alternative options for this co-culture model to study paracrine interactions between adipocytes and immune cells.

To standardize the contribution of adipocytes in immune cell co-culture models discussed herein, we utilize the murine 3T3-L1 pre-adipocyte cell line, which requires differentiation into lipid-laden mature adipocytes prior to co-culture with immune cells. For immune cells, we purified specific populations from a mouse spleen (to mimic the immune cells' first interaction with adipocytes), although primary immune cell populations can be isolated from any desired tissue site. To further recapitulate the obese AT microenvironment within the co-culture model, adipocytes and immune cells are stimulated with lipopolysaccharide (LPS) or endotoxin at the concentration reported in obese humans and rodents [15–17]. Finally, in this co-culture model, the immune cells (specifically macrophages and $CD8^+$ and $CD4^+$ T cell subsets) are co-cultured with adipocytes at a cellular ratio that recapitulates the percentage of each immune cell subset with the AT stromal vascular fraction [7–9, 13, 18]. This increases the translational relevance of this experimental approach, which was not included in earlier macrophage/adipocyte co-culture studies that demonstrated proof of concept for the relevance of paracrine interactions between these cell types in driving AT dysfunction [18–21].

Therefore, the methodologies provided herein outline the procedures for establishing adipocyte-primary immune cell co-cultures including (1) the growth and differentiation of 3T3-L1 pre-adipocytes into mature adipocytes; (2) the isolation of splenic immune cell populations wherein multiple immune cell types can be purified from a single splenic mononuclear cell population using positive selection with magnetic MicroBeads; and (3) the establishment of and sample collection from adipocyte/immune cell co-cultures under various conditions, namely when cells are in direct physical contact (cell contact-dependent co-culture) and when cells are separated by a trans-well semipermeable membrane (cell contact-independent, soluble mediator-driven co-culture). Furthermore, we outline different co-culture stimulation conditions that have been employed with adipocytes co-cultured with macrophages and T-cell ($CD4^+$ and $CD8^+$) subsets to demonstrate the scope of experimental manipulations of the co-culture model that could be utilized.

2 Materials

2.1 Equipment

1. Pipette controller/pipetting aid (compatible with 1, 5, 10 mL polystyrene, disposable serological pipettes).
2. 10, 200, 1000 μL adjustable single-channel pipettes.
3. 37 °C water bath.
4. Centrifuge.
5. Humidified incubator maintained at 37 °C and 5% CO_2.
6. Aspiration/vacuum waste collection system for cell culture.
7. Hemocytometer and coverslip.
8. Light microscope.
9. Laminar flow hood and aseptic work area.
10. Manual magnetic cell separator.

2.2 Cell Culture

1. 1, 5, 10 mL polystyrene serological pipettes, sterile, non-pyrogenic/endotoxin free, noncytotoxic.
2. 10, 200, 1000 μL polypropylene, sterile single-channel pipette tips.
3. T-25 culture flask: T-25 polystyrene tissue culture flask, 25 cm^2 growth area, sterile, non pyrogenic/endotoxin free, noncytotoxic, DNase/RNase/DNA free, canted neck, filter cap.
4. T-75 culture flask: T-75 polystyrene tissue culture flask, 75 cm^2 growth area, sterile, non-pyrogenic/endotoxin free, noncytotoxic, DNase/RNase/DNA free, canted neck, filter cap.

5. 15 mL Screw-cap, polypropylene tube, 120 × 17 mm, conical base, sterile, non-pyrogenic/endotoxin free, noncytotoxic (15 mL centrifuge tube).
6. 6-well culture plate: 6-well polystyrene cell culture plates, flat base, 8.87 cm^2 growth area, sterile, non-pyrogenic/endotoxin free, noncytotoxic, DNase/RNase/DNA free.
7. Transwell permeable supports: Sterile, polystyrene 6-well polystyrene cell culture plates, flat base (8.87 cm^2 growth area, sterile, non-pyrogenic/endotoxin free, noncytotoxic, DNase/RNase/DNA free) with a 24 mm insert and 0.4 μm polyester membrane.
8. Sterile 70 μm polypropylene, nylon mesh cell strainer.
9. Culture dish: Round polystyrene culture dish, 55 cm^2 surface area, 100 mm (D) × 20 mm (H), 11.0–16.5 mL volume, sterile, tissue culture-treated surface treatment.
10. Sterile 10 mL slip-tip syringe.
11. Sterile 1.5 mL polypropylene microcentrifuge tube, snap cap with flat top (1.5 mL microcentrifuge tube; sterile, DNase/RNase free).
12. MACS MS columns and plunger for immune cell elution from the column.
13. Cell scraper: Sterile polystyrene cell scraper for cell culture flasks, 20 mm blade width, 250 mm overall length, non-pyrogenic/endotoxin free, noncytotoxic.

2.3 Cell Line

1. Mouse 3T3-L1 preadipocytes (American Type Culture Collection, CL-173).
2. RAW264.7 cells (American Type Culture Collection, TIB-71™).

2.4 Cell Culture Media (See Note 1)

1. Basic medium: Dulbecco's modified Eagle's medium (DMEM) without sodium pyruvate supplemented with 4 mM L-glutamine, 4500 mg/L glucose, 10% (*v*/*v*) fetal bovine serum (FBS, low endotoxin, sterile filtered), 1% (*v*/*v*) penicillin-streptomycin (*see* **Note 2**).
2. Differentiation medium: Basic medium (contents outlined above) supplemented with 1 μM dexamethasone, 10 μg/mL insulin, 0.5 mM 3-isobutyl-1-methylxanthine (IBMX).
3. Maintenance medium: Basic medium (**item 1** above) supplemented with 10 μg/mL insulin.
4. Serum-free basic medium: Dulbecco's modified Eagle's medium (DMEM) without sodium pyruvate supplemented with 4 mM L-glutamine, 4500 mg/L glucose, 1% (*v*/*v*) penicillin-streptomycin.

2.5 Cell Culture Reagents

1. 70% (*v/v*) Ethanol.
2. 1× Phosphate-buffered saline (PBS) without calcium or magnesium, pH 7.4.
3. 0.25% (*w/v*) Trypsin-ethylenediaminetetraacetic acid (EDTA) solution, with phenol red.
4. 0.4% (*w/v*) Trypan blue solution diluted 1:10 in 1× PBS.
5. MACS MicroBeads: Magnetic CD8a (Ly-2) MicroBeads, mouse for positive selection of mouse $CD8a^+$ T cells from single-cell suspensions of lymphoid tissues.
6. MACS MicroBeads: Magnetic CD4 (L3T4) MicroBeads, mouse for positive selection of mouse $CD4^+$ T cells from single-cell suspensions of lymphoid tissues.
7. MACS MicroBeads: Magnetic CD11b (Mac-1α) MicroBeads, human and mouse for the positive selection of mouse $CD11b^+$ cells from single-cell suspensions of lymphoid tissues.
8. AutoMACS Running Buffer: MACS Separation Buffer (contains bovine serum albumin, EDTA, phosphate-buffered saline, 0.09% azide, pH 7.2, 0.2 μm filtered).
9. Cell lysis buffer: RNA/DNA/protein purification plus kit supplemented with 10 μL/mL 2-mercaptoethanol.

3 Methods

3.1 Thawing 3T3-L1 Pre-adipocytes

Steps 3–7 should be carried out under strict aseptic conditions.

1. When initially starting to culture the mouse 3T3-L1 preadipocytes, the cell line stock will be frozen. Thaw the vial by gentle agitation in a 37 °C water bath. To reduce the possibility of contamination, keep the O-ring and cap out of the water. Thawing should be rapid (approximately 2 min).
2. Remove the vial from the water bath as soon as the contents are thawed, and decontaminate by dipping in or spraying with 70% ethanol.
3. Transfer the vial contents to a 15 mL centrifuge tube containing 9.0 mL of the basic medium and centrifuge at approximately 400 × *g* for 5 min at room temperature.
4. Decant the supernatant without disturbing the cell pellet. Keep the centrifuge tube inverted and gently blot to remove as much supernatant as possible.
5. Resuspend the cell pellet with 1 mL of the basic medium and dispense into the T-25 culture flask containing 4 mL of the basic media (*see* **Note 3**).

6. Gently slide the T-25 culture flask back and forth, and side to side, to evenly distribute the cells across the culture surface.
7. Incubate the cells until they reach 70% confluence (i.e., the cells have divided enough to cover 70% of the culture flask surface area; after approximately 2 days) and are ready to be subcultured into T-75 culture flasks to continue dividing (*see* **Note 4**).

3.2 Subculturing 3T3-L1 Pre-Adipocytes for Continued Growth

1. Invert the culture flask so that the monolayer of adherent pre-adipocytes is on top and the medium is at the bottom.
2. Gently aspirate the medium from the bottom of the culture flask (avoiding the cell monolayer on the top; *see* **Note 5**).
3. Wash the cells by adding 5 mL of PBS to the bottom of the culture flask (avoiding the cell monolayer on the top; *see* **Note 6**).
4. Return the culture flask to its original position (with the cell monolayer on the bottom) and gently tilt the flask back and forth, and from side to side, to wash the cell monolayer with PBS.
5. Invert the culture flask again and gently aspirate the PBS from the bottom of the culture flask (avoiding the cell monolayer on the top). When finished, return the culture flask to its original position (with the cell monolayer on the bottom).
6. Repeat the washing with PBS (**steps 4** and **5**) one more time.
7. Slowly add trypsin to the culture flask, directly onto the cell monolayer (*see* **Note 7**). Add 1.5 mL trypsin to a T-25 flask or 2.5 mL trypsin to a T-75 culture flask. Gently tilt the culture flask back and forth, and from side to side, to ensure that the trypsin covers the cell monolayer.
8. Incubate the cells with trypsin for 5 min (*see* **Note 8**).
9. Deactivate the trypsin by adding the basic medium to the bottom of the culture flask, directly onto the cell monolayer. Add 3.5 mL of the basic media to a T-25 culture flask or 7.5 mL of the basic media to a T-75 culture flask.
10. Holding the culture flask at a 45° angle, use a serological pipette to rinse the bottom surface of the culture flask and direct the detached cells to one of the bottom corners of the flask. Repeat the rinse approximately five times with appreciable force in order to collect as many detached cells as possible.
11. Transfer the entire volume of trypsin + basic media containing the detached cells (5 mL from a T-25 culture flask or 10 mL from a T-75 culture flask) to a 15 mL centrifuge tube and centrifuge at 400 × *g* for 5 min at room temperature.

12. Decant the supernatant without disturbing the cell pellet. Keep the centrifuge tube inverted and gently blot it on an absorptive surface to remove as much supernatant as possible.
13. Resuspend the cell pellet in 0.5–2 mL basic medium, depending on the size of the cell pellet (*see* **Note 9**).
14. Add 10 μL of the resuspended cells to 10 μL of trypan blue solution to a 1.5 mL microcentrifuge tube and mix by pipetting.
15. Add 10 μL of the resuspended cells + trypan blue mixture to one chamber of a hemocytometer under a coverslip.
16. Using a light microscope, focus on the grid lines of the hemocytometer with a 10× objective. Count the live, unstained cells in each of the four quadrants (i.e., each set of 16 squares; *see* **Note 10**). Alternatively, cell number may be counted using an automated cell counter. If required, dead, stained cells can also be counted separately for a viability estimate.
17. To calculate the viable cell density (i.e., the number of viable cells/mL): Calculate the average number of viable cells from the four quadrants of the hemocytometer. Multiply by 10,000. Multiply by 2 to correct for the 1:2 dilution of the resuspended cells in trypan blue solution.
18. If required, calculate the percentage of viable cells: Calculate the number of dead cells/mL (as in **step 17**). Add the number of viable cells/mL and the number of dead cells/mL to obtain a total cell count/mL. Divide the number of viable cells/mL by the total cell count/mL. Multiply by 100 to obtain a percentage.
19. Calculate the volume of resuspended cells that would yield 300,000 cells (*see* **Note 11**).
20. Add the calculated volume of cells to 10 mL basic media in a T-75 culture flask (*see* **Note 3**). Gently slide the T-75 culture flask back and forth, and side to side, to evenly distribute the cells across the culture surface.
21. Incubate the cells until they reach 70% confluence (after approximately 2 days) and are ready to be subcultured into new T-75 culture flasks, or into 6-well culture plates for future experiments (*see* **Note 12**).

3.3 Subculturing 3T3-L1 Pre-adipocytes for Experimental Use

3T3-L1 pre-adipocytes must be subcultured into 6-well culture plates 15 days before experiential use (e.g., for co culture experiments).

1. Follow Subheading 3.2, **steps 1–16**.
2. Calculate the volume of resuspended cells that would yield 25,000 cells (*see* **Note 11**).

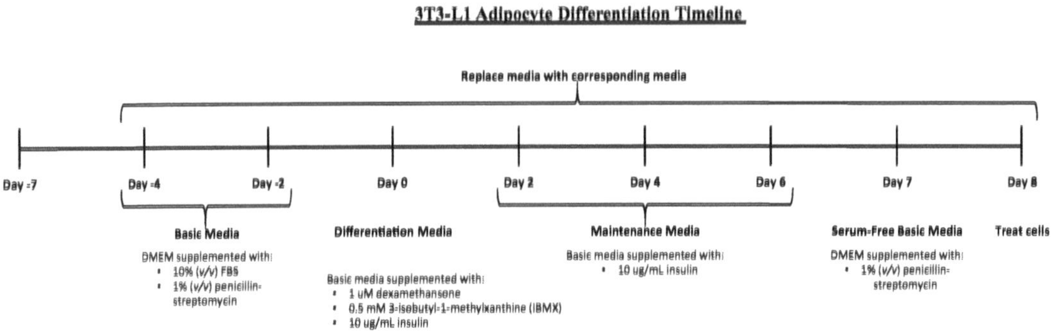

Fig. 1 3T3-L1 adipocyte differentiation timeline. Summary of the 3T3-L1 pre-adipocyte differentiation procedure into lipid-laden mature adipocytes for use in subsequent adipocyte/immune cell co-culture experiments

3. Add the calculated volume of cells to each well of a 6-well culture plate containing 2 mL basic medium (*see* **Note 3**). Gently slide the 6-well culture plate back and forth and side to side to evenly distribute the cells across the culture surface.
4. Incubate the cells. This time point is denoted as "day 7" before the cells are chemically differentiated into adipocytes (Fig. 1).
5. On day 4 (3 days after subculturing the cells into 6-well culture plates), replace the media with fresh basic medium. Tilt the 6-well culture plate on a 30° angle so that the media collects in the bottom half of each well.
6. Gently aspirate the media from the bottom side of each well to avoid disturbing the cell monolayer (*see* **Note 5**).
7. Return the 6-well culture plate to its original position and gently add 2 mL of fresh basic medium along the side of each well to avoid disturbing the cell monolayer.
8. On day 2 when the cells are 95–100% confluent with approximately 2 million cells in each culture well, replace the media with fresh basic medium again (as in **step 5**).

3.4 3T3-L1 Adipocyte Differentiation and Maintenance

3T3-L1 pre-adipocytes must be chemically differentiated 8 days prior to experimental use.

1. On day 0, follow Subheading 3.3, **step 5**, with a fresh quantity of differentiation medium (*see* **Note 13**).
2. On days 2, 4, and 6, follow Subheading 3.3, **step 5**, with some fresh maintenance medium (*see* **Note 14**).
3. On day 7, approximately 12 h prior to experimental use of the adipocytes on day 8, follow Subheading 3.3, **step 5**, with some fresh serum-free medium (*see* **Note 15**).

3.5 The Purification of Primary Splenic Immune Cell Populations

Multiple immune cell populations can be purified from a single mouse spleen for use in adipocyte/immune cell co-culture experiments. The isolation of a purified immune cell population for use in a co-culture is a two-step process. The first step is to generate a splenic mononuclear cell population, thereby removing nonimmune cells, such as red blood cells. The second step is the purification of specific immune cell populations from the mononuclear cell population.

3.5.1 The Preparation of a Splenic Mononuclear Cell Population

All steps should be carried out under strict aseptic conditions.

1. Aseptically remove the spleen and place it in a 15 mL centrifuge tube containing 5 mL basic medium. Store on ice until all spleens have been collected (*see* **Note 16**).
2. Place a 70 μm cell strainer in a 100 mm culture dish. Pour out the basic media and spleen into the cell strainer and ensure that the cell strainer surface is wet with basic medium.
3. Remove the plunger from a 10 mL syringe. Use the rubber end of the plunger to crush the spleen and push the small pieces of splenic tissue through the cell strainer (*see* **Note 17**).
4. Tilt the culture dish on a 45° angle to collect the 5 mL of splenic cell suspension at the bottom. Pipette the splenic cell suspension up and down to further disassociate the cells.
5. Add the 5 mL of splenic cell suspension back to the 15 mL centrifuge tube stored on ice.
6. Rinse the cell strainer with 5 mL of fresh basic medium over the culture dish to recover any remaining splenic cells. Continue to rinse the cell strainer until all the splenic tissue has passed through the cell strainer. Discard the cell strainer when finished, using one cell strainer per spleen.
7. Tilt the culture dish on a 45° angle and rinse the culture surface with the 5 mL of basic medium containing any remaining splenic cells.
8. Add the 5 mL of basic media containing any remaining splenic cells to the 15 mL centrifuge tube containing the splenic cell suspension.
9. Centrifuge at 350 × *g* for 5 min at room temperature.
10. Decant the supernatant and resuspend the cell pellet in 5 mL basic medium.
11. Slowly add 5 mL of Lympholyte-M to the bottom of the 15 mL centrifuge tube: insert a 5 mL sterile pipette containing the Lympholyte-M to the bottom of the tube. Lift the pipette tip up off the bottom of the tube by 2 cm and slowly dispense the 5 mL of Lympholyte-M. Two phases will develop in the

solution; the bottom phase will be clear (containing Lympholyte-M) and the top phase will be the color of basic medium (pink/red) (*see* **Note 18**).

12. Centrifuge at 800 × *g* for 22 min at room temperature (*see* **Note 19**).
13. There will be two phases visible still and the mononuclear cells will be visible at the interface between the clear and pink/red layers in the tube. Collect the top pink/red layer of liquid and the mononuclear cells at the interface (approximately 6 mL total). Try to avoid collecting the clear liquid from the bottom layer.
14. Transfer the mononuclear cells to a new 15 mL centrifuge tube.
15. Add basic medium to the 15 mL centrifuge tube to achieve 12 mL total volume.
16. Centrifuge at 350 × *g* for 5 min at room temperature.
17. Decant the supernatant and resuspend the cell pellet in 4 mL of MACS buffer (*see* **Note 20**). The mononuclear cell population is now ready for the purification of specific immune cell subsets using microbeads (*see* **Note 21**).

3.5.2 The Purification of Immune Cell Populations Using MACS MicroBeads

This section describes the purification of specific immune cell populations by positive selection, based on the surface marker expression of $CD4^+$, $CD8^+$, and $CD11b^+$ using MACS MicroBeads. Magnetic MicroBeads are available for the purification of other immune cell populations. From a single splenic mononuclear cell population, the protocol user can purify a single immune cell population, or multiple immune cell populations in succession. The purity of the positively selected immune cell populations should be confirmed using flow cytometry. We have successfully used this protocol to obtain immune cell purities >90% for $CD11b^+$ cells [7], $CD8^+$ cells [8–11], and $CD4^+$ cells [12].

1. Calculate the total number of viable cells: Add 10 μL of the resuspended mononuclear cells to 90 μL of trypan blue solution and mix.
2. Add 10 μL of the resuspended mononuclear cells + trypan blue mixture to one chamber of a hemocytometer under a coverslip.
3. Using a microscope, focus on the grid lines of the hemocytometer with a 10× objective. Count the live, unstained cells in each of the four quadrants (i.e., each set of 16 squares; *see* **Note 10**). If required, dead, stained cells can also be counted separately for a viability estimate.
4. Calculate the average number of viable cells from the four outer quadrants of the hemocytometer (*see* **Note 22**).

5. Multiply by 10,000.
6. Multiply by 10 to correct for the 1:10 dilution of the resuspended mononuclear cells in trypan blue solution.
7. Multiply by 4 to correct the volume of MACS buffer that the mononuclear cells are resuspended in.
8. If required, calculate the percentage of viable cells: Calculate the total number of dead cells (as Subheading 3.5.2, **step 1**).
9. Add the number of viable cells and the number of dead cells to obtain a total cell count.
10. Divide the number of viable cells by the total cell count. Multiply by 100 to obtain a percentage.
11. Having determined the total number of mononuclear cells, calculate the volume of MicroBeads and MACS buffer that would be required to produce a ratio of 10 μL of MicroBeads for every 90 μL MACS buffer containing 10^7 cells (i.e., add 10 μL of MicroBeads and 90 μL for every 10^7 cells).
12. Add 6 mL of MACS buffer to the mononuclear cell suspension and centrifuge at 300 × *g* for 10 min at room temperature.
13. Decant the supernatant and resuspend the mononuclear cells in the volume of MACS buffer calculated in **step 3** above.
14. Add the volume of MicroBeads calculated in **step 3** above to the mononuclear cell suspension. Gently mix the mononuclear cell and MicroBead suspension (*see* **Note 23**).
15. Incubate the mononuclear cell and MicroBead suspension for 15 min at 4 °C. Gently mix the suspension after 7 min.
16. Add the MACS buffer to the mononuclear cell and MicroBead suspension to achieve a total volume of 6 mL.
17. Centrifuge the mononuclear cell and MicroBead suspension at 300 × *g* for 10 min at room temperature.
18. Decant the supernatant and resuspend the mononuclear cells in 500 μL of MACS buffer.
19. Place a MS MACS column in the magnetic MACS separator.
20. Place a new 15 mL centrifuge tube under the MACS column to collect the flow-through. Since the immune cell population of interest is being purified by positive selection, the cells will be retained in the MACS column and all remaining immune cell populations within the mononuclear cell suspension will flow through the MACS column into the tube underneath. These unlabeled cells (i.e., surface marker of interest negative cells) can be incubated with MicroBeads specific for another immune cell surface marker (as described below in **steps 24–29**; *see* **Note 24**).
21. Rinse the MACS column with 500 μL of MACS buffer.

22. Once the entire MACS buffer has passed through the column, add 500 μL of the mononuclear cell suspension to the MACS column (*see* **Note 25**).
23. Add 1.5 mL of the MACS buffer to the 15 mL centrifuge tube that held the mononuclear cell suspension and rinse the sides of the tube to collect any remaining mononuclear cells.
24. Once the first 500 μL of the mononuclear cell suspension has passed through the MACS column, add 500 μL of the MACS buffer containing any remaining mononuclear cells to the MACS column.
25. Repeat **step 16** two additional times so that the entire 1.5 mL volume of the MACS buffer containing any remaining mononuclear cells has passed through the MACS column.
26. If required, store the 15 mL centrifuge tube containing the flow-through of unlabeled/negative mononuclear cells on ice to await further purification of a different immune cell population using MicroBead-based positive selection.
27. Elute the positively selected immune cell population: Ready 5 mL of the MACS buffer in a new 15 mL centrifuge tube.
28. Remove any remaining droplets that form at the base of the MACS column to avoid any unlabeled/negative cells being added to the positively selected immune cell population during the subsequent steps.
29. Remove the MACS column from the magnetic field and place it in the 15 mL centrifuge tube containing the 5 mL of MACS buffer (*see* **Note 26**).
30. Add 1 mL of the MACS buffer to the MACS column and use the plunger to force 1 mL of the MACS buffer through the column to collect the positively selected immune cell population in the 15 mL centrifuge tube.
31. Centrifuge the positively selected immune cell population at 300 × *g* for 10 min at room temperature.
32. Decant the supernatant and resuspend the cell pellet in 300 μL basic medium.
33. Calculate the total number of viable cells (as in **step 1** above).
34. Calculate, and then add the volume of basic medium required to produce a viable immune cell density of 2000 cells/μL to the immune cell resuspension. Store the cells on ice for co-culture with adipocytes (as outlined in Subheading 3.4).

The following steps outline the procedure of purifying additional immune cell populations (based on the expression of a different cell surface marker, unique to the cell population of interest).

35. Centrifuge the 15 mL centrifuge tube containing the flow-through of unlabeled/negative mononuclear cells at 350 × *g* for 5 min at room temperature.
36. Decant the supernatant and resuspend the cell pellet in 4 mL of the MACS buffer (*see* **Note 20**).
37. Repeat Subheading 3.5.2, **steps 1–23**, using the appropriate MicroBeads specific for a new immune cell population.

3.6 Adipocyte and Immune Cell Co-culture: Direct Cell Contact

1. Calculate the volume of immune cell suspension that should be added to each adipocyte monolayer of each well of a 6-well culture plate to achieve the desired ratio of adipocytes to immune cells. The final volume in each adipocyte/immune cell co-culture well of a 6-well culture plate should be 2 mL and will contain fresh basic medium and the following components: immune cells (*see* **Note 27**).
2. Stimuli or experimental treatments (*see* **Note 28**).
3. Remove the basic medium from each well of the 6-well culture plate.
4. Add the required volume of fresh basic medium (calculated in **step 1**) to each culture well without disrupting the adipocyte monolayer on the bottom of the culture well.
5. Add the required volume of the immune cell suspension to obtain the desired ratio of adipocytes:immune cells (calculated in **step 1**) to each culture well without disrupting the adipocyte monolayer on the bottom of the culture well (as described in Subheading 3.3, **step 6**). The two cell types (adipocytes and immune cells) will now be in direct cell contact. Gently slide the 6-well culture plate back and forth, and side to side, to help distribute the immune cells across the culture surface of each well.
6. Add the required volume of stimuli or experimental treatments to each culture well and gently slide the 6-well culture plate back and forth, and side to side, to help distribute the stimuli or experimental treatments across the culture surface of each well.
7. Incubate the adipocyte and immune cell co-cultures for the desired length of time (*see* **Note 29**).
8. At the end of the incubation period, harvest the co-culture samples, which can include the following: Co-culture conditioned media (CCM): Tilt the 6-well culture plate on a 45° angle. Without disturbing the cells on the bottom of the culture well, collect 1.0–1.5 mL of CCM into a microcentrifuge tube. Centrifuge the CCM at 1000 × *g* for 10 min at 4 °C to separate any immune cells. Transfer the supernatant to a new microcentrifuge tube, which can be stored at −80 °C to await analysis (*see* **Note 30**).

9. Co-culture cell lysate: Tilt the 6-well culture plate on a 45° angle. Without disturbing the cells on the bottom of the culture well, gently aspirate the remaining media. Return the 6-well culture plate to its original position. Add the volume of lysis buffer for the duration of time indicated by the manufacturer. Tilt the 6-well culture plate on a 45° angle again and use a cell scraper to collect the lysed cells at the bottom of the culture well. Collect the lysed cells into a microcentrifuge tube, which can be stored at −80 °C to await analysis (*see* **Note 31**).

3.7 Adipocyte and Immune Cell Co-culture: Indirect Cell Contact via Trans-well Separation

1. Repeat Subheading 3.6, **steps 1–3**.
2. Place a trans-well-permeable support into each culture well.
3. Add the required volume of the immune cell suspension to obtain the desired ratio of adipocytes to immune cells (calculated in **step 1**) inside the trans-well-permeable support which physically separates the immune cells from the adipocyte monolayer below, thereby producing cell contact-independent co-cultures (*see* **Note 32**). Gently slide the 6-well culture plate back and forth, and side to side, to help distribute the immune cells across the trans-well culture surface in each well.
4. Repeat Subheading 3.6, **steps 5** and **6**.
5. At the end of the incubation period, harvest the co-culture samples, which can include the following: CCM: Tilt the 6-well culture plate containing the trans-well-permeable supports on a 45° angle. Without disturbing the adipocyte monolayer, collect 1.0–1.5 mL of CCM from below the trans-well-permeable support and transfer to a microcentrifuge tube. Centrifuge the CCM at 1000 × *g* for 10 min at 4 °C to separate any immune cells. Transfer the supernatant to a new microcentrifuge tube, which can be stored at −80 °C to await analysis (*see* **Note 30**).
6. Immune cell lysate: Tilt the 6-well culture plate containing the trans-well-permeable supports on a 45° angle. Collect the approximately 0.5 mL of CMM containing the immune cell suspension from inside the trans-well-permeable support and transfer it to a microcentrifuge tube. Over a culture dish, rinse the inside of the trans-well-permeable support with 0.5 mL fresh basic medium and transfer to the microcentrifuge tube. Centrifuge the immune cell suspension at 1000 × *g* for 10 min at 4 °C. Gently aspirate the supernatant without disrupting the cell pellet. Add the volume of lysis buffer to the microcentrifuge tube for the duration indicated by the manufacturer. Vortex the lysed cells and store at −80 °C until analysis.
7. Adipocyte lysate: Tilt the 6-well culture plate on a 45° angle. Without disturbing the adipocyte monolayer, gently aspirate any remaining media. Return the 6-well culture plate to its

original position. Add the volume of lysis buffer for the duration indicated by the manufacturer. Tilt the 6-well culture plate on a 45° angle again and use a cell scraper to collect the lysed cells at the bottom of the culture well. Collect the lysed cells into a microcentrifuge tube, which can be stored at −80 °C to await analysis (*see* **Note 31**).

4 Notes

1. The necessary amount of medium should be aliquoted before use and warmed to 37 °C in a hot water bath. Decontaminate the container by spraying it with 70% ethanol.
2. FBS must be aseptically filtered to remove any precipitates before it is added to the basic medium.
3. Prior to the addition of cells to a culture flask or plate, the appropriate amount of basic media should be added to the culture flask or plate and incubated for at least 15 min to allow the medium to reach its normal pH (7.0–7.6) and temperature of 37 °C.
4. Cells must be subcultured when they reach 70% confluence in order to avoid "contact inhibition," wherein cells growing in a monolayer stop dividing when in contact with other cells.
5. 3T3-L1 pre-adipocytes and adipocytes are adherent cells, so care must be taken not to disrupt the cell monolayer when aspirating the media.
6. Basic medium contains FBS, which can deactivate trypsin. Therefore, 3T3-L1 pre-adipocytes are rinsed with 1X PBS to remove any residual FBS from the cell surface prior to the addition of trypsin.
7. Trypsin is a serine protease that detaches adherent cells from the culture surface.
8. Trypsin works optimally at 37 °C.
9. If the volume of basic media to resuspend the cells is unclear, resuspend in 0.5 mL and count the live, unstained (with trypan blue solution) cells. If the cell count exceeds 70/quadrant of the hemocytometer, and therefore may be inaccurate due to human error, add 0.5 mL basic medium to the resuspended cells and count the live, unstained cells again. Continue this process until the cell count does not exceed 70/quadrant. Alternatively, increase the volume of trypan blue solution that contains the 10 μL of resuspended cells prior to addition to the hemocytometer and increase this correction for the viable cell density calculation accordingly.

10. When counting cells, employ a system whereby cells are only counted when they are set within a square or on the right-hand or bottom boundary line, and apply this system to every cell count of the experiment.
11. Cells should be added to the culture surface at a density of 3000–5000 cells/cm^2. Since these cells divide at a predictable rate they will therefore reach the desired confluence at a predictable time point.
12. Cells should not be subcultured more than five times to reduce the possibility of phenotypic variations, genetic drift, and contamination as much as possible.
13. Differentiation medium contains dexamethasone, which is an anti-inflammatory glucocorticoid that plays a role in cell survival, cell signaling, and adipogenic gene expression; IBMX, which works to decrease cell proliferation and increase adipogenic gene expression; and insulin, which promotes lipid accumulation.
14. Maintenance medium contains insulin, which promotes lipid accumulation.
15. Serum-free medium ensures cell quiescence prior to experimental use on day 8 by eliminating all potential stimuli in the maintenance medium such as insulin or growth factor-containing FBS.
16. Aseptic spleen removal requires clean surgical tools soaked in 70% ethanol.
17. The spleen has a capsule surrounding the tissue; therefore, it will require gentle pressure to break through the capsule and release the contents into the cell strainer. It is also important to lift the cell strainer up off the bottom of the culture dish (2–3 cm) to better allow the splenic contents to pass through the cell strainer and into the basic media in the culture dish. Do not touch the bottom of the cell strainer directly with your gloved fingers. Instead, gently tap the cell strainer against the bottom of the petri dish in the basic medium to help recover splenic cells.
18. Lympholyte-M is light sensitive, so this step must be performed in minimal light conditions. Lympholyte-M is stored at 4 °C but should be allowed to warm to room temperature for experimental use.
19. Once the centrifuge stops, the mononuclear cells enriched at the interface will start to move through the lower phase of the solution and will be lost, so it is important to work quickly during this step to collect the mononuclear cells.
20. MACS buffer and mononuclear cell suspension should be stored on ice throughout the protocol.

21. There are different techniques that can be employed at this step of the protocol to purify specific immune cell populations (identified by cell surface markers) using either positive or negative selection approaches for later use in adipocyte co-culture experiments. Possible methodologies include a cell sorting of the desired immune cell population by flow cytometry, or MicroBead/column purification techniques.
22. If the mononuclear cell density observed in the hemocytometer is too high, the cell counts will not be accurate; therefore, adjust the volume of trypan blue solution that the 10 μL of mononuclear cells are mixed as necessary and correct for this dilution in the calculation as appropriate.
23. The MicroBeads are light sensitive, so this step must be performed in minimal light conditions.
24. Positively selected cells will be retained within the MACS column in the magnetic field. All unlabeled (i.e., negative) cells will pass through the column to the tube underneath. By collecting the negative cells, it is possible to purify multiple immune cell populations from a single splenic mononuclear cell population.
25. The maximum volume to add to the MACS column is 500 μL.
26. Once the MACS column is removed from the magnetic field, it is important to work quickly to elucidate the positively selected immune cell population from the MACS column.
27. The desired cellular ratio for adipocyte/immune cell co-culture will vary based on the experimental design. Our research team developed adipocyte/immune cell co-culture models to recapitulate the cellular ratio of specific immune cell populations reported in obese adipose tissue including 17% $CD11b^+$ cells (i.e., $CD11b^+$ macrophages) [7], 5% $CD4^+$ T cells [12], and 10% $CD8^+$ T cells [8–11]. The ratio of adipocytes to immune cells will determine the volume of basic media and immune cell suspension required to set up the co-cultures. For example, if the desired co-culture ratio is 10:1 adipocytes to immune cells [8, 9], then a single culture well of a 6-well culture plate containing approximately 2 million differentiated adipocytes should be co-cultured with 200,000 purified immune cells. Since the immune cells were adjusted to a density of 2000 cells/μL, add 100 μL (200,000 cells divided by 2000 cells/μL) of the immune cell suspension to 1.9 mL of fresh basic media in each culture well.
28. Any soluble stimuli or experimental treatment could be added to the adipocyte/immune cell co-cultures. We have used a variety of adipocyte/immune cell co-culture stimulation conditions [9]. Additionally, in the experimental design, it is recommended to include the following controls: unstimulated

co-cultures (co-cultured cells in basic media only, negative control) and individually cultured adipocytes and immune cell populations in the appropriate stimulation conditions.

29. Adipocyte-immune cell co-cultures are typically incubated for 24 h [7–12] wherein immune cell viability exceeds 89%; however, the co-culture incubation period could be extended but it is not recommended to exceed 48 h as cell viability decreases with longer co-culture durations.
30. The volume of CCM collected is conservative to minimize contamination of the CCM with cells and to avoid disrupting the cells in culture. The CCM will contain the secreted mediators produced from the co-culture of adipocytes and immune cells and can be used in subsequent experiments using either primary cell types or cell lines.
31. Both RNA and protein can be isolated from lysed cells of a single co-culture well by using the commercially available cell lysis buffer that is included in specialized RNA/protein co-isolation kits, e.g., the RNA/DNA/protein purification plus kit.
32. The permeable membrane of the trans-well permits secreted proteins produced from the co-cultured cells to pass freely between the adipocyte (lower) and immune cell (upper) layers of the co-culture system (Fig. 2); however, the 0.4 μM pore size is too small for the co-cultured cells to make physical contact with each other; hence, this co-culture condition provides insight into the cell contact-independent mechanisms of communication (or paracrine interactions) between the co-cultured cell types.

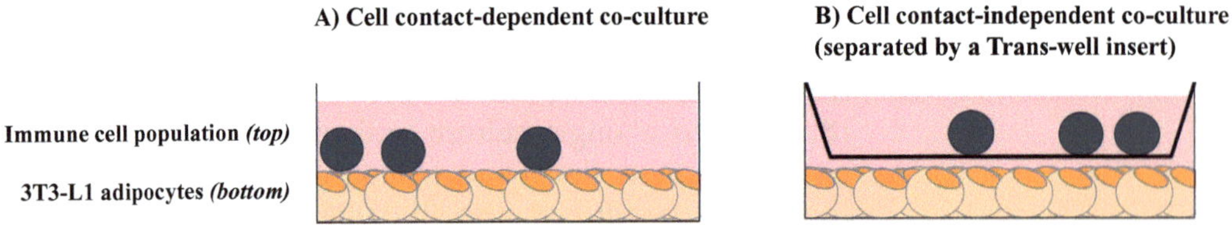

Fig. 2 Diagram of adipocyte/immune cell co-cultures that are (**a**) cell contact dependent (i.e., co-cultured cells are in direct physical contract) and (**b**) cell contact independent (i.e., adipocytes and immune cells are separated by a trans-well insert) (depicted by a solid black line between the co-cultured cell types) that permits the passage of soluble mediators but prevents physical contact between co-cultured adipocytes (bottom monolayer in the culture well) and immune cells (top layer of cells in the culture well)

Acknowledgments

Jennifer M. Monk and Danyelle M. Liddle contributed equally to this work.

References

1. Liddle DM, Hutchinson AL, Wellings HR, Power KA, Robinson LE, Monk JM (2017) Integrated immunomodulatory mechanisms through which long-chain n-3 polyunsaturated fatty acids attenuate obese adipose tissue dysfunction. Nutrients 9(12):pii:E1289
2. McLaughlin T, Liu LF, Lamendola C, Shen L, Morton J, Rivas H, Winer D, Tolentino L, Choi O et al (2014) T-cell profile in adipose tissue is associated with insulin resistance and systemic inflammation in humans. Arterioscler Thromb Vasc Biol 34:2637–2643
3. Winer S, Winer DA (2012) The adaptive immune system as a fundamental regulator of adipose tissue inflammation and insulin resistance. Immunol Cell Biol 90:755–762
4. Huh JY, Park YJ, Ham M, Kim JB (2014) Crosstalk between adipocytes and immune cells in adipose tissue inflammation and metabolic dysregulation in obesity. Mol Cells 37:365–371
5. Nishimura S, Manabe I, Nagasaki M, Eto K, Yamashita H, Ohsugi M, Otsu M, Hara K, Ueki K et al (2009) CD8+ effector T cells contribute to macrophage recruitment and adipose tissue inflammation in obesity. Nat Med 15:914–920
6. Lolmède K, Duffaut C, Zakaroff-Girard A, Bouloumié A (2011) Immune cells in adipose tissue: key players in metabolic disorders. Diabetes Metab 37:283–290
7. De Boer AA, Monk JM, Liddle DM, Hutchinson AL, Power KA, Ma DW, Robinson LE (2016) Fish-oil-derived n-3 polyunsaturated fatty acids reduce NLRP3 inflammasome activity and obesity-related inflammatory cross-talk between adipocytes and CD11b(+) macrophages. J Nutr Biochem 34:61–72
8. Monk JM, Liddle DM, De Boer AA, Brown MJ, Power KA, Ma DW, Robinson LE (2015) Fish-oil-derived n-3 PUFAs reduce inflammatory and chemotactic adipokine-mediated cross-talk between co-cultured murine splenic CD8+ T cells and adipocytes. J Nutr 145:829–838
9. Monk JM, Liddle DM, Brown MJ, Zarepoor L, De Boer AA, Ma DW, Power KA, Robinson LE (2016) Anti-inflammatory and anti-chemotactic effects of dietary flaxseed oil on CD8(+) T cell/adipocyte-mediated cross-talk. Mol Nutr Food Res 60:621–630
10. Liddle DM, Monk JM, Hutchinson AL, Ma DWL, Robinson LE (2015) Fish oil-derived n-3 polyunsaturated fatty acids mitigate lean and obese CD8$^+$ T cell-adipocyte inflammatory cross-talk, in part through a TNFα-dependent mechanism. Appl Physiol Nutr Metab 40:516
11. Liddle DM, Monk JM, Hutchinson AL, Power KA, Robinson LE (2016) Adipocyte dysfunction induced by inflammatory cross-talk with CD8$^+$ T cells is mitigated by fish oil-derived n-3 polyunsaturated fatty acids, in part through a TNFα-dependent mechanism. Appl Physiol Nutr Metab 41:S23
12. Liddle DM, Hutchinson AL, Monk JM, Ma DWL, Robinson LE (2016) N-3 polyunsaturated fatty acids mitigate adipocyte-CD4$^+$ T cell inflammatory cross-talk in co-culture models of obese adipose tissue. Appl Physiol Nutr Metab 41:S24
13. De Boer AA, Monk JM, Robinson LE (2014) Docosahexaenoic acid decreases pro-inflammatory mediators in an in vitro murine adipocyte macrophage co-culture model. PLoS One 9:e85037
14. De Boer AA, Monk JM, Liddle DM, Power KA, Ma DW, Robinson LE (2015) Fish oil-derived long-chain n-3 polyunsaturated fatty acids reduce expression of M1-associated macrophage markers in an ex vivo adipose tissue culture model, in part through adiponectin. Front Nutr 2:31
15. Cani PD, Amar J, Iglesias MA, Poggi M, Knauf C, Bastelica D, Neyrinck AM, Fava F, Tuohy KM et al (2007) Metabolic endotoxemia initiates obesity and insulin resistance. Diabetes 56:1761–1772
16. Creely SJ, McTernan PG, Kusminski CM, Fisher F, Da Silva NF, Khanolkar M, Evans M, Harte AL, Kumar S (2007) Lipopolysaccharide activates an innate immune system response in human adipose tissue in obesity and type 2 diabetes. Am J Physiol Endocrinol Metab 292:E740–E747
17. Laugerette F, Furet JP, Debard C, Daira P, Loizon E, Géloën A, Soulage CO, Simonet C, Lefils-Lacourtablaise J et al (2012) Oil composition of high-fat diet affects metabolic

inflammation differently in connection with endotoxin receptors in mice. Am J Physiol Endocrinol Metab 302:E374–E386

18. Lumeng CN, Bodzin JL, Saltiel AR (2007) Obesity induces a phenotypic switch in adipose tissue macrophage polarization. J Clin Invest 117:175–184

19. Xie L, Ortega MT, Mora S, Chapes SK (2010) Interactive changes between macrophages and adipocytes. Clin Vaccine Immunol 17:651–659

20. Oliver E, McGillicuddy FC, Harford KA, Reynolds CM, Phillips CM, Ferguson JF, Roche HM (2012) Docosahexaenoic acid attenuates macrophage-induced inflammation and improves insulin sensitivity in adipocytes-specific differential effects between LC n-3 PUFA. J Nutr Biochem 23:1192–1200

21. Suganami T, Nishida J, Ogawa Y (2005) A paracrine loop between adipocytes and macrophages aggravates inflammatory changes: role of free fatty acids and tumor necrosis factor alpha. Arterioscler Thromb Vasc Biol 25:2062–2068

Chapter 10

A Gene Expression Analysis of M1 and M2 Polarized Macrophages

Nour Eissa, Hayam Hussein, and Jean-Eric Ghia

Abstract

Macrophages are professional innate immune cells that are broadly disseminated throughout the body, shape various innate and adaptive immune responses, and play crucial roles in inflammation, homeostasis, wound healing, and tissue remodelling. According to their surrounding microenvironments, macrophages can differentiate themselves in different phenotypes. Over the last two decades, gene expression profiling has been used to decipher new transcripts associated with macrophage phenotypes. This chapter outlines protocols used to isolate and culture murine macrophages and how they can be "polarized" to obtain a specific phenotype. Furthermore, we describe a protocol for gene expression profiling using a quantitative real-time polymerase chain reaction (qPCR), a high-standard technology in the field of gene expression.

Key words Macrophage polarization, qPCR, Murine macrophages, Innate immunity, Gene expression

1 Introduction

Macrophages are a highly heterogeneous population of cells that exhibit a continuum of activation states. They play a critical role in homeostasis and tissue remodelling, orchestrate various responses to antigens, and are one of the key players in inflammation and host defence [1]. Macrophages have extraordinary plasticity to adjust their phenotype according to their surrounding microenvironment. The wide range of macrophage phenotypes is often oversimplified into two functional groups: group 1, "inflammatory, M1 macrophages" (producing a low amount of interleukin (IL)-10) that are involved in the pro-inflammatory response, and group 2, "wound healing, M2 macrophages" (producing a low to a high amount of IL-10) that are involved in anti-inflammatory mechanisms, but also fibrosis and wound healing. Recently a new subset of macrophages producing a high level of IL-10 has been discovered and referred to "regulatory macrophages: M_{regs}" [2]. M1 macrophages express and release pro-inflammatory cytokines such as a monocyte

Suresh Mishra (ed.), *Immunometabolism: Methods and Protocols*, Methods in Molecular Biology, vol. 2184,
https://doi.org/10.1007/978-1-0716-0802-9_10, © Springer Science+Business Media, LLC, part of Springer Nature 2020

chemoattractant protein (MCP)-1, IL-6, a tumour necrosis factor (TNF), IL-12, and nitric oxide (NO). Conversely, M2 macrophages express a high level of the mannose receptor (MR), arginase-1 (Arg-1), IL-10, and transforming growth factor (TGF)-β [3]. Macrophages' polarization depends on specific combinations of cytokines within the microenvironment. For example, naïve macrophages can be polarized to an M1 macrophage phenotype using lipopolysaccharide (LPS) which triggers M1 activation and enhances pro-inflammatory cytokine production, and to an M2a macrophage phenotype using a combination of IL-4/IL-13 [4–6]. Evidence suggests that macrophages maintain considerable plasticity between activation states, which make them a unique therapeutic target for inflammatory diseases [5–8]. M2 macrophages are also subclassified into M2a, M2b, M2c, and M2d sub-phenotypes, depending on the nature of stimuli. M2a activation is a response to IL-4 and IL-13, M2b to immune complexes and bacterial LPS, M2c to glucocorticoids and TGF-beta, and M2d to IL-6 and adenosines [9].

Gene expression profiling is a powerful approach to investigate and understand the molecular pathways and gene signatures associated with the different phenotypes of macrophages [10]. Furthermore, gene expression analysis is widely used to support protein expression profiles from an omics assay. During the last two decades, quantitative real-time PCR (qPCR) has become one of the standard techniques for investigating and assessing the differences in macrophage gene expression levels [11]. Each macrophage phenotype has a determined gene signature associated with a specific activation status. Therefore, nowadays, gene expression analysis is a critical tool to identify and characterize the different macrophage phenotypes.

In this chapter, we describe protocols to obtain a cell culture of in vitro murine macrophages, to "polarize" macrophages to an M1 or M2 phenotype, and to characterize them using the qPCR tool.

2 Materials

Prepare all media and solutions under a biological safety cabinet in an aseptic environment to avoid microbial contamination. All qPCR materials are molecular biology-grade materials and RNase free. Conscientiously follow all waste disposal guidelines when disposing of waste materials.

2.1 Isolation of Murine Macrophages

1. 10 mL Syringes.
2. 12-, 24-, or 48-well cell culture plates.
3. 20–25 G Needles.
4. 50 mL Conical centrifuge tubes.

5. 70% Ethanol.
6. Biological safety cabinet.
7. Cell culture incubator.
8. Centrifuge.
9. Growth media: Complete RPMI 1640 or DMEM/F12 media supplemented with 100 U/mL penicillin, 100 μg/mL streptomycin, and 5% endotoxin-free heat-inactivated fetal bovine serum (FBS).
10. Different-sized pipettes.
11. Gas evaporator, O_2, isoflurane anesthetic gas, and an anesthesia machine.
12. Hemocytometer and/or automated cell counter.
13. Ice.
14. Ketamine/xylazine cocktail (100 mg/kg)/xylazine (10 mg/kg).
15. Mouse dissection set.
16. Pathogen-free BALB/c and C57Bl/6 mice can be used (*see* **Note 1**).
17. Personal protection: gloves, lab coat, mask, etc. (according to requirements of your facility).
18. Plastic dishes.
19. Recombinant macrophage colony-stimulating factor (M-CSF).
20. Sterile 1× phosphate-buffered saline (PBS).

2.2 Activation of Macrophages

1. RPMI 1640 or DMEM/F12 cell culture medium.
2. CO_2 incubator.
3. Growth medium: Complete RPMI 1640 or DMEM/F12 medium supplemented with 100 U/mL penicillin, 100 μg/mL streptomycin, and 5% endotoxin-free heat-inactivated fetal bovine serum (FBS).
4. Interferon gamma (IFNγ) recombinant protein.
5. Interleukin (IL)-13 recombinant protein.
6. Interleukin (IL)-4 recombinant protein.
7. Lipopolysaccharide (LPS).
8. Water bath.

2.3 Gene Expression Analysis of M1/M2a Macrophages

1. cDNA synthesis kits.
2. Cell scraper.
3. 2 mL Centrifuge tubes.
4. Chloroform.

5. 75% Ethanol.
6. 99% Ethanol.
7. Biological safety cabinet.
8. Nanodrop spectrophotometer.
9. Primers.
10. qPCR machine.
11. RNaseAWAY®.
12. RNase-free water.
13. SYBR green.
14. Thermocyclers.
15. Trizol.

3 Methods

3.1 Isolation of Murine Macrophages

In this chapter, we focus on the murine peritoneal macrophages, bone marrow-derived macrophages, and alveolar macrophages, which are extensively used to study and understand the immunobiology and pathophysiology of several diseases. Other macrophages, such as the THP-1-derived macrophages or RAW264.7 cells, can also be used.

3.1.1 Isolation of Murine Resident Peritoneal Macrophages

The murine peritoneal cavity provides an accessible location for collecting about 1×10^6 resident peritoneal macrophages [12]. The injection of eliciting agents into the peritoneal cavity can increase the macrophages' numbers, but with some restrictions (*see* **Note 2**). The following steps show briefly a commonly used protocol to isolate and extract resident peritoneal macrophages:

1. Pathogen-free mice can be sacrificed by inhalation of isoflurane (3–4%) or other gases followed by a quick cervical dislocation, by a skilled and trained technician to avoid pain and excessive hemorrhage in the peritoneal cavity. Euthanasia protocol should follow institutional animal care guidelines.
2. Clean and sterilize the mouse abdomen with 70% ethanol for 2–20 s, then pitch up the skin using forceps and make a small incision with a scissor to expose the peritoneum.
3. Inject 5–10 mL of cold PBS in the peritoneal cavity using 25 G needle and shake the mouse for 20 s. Be careful not to penetrate any internal organ or macrophages can get contaminated.
4. Using 20–22 G needle and a 10 mL syringe, aspirate the macrophages' cell suspension from the peritoneal cavity. The abdominal/visceral fat might block the needle. To resolve this hitch, remove the needle and introduce it again.

5. Remove the needle from the syringe and dispense the macrophages' cell suspension in a 50 mL tube containing 20 mL of ice-cold phosphate-buffered saline (PBS) positioned on ice. Cells should be kept cold throughout the process, and an average of 3–5 mice can be sacrificed when using the suspension pooled in one tube.
6. Using a 4 °C refrigerated centrifuge, centrifuge the cell suspension at 350–400 × *g* for 5 min. Discard the supernatant and resuspend cell pellets with cold complete RPMI 1640 or DMEM/F12 supplemented with 100 U/mL penicillin, 100 μg/mL streptomycin, and 5% of endotoxin-free heat-inactivated fetal bovine serum (FBS) (*see* **Notes 3** and **4**). Count the cells using a hemocytometer or automated cell counter, and plate on plastic plates (*see* **Note 5**) at 1–3 × 10^6 cells/mL, and incubate for 2 h at 37 °C. For a 24-well plate, this number should be doubled.
7. Gently, non-adherent cells are removed by washing three times with warm 1x PBS at 37 °C and then add RPMI 1640 or DMEM/F12 supplemented with 5% of FBS. At that time, >90% of adherent cells should be macrophages.
8. The macrophages are now ready for polarization or total RNA isolation for gene expression analysis.

3.1.2 Culture of Murine Bone Marrow-Derived Macrophages (BMDM)

Bone marrow cells are collected and cultured for 7 days in a medium containing recombinant macrophage colony-stimulating factor (M-CSF, 10 ng/mL) [12, 13]. This protocol can yield >90% pure adherent bone marrow-derived macrophages [14] and 2–3 × 10^7 macrophages using two femurs or higher by using the tibias [15]. As an alternative to recombinant M-CSF, 20% (v/v) L929 conditioned medium can be used to stimulate macrophages' differentiation [16]. Here, we describe the culture of BMDM using recombinant M-CSF.

1. Pathogen-free BALB/c and C57Bl/6 mice can be sacrificed by inhalation of isoflurane (3–4%) or other gases followed by a quick cervical dislocation. Euthanasia protocol should follow animal care guidelines approved by your research institute.
2. Using a laminar flow hood (*see* **Note 6**), prepare the abdomen in an aseptic manner by rubbing it with 70% ethanol. With a scissor, make an incision in the abdomen, peel skin, and remove the muscles from the bones to be able to uncover the femur head. Cut off the hind legs at the hip joint without breaking the femur. Collect femurs in a plastic dish containing sterile 1x PBS. The removal of excess muscle from the legs is necessary by holding the end of the bone with forceps and using scissors to push muscle away.

3. Sterilize the bones by soaking in alcohol for 10–20 s and then keep them in ice-cold sterile 1× PBS.
4. Cut the leg bones proximal to each joint with a sharp scissor and then flush the bones with cold sterile RPMI 1640 or DMEM/F12 using 24–25 G needle attached to a 10 mL syringe in a sterile 50 mL conical centrifuge tube on ice. Repeat the flush 3–5 times until bone cavity colours change to white.
5. With a 10 mL pipette, do up and downs to disperse the cell aggregates.
6. At room temperature, centrifuge cells for 10 min at 400 × *g*.
7. Discard the supernatant, and then resuspend the cell pellets in macrophage complete RPMI 1640 by pipetting up and down.
8. Using a hemocytometer or an automated cell counter, seed 3–5 × 10^6/mL per sterile plastic Petri dish in 10 mL of growth media and add recombinant M-CSF at a concentration of 10 ng/mL.
9. Incubate Petri dishes in a humidified incubator with 5% CO_2 at 37 °C.
10. On day 3, add another 5 mL of growth media.
11. On day 7, remove cell culture supernatants and wash adherent cells with 5 mL of warm 1× PBS 37 °C. By using a cell scraper, gently scrape the cells off the plate, and then pool the dislodged cells in a 50 mL conical tube.
12. Centrifuge cells for 10 min at 400 × *g*, 4 °C, and remove the supernatant.
13. Resuspend cells in 5 mL growth media, and count cells using a hemocytometer or an automated cell counter.
14. At this point, the cells should be 100% macrophages with a yield of 7–15 × 10^7 macrophages per mouse. Typically, the cells are plated in growth media at 1 × 10^6 cells/well in 6-well tissue culture plates, or 2 × 10^5 cells/well in 24-well tissue culture plates, or 1 × 10^5 cells/well in 48-well tissue culture plates.
15. The macrophages are now ready for polarization or total RNA isolation for gene expression analysis.

3.1.3 Isolation of Alveolar Macrophages

Alveolar macrophages are key players of host defence and resolution of lung inflammation [17]. In this protocol, a method for isolating murine alveolar macrophages is described. Following the described protocol, a mouse can yield 3–5 × 10^5 alveolar macrophages:

1. Anesthetize mice by an intraperitoneal injection of a cocktail of 100–200 μL of ketamine (100 mg/kg)/xylazine (10 mg/kg) according to the weight of the animal.

2. Cervical dislocation is not recommended since it can damage the trachea and result in contaminating the sample with blood. Also, the inhalation of isoflurane or other gases is not indicated as a viable option as it might change the alveolar macrophage response.
3. Scrub the neck and abdomen with 70% ethanol, and then surgically expose the trachea on the ventral side of the neck.
4. Insert an 18 G needle connected to a 5 mL syringe containing 4 mL of pre-warmed (37 °C) complete RPMI 1640 medium with 5% FBS into the tracheal lumen just below the larynx. Tie surgical sutures around the needle. Then inject a sterile complete RPMI 1640 medium with 5% FBS into the lungs and apply gentle massage of the lungs.
5. Gently aspirate the fluid into the syringe, reinfuse it back into the lung 3–5 times and then collect the final lung fluid into a 50 mL conical polypropylene tube on ice.
6. Centrifuge the cell suspension for 10 min at 400 × *g* at 4 °C.
7. Discard the supernatant and resuspend the cell pellet in growth media for counting.
8. Plate the cells and incubate in a humidified incubator with 5% CO_2 at 37 °C for 2 h, and then wash the non-adherent cells gently with 1× PBS at 37 °C.
9. Add growth media to the adherent macrophages.
10. The macrophages are now ready for polarization, or total RNA isolation for gene expression analysis.

3.2 Activation of Macrophages

One of the unique features of macrophages is their capability to react to environmental stimuli and subsequently to change their phenotype and physiology; this is referred to as "activation" [18]. There are several ways to activate the different subsets of macrophages [19]; for clarity, in this chapter, we only describe the main two phenotypes of macrophages (M1 and M2a), each with distinct physiology.

3.2.1 Polarization of Classically Activated M1 Macrophages

Classically activated macrophages with the designation M1 are the most comprehensively defined activated macrophage phenotypes and are one of the critical immune effector cells for host defence and inflammatory responses [19]. A combination of two signals can result in M1 polarization: the first signal is known as a "priming" step and is induced by IFNγ, which is produced by natural killer (NK) cells and Th1 cells. The second signal can be portrayed by a Toll-like receptor (TLR) ligand, such as LPS [20]. Here, we describe the following steps for in vitro M1 polarization, using naïve peritoneal macrophages or BMDM. Overall, macrophages are typically primed with IFNγ overnight and are stimulated with LPS the next morning.

1. Isolate the peritoneal, BMDM, or alveolar macrophages, as described in the previous sections.
2. Macrophage priming: Add 150 U/mL of IFNγ [21] to the cultured macrophages in a complete medium and incubate for 6–12 h. Controls should include naïve cells that are not primed.
3. Depending on the origin of the macrophages in question, optimal concentrations of IFNγ should be established in each laboratory, ranging from 50 to 250 U/mL.
4. Wash the cells with a warm and complete medium.
5. Macrophage stimulation: Add LPS (1–100 ng/mL) to the growth media and incubate for 4–6 h. Then harvest the cells for gene expression analysis. The optimal range of LPS should be determined accordingly (*see* **Note 7**).

3.2.2 Polarization of Alternatively Activated M2a Macrophages

Here, we describe the polarization of naïve macrophages into M2a macrophages by adding IL-4, IL-13, or both cytokines together.

1. Isolate the peritoneal, BMDM, or alveolar macrophages, as described in the previous sections.
2. Add IL-4 or IL-13 (optimal concentration, 10–20 ng/mL) and incubate overnight. This should be sufficient to generate the M2 macrophages and induce arginase expression.
3. Harvest the M2 cells for gene expression analysis.

3.3 Gene Expression Analysis of M1/M2a Macrophages

The extraction of RNA is a critical step and should be done in an RNase-free environment under a molecular biology safety cabinet (*see* **Note 8**), according to the following steps:

3.3.1 RNA Extraction and Quantification

1. Use RNaseAWAY® (and 100% ethanol) to clean all tools, pipettes, and benches before starting.
2. Aspirate the culture medium from the adherent cells generated previously.
3. Add 1 mL of TRIzol reagent per well in a 24-well plate (in the 48-well plate, add 0.5 mL TRIzol).
4. Use a cell scraper to help detach the adherent cells.
5. Incubate the homogenate for 3–5 min at room temperature.
6. Collect the cells in TRIzol in 2 mL centrifuge tubes.
7. Add 0.2 mL of chloroform per 1 mL of TRIzol® (per every 100 μL of TRIzol®, use 20 μL of chloroform). Close sample tubes securely.
8. Vortex tubes vigorously for 15–20 s.

9. Incubate for 5 min at room temperature.
10. Centrifuge for 15 min at 4 °C (at no more than 12,000 × *g*). Following centrifugation, the mixture separates into a lower red phenol-chloroform phase, interphase, and a colourless upper aqueous phase. RNA remains exclusively in the aqueous phase (the volume of the aqueous phase is about 60% of the used volume of TRIzol®) (*see* **Note 9**).
11. Transfer the aqueous phase (top layer) to a new tube and measure its volume.
12. Precipitate the RNA from the aqueous phase by mixing it with 500 μL 99% ethanol (0.5 mL: 1 mL of TRIzol); then vortex moderately for 10–20 s.
13. Incubate for 10 min at room temperature and continue.
14. Centrifuge for 15 min at 4 °C (at no more than 12,000 × *g*).
15. After centrifugation, the RNA precipitate forms a gel-like pellet on the side and bottom of the tube. Carefully remove the supernatant without disturbing the pellet. Add 1 mL of 75% ethanol (in nuclease-free water) per 1 mL of TRIzol used in the isolation step.
16. Vortex gently and centrifuge for 5 min at 4 °C (at no more than 7500 × *g*).
17. Carefully remove the ethanol wash without disturbing the pellet. Remove all residual ethanol by centrifuging again briefly and removing ethanol with a fine pipette. The complete removal of all ethanol is necessary for the RNA.
18. Air-dry the RNA pellet (leave it for 2–3 min) on ice with Kimwipe on top of an open tube (*see* **Note 10**).
19. Dissolve the pellet in RNase-free water by passing the solution a few times through a pipette tip, or by more vigorous vortexing (30–100 μL depending on the pellet size).
20. Vortex and place tube back on the ice for 1 min.
21. Centrifuge for 5 min at 4 °C (not more than 7500 × *g*) to remove insoluble material.
22. Transfer supernatant to a final 1.5 mL tube.
23. Add 1.5 μL of isolated RNA in Nanodrop, and then read.
24. Take OD at 260 and 280 nm to determine sample concentration and purity. The A260/A280 ratio should be around 1.8–2.2.
25. Write final concentration and dates on the side of the tube, and then store at −80 °C.

3.3.2 cDNA Synthesis

There are several available commercial kits to synthesize cDNA from RNA by simply following the manufacturer's instructions. In our experiments, reverse transcription is carried out using the SuperScript VILO cDNA Synthesis Master Mix, according to the manufacturer's instructions, in a thermocycler at 25 °C for 10 min, followed by 42 °C for 60 min, and finally 85 °C for 5 min. cDNA samples are then cooled to 4 °C and stored at −20 °C for qPCR analysis.

3.3.3 M1/M2 Macrophage Primers

Several specific genes are associated with M1 macrophage activation and show high mRNA expression levels: *Il6*, *Il1 β*, *TNF-α*, inducible nitric oxide synthase (*iNos*), macrophage chemoattractant protein (*Mcp*)-1, macrophage inflammatory proteins (*Mip*)-1α, and (*Mip*)-1β [5, 8, 20–22] are examples. The following specific genes are associated with M2 macrophage activation and show high mRNA expression levels: *Il10*, *Arg-1*, chitinase-like proteins (*Ym1*), and inflammatory zone proteins (*Fizz1*) [6, 8, 11].

Primer design is a critical step in gene expression analysis [22, 23]. The primers can be designed from nucleotide sequences identified using the NCBI BLAST (http://blast.ncbi.nlm.nih.gov/Blast.cgi). The tested and validated primers for M1 and M2 markers are listed in Table 1. The ideal primer should have a:

1. Melting temperature (T_m) of 58–62 °C
2. GC content of 45–55%
3. Length of 18–22 bp
4. Amplicon size between 75 and 175 bp

As recommended by the Minimum Information for Publication of Quantitative Real-Time PCR Experiments (MIQE) guidelines [24], qPCR efficiencies in the exponential phase should be calculated for each primer pair using standard curves (a 5-point, fivefold serial dilution of pooled cDNA that includes equal amounts from the sample set). The mean Ct values for each serial dilution should be plotted against the logarithm of the cDNA dilution factor, and computed according to the equation $E = 10[-1/\text{slope}]$ [24], where the slope is the gradient of the linear regression line.

3.3.4 Quantitative Real-Time Polymerase Chain Reaction (qPCR)

The Minimum Information for Publication of Quantitative Real-Time PCR Experiments (MIQE) guidelines [24] should be considered in all the gene expression analysis.

1. The qPCR reactions can be carried out in any suitable real-time system using SYBR green, e.g., Power SYBR green (Life Technologies), in a final volume of 20 μL reactions.
2. The PCR conditions are as follows: 95 °C for 10 min, followed by 40 cycles at 95 °C for 15 s and at 60 °C for 60 s.

Table 1
Primer sequence for M1 and M2 macrophage markers

	Gene	Forward	Reverse
M1 markers	*Inos*	GTTCTCAGCCCAACAATACAAGA	GTGGACGGGTCGATGTCAC
	Mip1a	TGTACCATGACACTCTGCAAC	CAACGATGAATTGGCGTGGAA
	Mip1b	TTCCTGCTGTTTCTCTTACACCT	CTGTCTGCCTCTTTTGGTCAG
	Il1b	GCAACTGTTCCTGAACTCAACT	ATCTTTTGGGGTCCGTCAACT
	Il-6	TAGTCCTTCCTACCCCAATTTCC	TTGGTCCTTAGCCACTCCTTC
	Tnfa	CCCTCACACTCAGATCATCTTCT	GCTACGACGTGGGCTACAG
	Mcp-1	TTAAAAACCTGGATCGGAACCAA	GCATTAGCTTCAGATTTACGGGT
	CD80	TCGGCGCAGTAATAACAGTC	GTTTCTCTGCTTGCCTCATTTC
	CD86	TTACGGAAGCACCCACGATG	ACTACCAGCTCACTCAGGCT
M1 cytokines	*Il12p40*	TGGTTTGCCATCGTTTTGCTG	ACAGGTGAGGTTCACTGTTTCT
	Il23p19	AATAATGTGCCCCGTATCCAGT	GCTCCCCTTTGAAGATGTCAG
M2 markers	*Il10*	GCTCTTACTGACTGGCATGAG	CGCAGCTCTAGGAGCATGTG
	Fizz1	AAGCCTACACTGTGTTTCCTTTT	GCTTCCTTGATCCTTTGA TCCAC
	Ym1	CAGGTCTGGCAATTCTTCTGAA	GTCTTGCTCATGTGTGTAAG TGA
	Arg-1	TTGGGTGGATGCTCACACTG	GTACACGATGTCTTTGGCAGA
M2 cytokine	*Tgfb*	CTCCCGTGGCTTCTAGTGC	GCCTTAGTTTGGACAGGATCTG
	Il1ra	GCTCATTGCTGGGTACTTACAA	CCAGACTTGGCACAAGACAGG
Reference gene	*Tbp*	ACCGTGAATCTTGGCTGTAAAC	GCAGCAAATCGCTTGGGATTA

3. To test the specificity of each primer, the melting curve analysis should be included (65–95 °C, with fluorescence measured every 0.5 °C).
4. The absence of contamination from either the genomic DNA amplification or the primer formation should be ensured using two types of controls: the first without reverse transcriptase (no-RT control, one for each RNA) and the second with no DNA template (NTC control, one for each primer pair).
5. All qPCRs should be run in duplicate; the average standard deviation within duplicates of all samples studied is <0.6 cycles.
6. Data analysis: the relative M1 and M2 genes' expression is determined by calculating differences in the comparative threshold cycle ($\Delta\Delta Ct$), using a reference gene such as the Tata-box binding protein (TBP), and the difference in threshold cycle (ΔCt) methods [25–27].

4 Notes

1. Mice obtained from an external source (e.g., a commercial supplier) should be housed for at least 7 days following arrival to acclimate them from stress induced by the transit or transfer.
2. Thioglycolate medium is frequently used as an eliciting agent to increase the number of peritoneal macrophages. The above protocol can be applied 4 days from the injection of 1 mL of 3% thioglycolate medium into the peritoneal cavity. Thioglycolate should be used carefully as it can dramatically alter the physiological properties of the harvested macrophages [12].
3. Macrophages are tremendously sensitive to environmental signals, and therefore all reagents must be of high quality and endotoxin-free to avoid aberrant stimuli.
4. Heat-inactivated FBS that is low in endotoxin should be used; however, higher quantities of FBS in the culture medium will influence macrophage dynamics.
5. As macrophages tend to stick to most surfaces, in our experiments, for maximum cell recovery, polypropylene plastic wares were used.
6. All solutions and equipment must be sterile, and proper sterile procedures should be employed.
7. The concentration of LPS can be different according to the source of LPS and its purity.
8. An RNase-free environment is critical during the gene expression experiments.
9. Be advised that temperatures >8 °C may cause some DNA to partition in the aqueous phase.
10. It is important not to let the RNA pellet dry completely, as this will greatly decrease its solubility.

Acknowledgments

This chapter is supported by grants from the Canada Foundation for Innovation, Crohn's and Colitis Canada, Research Manitoba, the Children's Hospital Research Institute of Manitoba, the Natural Sciences and Engineering Research Council, and finally the Canadian Institutes of Health Research, to Jean-Eric Ghia. Nour Eissa is supported by the Canadian Institutes of Health Research (CIHR) (Grant# 395678), Children's Hospital Research Institute of Manitoba, Health Science Centre Foundation (HSCF)-Mindel, and the Tom Olenick Research Excellence Award in Immunology and the MITACS Accelerate Program.

References

1. Wynn TA, Chawla A, Pollard JW (2013) Macrophage biology in development, homeostasis and disease. Nature 496(7446):445
2. DeNardo DG, Ruffell B (2019) Macrophages as regulators of tumour immunity and immunotherapy. Nat Rev Immunol 19:369–382
3. Okabe Y, Medzhitov R (2016) Tissue biology perspective on macrophages. Nat Immunol 17 (1):9
4. Eissa N, Hussein H, Kermarrec L, Ali AY, Marshall A, Metz-Boutigue M-H et al (2018) Chromogranin-a regulates macrophage function and the apoptotic pathway in murine DSS colitis. J Mol Med 96(2):183–198
5. Eissa N, Hussein H, Kermarrec L, Elgazzar O, Metz-Boutigue M-H, Bernstein CN, Ghia J-E (2017) Chromofungin (CHR: CHGA47-66) is downregulated in persons with active ulcerative colitis and suppresses pro-inflammatory macrophage function through the inhibition of NF-κB signaling. Biochem Pharmacol 145:102–113
6. Eissa N, Hussein H, Kermarrec L, Grover J, Metz-Boutigue M-HE, Bernstein CN, Ghia J-E (2017) Chromofungin ameliorates the progression of colitis by regulating alternatively activated macrophages. Front Immunol 8:1131
7. Eissa N, Hussein H, Hendy GN, Bernstein CN, Ghia J-E (2018) Chromogranin-A and its derived peptides and their pharmacological effects during intestinal inflammation. Biochem Pharmacol 152:315–326
8. Eissa N, Hussein H, Mesgna R, Bonin S, Hendy G, Metz-Boutigue M-H et al (2018) Catestatin regulates epithelial cell dynamics to improve intestinal inflammation. Vaccine 6 (4):67
9. Rőszer T (2015) Understanding the mysterious M2 macrophage through activation markers and effector mechanisms. Mediat Inflamm 2015:1
10. Gautier EL, Shay T, Miller J, Greter M, Jakubzick C, Ivanov S et al (2012) Gene-expression profiles and transcriptional regulatory pathways that underlie the identity and diversity of mouse tissue macrophages. Nat Immunol 13(11):1118
11. Taylor S, Wakem M, Dijkman G, Alsarraj M, Nguyen M (2010) A practical approach to RT-qPCR—publishing data that conform to the MIQE guidelines. Methods 50(4):S1–S5
12. Gonçalves R, Mosser DM (2015) The isolation and characterization of murine macrophages. Curr Protoc Immunol 111 (1):14.11.11–14.11.16
13. Zhang X, Goncalves R, Mosser DM (2008) The isolation and characterization of murine macrophages. Current Protoc Immunol 83 (1):14.11.11–14.11.14
14. Cunnick J, Kaur P, Cho Y, Groffen J, Heisterkamp N (2006) Use of bone marrow-derived macrophages to model murine innate immune responses. J Immunol Methods 311 (1–2):96–105
15. Rios FJ, Touyz RM, Montezano AC (2017) Isolation and differentiation of murine macrophages. Editors: Rhian M. Touyz Ernesto L. Schiffrin. Methods Mol Biol 1527:297–309. https://doi.org/10.1007/978-1-4939-6625-7_23
16. De Nardo D, Kalvakolanu DV, Latz E (2018) Immortalization of murine bone marrow-derived macrophages. Editor: Germain Rousselet. Methods Mol Biol 1784:35–49. https://doi.org/10.1007/978-1-4939-7837-3_4
17. McQuattie-Pimentel AC, Budinger GS, Ballinger MN (2018) Monocyte-derived alveolar macrophages: the dark side of lung repair? Am J Respir Cell Mol Biol 58:5–6; American Thoracic Society
18. Glass CK, Natoli G (2016) Molecular control of activation and priming in macrophages. Nat Immunol 17(1):26
19. Gordon S (2007) The macrophage: past, present and future. Eur J Immunol 37(S1):S9–S17
20. Mosser DM, Edwards JP (2008) Exploring the full spectrum of macrophage activation. Nat Rev Immunol 8(12):958
21. Mosser DM, Zhang X (2008) Activation of murine macrophages. Current Protoc Immunol 83(1):14.12.11–14.12.18
22. Eissa N, Hussein H, Wang H, Rabbi MF, Bernstein CN, Ghia J-E (2016) Stability of reference genes for messenger RNA quantification by real-time PCR in mouse dextran sodium sulfate experimental colitis. PLoS One 11(5): e0156289
23. Eissa N, Kermarrec L, Hussein H, Bernstein CN, Ghia J-E (2017) Appropriateness of reference genes for normalizing messenger RNA in mouse 2, 4-dinitrobenzene sulfonic acid (DNBS)-induced colitis using quantitative real time PCR. Sci Rep 7:42,427

24. Bustin SA, Benes V, Garson JA, Hellemans J, Huggett J, Kubista M et al (2009) The MIQE guidelines: minimum information for publication of quantitative real-time PCR experiments. Clin Chem 55(4):611–622
25. Schmittgen TD, Livak KJ (2008) Analyzing real-time PCR data by the comparative CT method. Nat Protoc 3(6):1101–1108
26. Eissa N, Hussein H, Diarra A, Elgazzar O, Gounni AS, Bernstein CN, Ghia J-E (2019) Semaphorin 3E regulates apoptosis in the intestinal epithelium during the development of colitis. Biochem Pharmacol 166:264
27. Kermarrec L, Eissa N, Wang H, Kapoor K, Diarra A, Gounni AS et al (2019) Semaphorin 3E attenuates intestinal inflammation through the regulation of the communication between splenic CD11C+ and CD 4+ CD 25-T cells. Br J Pharmacol 176:1235

Chapter 11

Simultaneous, Quantitative Characterization of Protein ADP-Ribosylation and Protein Phosphorylation in Macrophages

Casey M. Daniels, Arthur Nuccio, Pauline R. Kaplan, and Aleksandra Nita-Lazar

Abstract

The posttranslational modifications (PTMs) ADP-ribosylation and phosphorylation are important regulators of cellular pathways, and while mass spectrometry (MS)-based methods for the study of protein phosphorylation are well developed, protein ADP-ribosylation methodologies are still in a rapidly developing stage. The method described in this chapter uses immobilized metal affinity chromatography (IMAC), a phosphoenrichment matrix, to enrich ADP-ribosylated peptides which have been cleaved down to their phosphoribose attachment sites by a phosphodiesterase, thus isolating the ADP-ribosylated and phosphorylated proteomes simultaneously. To achieve the robust, relative quantification of PTM-level changes we have incorporated dimethyl labeling, a straightforward and economical choice which can be used on lysate from any cell type, including primary tissue. The entire pipeline has been optimized to work in ADP-ribosylation-compatible buffers and with protease-laden lysate from macrophage cells.

Key words ADP-ribosylation, Poly(ADP-ribose), Mono(ADP-ribose), Mass spectrometry, Phosphorylation, Phosphoribosylation, LC-MS/MS, Posttranslational modifications, IMAC, Phosphoenrichment, Dimethyl labeling, Phosphodiesterase, SVP, Proteomics

1 Introduction

The mass spectrometry (MS)-aided identification of protein posttranslational modifications (PTMs) has transformed our understanding of biochemical networks within cells, illuminating pathways of communication and highlighting potential targets for drug development. As our technological ability progresses, new complexities in these signaling mechanisms are appreciated, spurring more development and progress. It is now widely accepted that all protein networks are regulated by a large variety of PTMs, in addition to inter-protein interactions and protein translocation events, and a thorough understanding of these networks requires analysis of changes at all of these levels. The method we describe

Suresh Mishra (ed.), *Immunometabolism: Methods and Protocols*, Methods in Molecular Biology, vol. 2184,
https://doi.org/10.1007/978-1-0716-0802-9_11,

here addresses a need for looking at simultaneous changes in two unique PTMs—ADP-ribosylation and phosphorylation—both globally and in a semi-targeted manner (following the affinity purification of proteins of interest). In addition to site identification, we present our adapted protocol for dimethyl labeling (DML) of ADP-ribosylated peptides for a robust, relative quantification.

Protein ADP-ribosylation exists in many forms: as a free polymer, a monomer (mono-ADP-ribose, or MAR), a linear polymer of 2–200 subunits (poly-ADP-ribose or PAR), or a branched polymer of potentially hundreds of subunits (also referred to as PAR). Additionally, there are ten known amino acid acceptors of this PTM, further increasing the complexity of study and requiring neutral or acidic reaction conditions for the maintenance of all ADP-ribosylation sites [1]. As such, the MS-aided identification of these PTMs has proven challenging, with global site identification only becoming possible in the last decade [2–4]. Our contribution of the phosphodiesterase-based method [5] has allowed us to extensively characterize the changing ADP-ribosylated and phosphorylated proteomes in primary human macrophages (not shown here), a feat which required the development and optimization of this method, as reported in this chapter.

As shown in Fig. 1, the execution of this method will simultaneously reveal phosphorylation and ADP-ribosylation sites following the immobilized metal affinity chromatography (IMAC) enrichment of both. This enrichment is enabled by the phosphodiesterase (snake venom phosphodiesterase, or SVP) digestion of mono- and poly-ADP-ribosylated residues to phosphoribose (pR), a 212 Dalton molecular tag that contains the terminal phosphate residue amenable to IMAC enrichment. All steps have been adjusted as needed to work at a neutral pH, for maintaining all ADP-ribosylation events, and to work specifically for a macrophage cell lysate, which appears to contain urea-resistant proteases (Fig. 2). This protease activity, which was not observed in lysates from previously used cell types (data not shown), mandated that lysates be digested to tryptic peptides as quickly as possible to avoid nonspecific protease activity, and the associated loss of bottom-up proteomic data. Since the original protocol called for the SVP digestion of ADP-ribose at the protein level, before the addition of LysC and trypsin to the reaction mixture, it was necessary to determine whether SVP could be digested by these proteases if the ADP-ribose digestion was performed simultaneously or right after protease digestion. Figure 3 reveals that SVP is indeed resistant to trypsin and Lys-C digestion in these reaction conditions (1 M urea, 1 mM $CaCl_2$, 50 mM Tris–HCl pH 7, 37 °C, 2 h) and can therefore be added to the protease reaction mixture after the proteins have been digested to peptides, allowing for the immediate protease digestion of the lysates, and perhaps an increased efficiency of SVP, as peptides may be more susceptible to enzymatic activity than proteins (not tested directly).

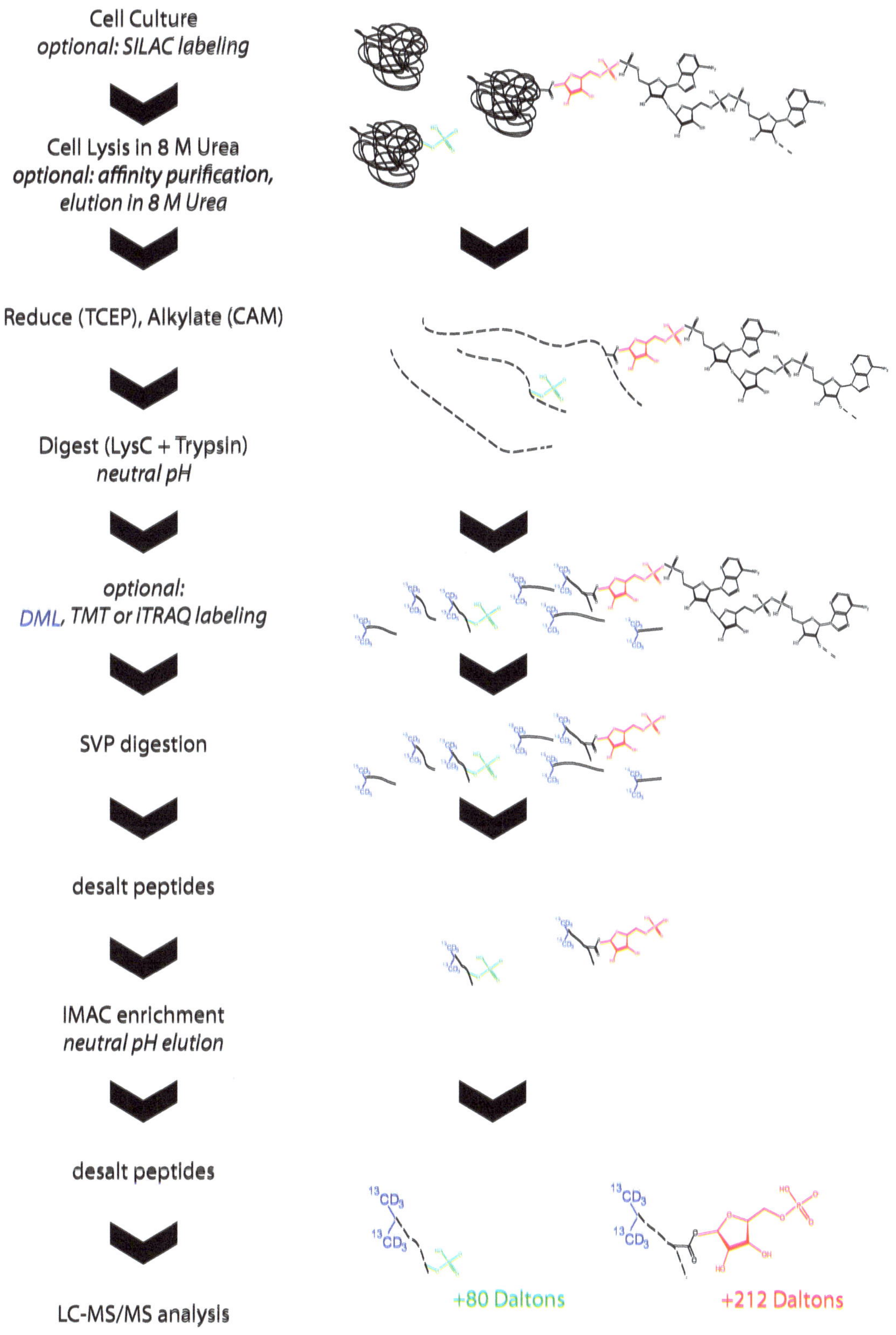

Fig. 1 Sample processing steps described in this protocol

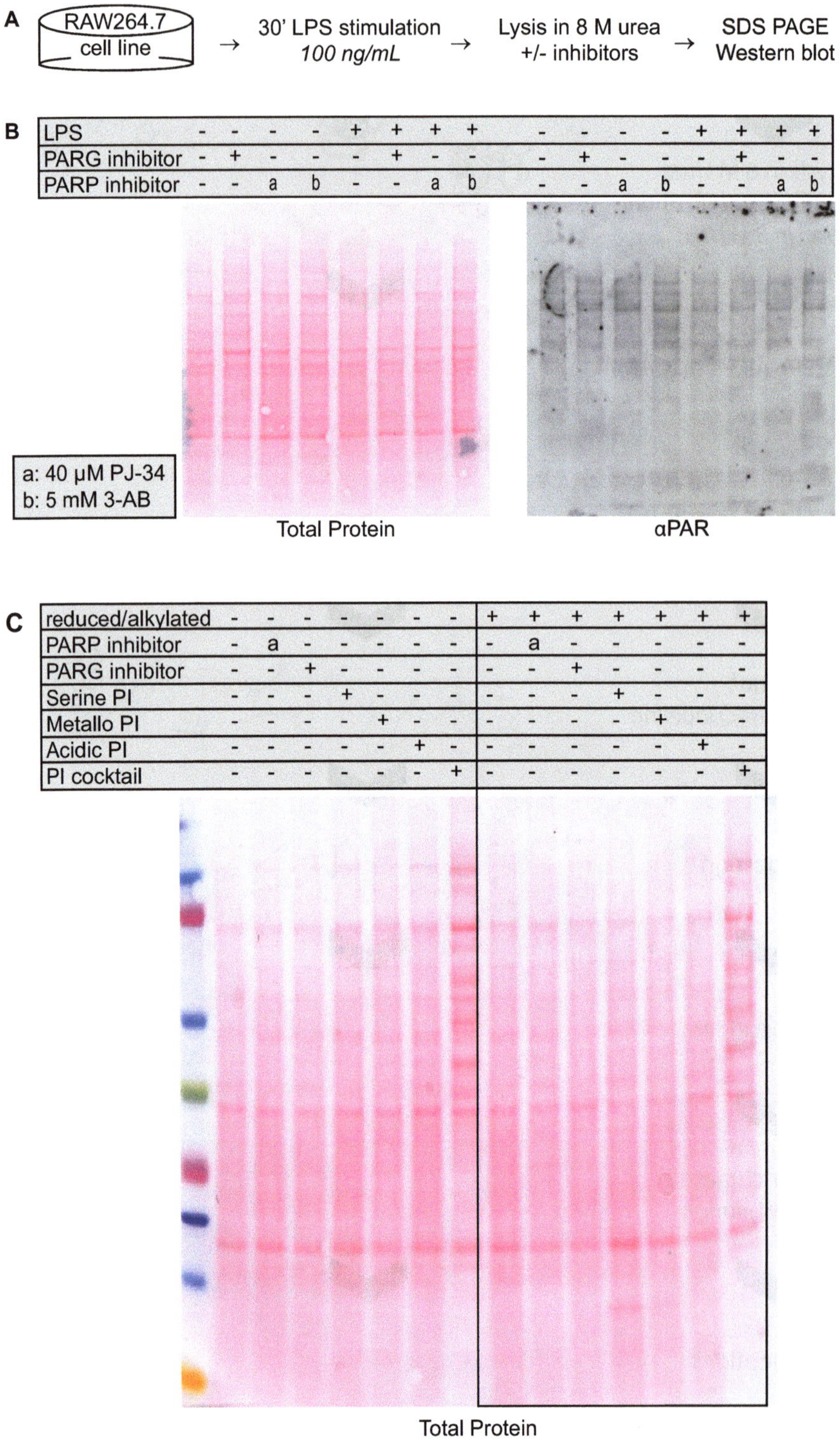

Fig. 2 Optimization of macrophage cell lysis conditions. RAW264.7 cells were treated with either vehicle or LPS before being lysed in 8 M urea with or without inhibitors (**a**) for analysis by western blot (**b**, **c**). Total protein levels are shown following Ponceau S staining, and protein poly(ADP-ribosylation) levels are shown

SVP purification from snake venom was also optimized for this protocol, as it was recognized that stabilizing SVP with glycerol is problematic due to its suppression of protein/peptide ionization [6]. Instead of glycerol, the reducing agent TCEP was used, in addition to an increased amount of BSA as a stabilizing protein. SVP purification and an assessment of the concentration are shown in Fig. 4.

To facilitate a robust, relative quantification of peptide levels in lysate from primary cells (monocyte-derived macrophages from human blood), dimethyl labeling (DML) was adapted for use with ADP-ribosylated peptides. DML involves the modification of peptide α- and ε-amine groups with light-, medium-, or heavy-labeled methyl groups, as shown in panel A of Fig. 5. As with the more common method for relative quantification—stable isotope-labeled amino acids in cell culture (SILAC)—peptide levels are ascertained from MS1 peptide information, abrogating the need for especially high-resolution chromatography and mass spectrometry (as required by other peptide labeling relative quantification methods, such as the Tandem Mass Tag [TMT™] or iTRAQ® commercial kits from Thermo and SCIEX, respectively). Unlike

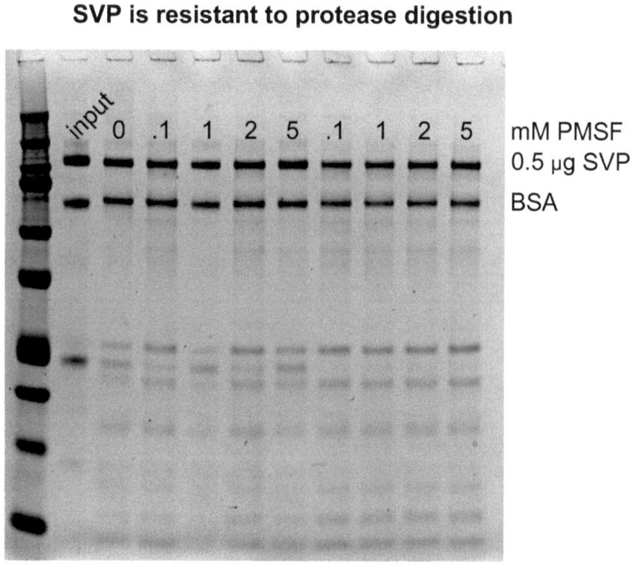

Fig. 3 SVP is resistant to protease digestion. The ability for SVP to withstand any protease activity encountered during combination protein/ADP-ribose digestion is shown

Fig. 2 (continued) following blotting with rabbit and mouse αPAR antibodies. PARP and PARG inhibitors have no effect on protein or poly(ADP-ribose) levels, while protease inhibitors appear to affect protein composition, suggesting that 8 M urea is sufficient for stopping PARG and PARG activity, but not all protease activity in macrophage whole-cell lysate

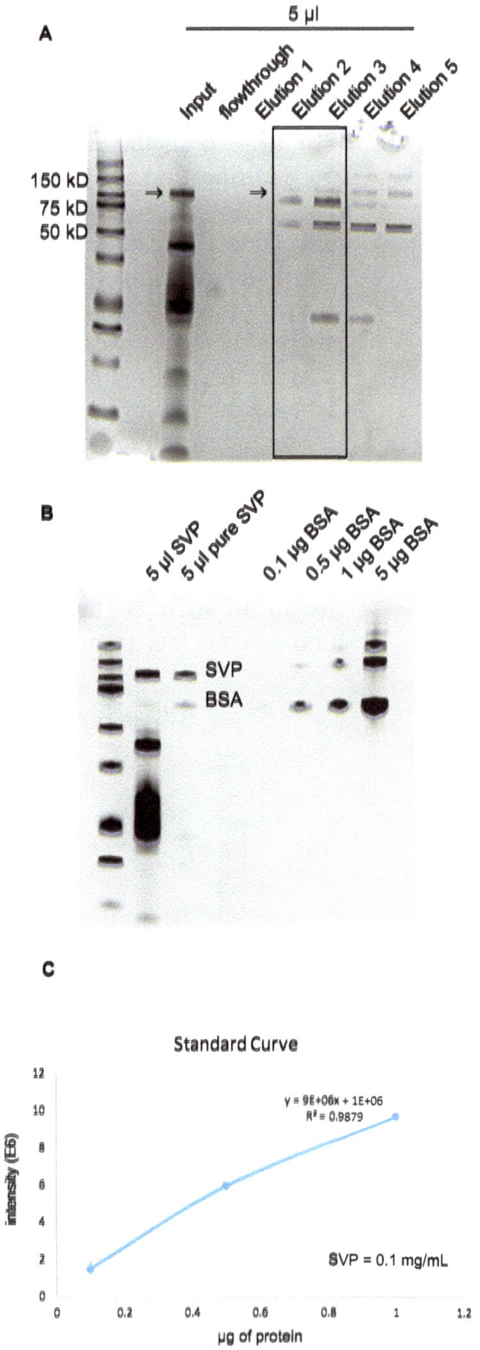

Fig. 4 Blue sepharose affinity purification of SVP for use in digestion of protein ADP-ribose to phosphoribose. Partially purified SVP was purchased from Worthington and further purified in-house using our established protocol. Purification of the 96 kDa SVP is shown in (**a**), and quantification of protein concentration is shown in (**b**) and (**c**)

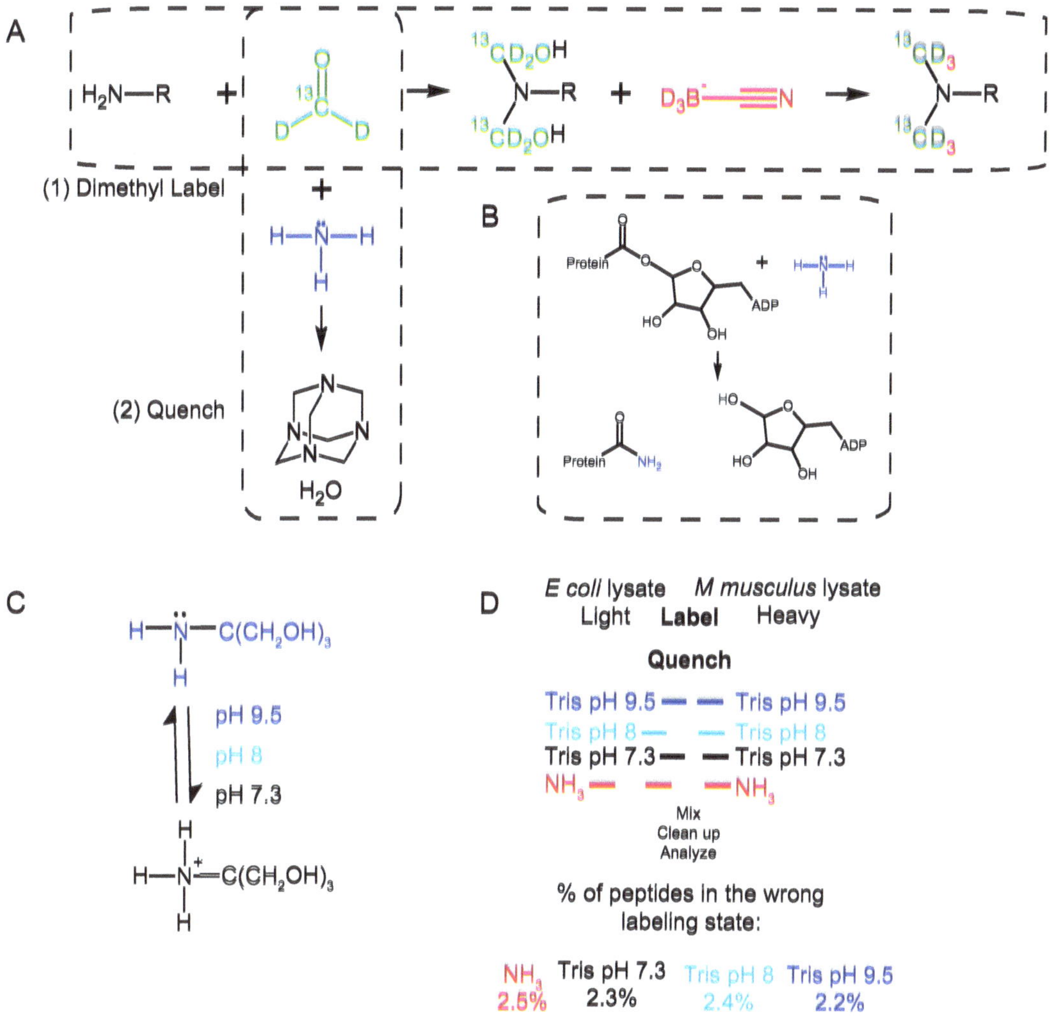

Fig. 5 Adaptation of the standard dimethyl labeling protocol for use with ADP-ribosylated peptides. Dimethyl labeling involves the exposure of peptides to isotopomers of formaldehyde and sodium cyanoborohydride to incorporate heavy labels into the dimethyl tags at the N-terminus and lysine groups of peptides (**a-1**). In order to prevent cross-labeling of peptide groups upon mixing, free formaldehyde is reacted with an amine-containing molecule such as ammonia (**a-2**). As shown in (**b**), however, exposure of ADP-ribosylated peptides to ammonia will result in loss of the PTM at acidic modification sites. The amine-containing Tris molecule, when used at a neutral or slightly basic pH, serves here as an alternative, ADPr-safe quenching solution (**c**) and (**d**)

SILAC, however, the labeling is done at the peptide level, making this approach universal and not dependent on cell cultures [7] or whole-animal labeling [8], a fact which has allowed our group to benefit from an MS1-level relative quantification of peptides from primary human immune cells. Finally, DML is considered highly cost effective [9, 10], as compared to alternative labeling methods for relative quantification.

The adaptation of a well-established DML protocol [9] for ADP-ribosylation site identification requires pH levels to be kept at or below 7. As DML is typically performed in a triethyl ammonium bicarbonate (TEAB) buffer, with a pH of 8.5, we lowered the pH of this buffer to 7 using formic acid, determining that labeling efficiency was still at or near 100% with this change. We also addressed the quenching step, which is typically performed with ammonia, a molecule known to release ADP-ribose modifications (Fig. 5, panel B, and [2]), therefore showing that it is possible to instead quench with neutral Tris–HCl (Fig. 5, panels C and D).

The pipeline presented in this chapter will allow researchers to analyze the phosphoproteome and ADP-ribosylated proteome of cells and proteins of interest, with sample preparations occurring in 2 days and application available for all cell and tissue types, including macrophages. The findings from such work will be welcome additions to the small but growing knowledge base of protein ADP-ribosylation and its cross talk with phosphorylation.

2 Materials

2.1 SVP Purification

1. Bovine serum albumin (BSA) standard (powder form).
2. Phosphodiesterase I from *Crotalus adamanteus* venom (Worthington or Sigma).
3. HiTrap blue sepharose column, 1 mL.
4. Loading/dialysis buffer: 10 mM Tris–HCl pH 7.5, 50 mM NaCl, 15 mM $MgCl_2$.
5. Elution buffer: 10 mM Tris–HCl pH 7.5, 50 mM NaCl, 15 mM $MgCl_2$, 150 mM potassium phosphate pH 7.5, 1 mM TCEP, 1 mg/mL BSA.
6. 1 mL and 10 mL syringes with a Luer-Lok™ adapter.
7. Blunt fill needle.
8. 20% Ethanol.
9. Dialysis tubing and clips, 30 kD MWCO.
10. Magnetic stir plate positioned in a cold room.
11. 2 L Glass beaker.
12. SDS-PAGE system (e.g., NuPAGE® from Thermo Fisher or Mini-PROTEAN® from Bio-Rad).
13. Heat block.
14. Sample rocker.
15. Coomassie stain.
16. Gel imager and analysis software (e.g., ImageJ, which is freely available from imagej.nih.gov).

2.2 Cell Culture and Lysis with Optional SILAC Labeling and/or Affinity Purification

1. Cell line of interest with appropriate growth media.
2. Chilled phosphate-buffered saline (PBS), pH 7.4.
3. Cell scraper.
4. Chilled collection tubes, 1.5–2 mL.
5. Chilled bath sonicator.
6. Chilled benchtop centrifuge.
7. Appropriate enrichment materials for affinity purification, if desired.
8. 8 M Urea in 50 mM Tris–HCl pH 7.3.
9. Kit for protein quantification.

2.3 Reduction, Alkylation, and Digestion of Proteins at Neutral pH

1. 50 mM Tris(2-carboxyethyl)phosphine (TCEP) in water.
2. 100 mM Chloroacetamide (CAM) in water.
3. Heat block with lid.
4. 1 mg/mL Trypsin/Lys-C mix, mass spec grade.
5. 1 M $MgCl_2$ solution.
6. 50 mM $CaCl_2$ solution.
7. 200 mM Triethylammonium bicarbonate (TEAB) buffer at pH 7. Adjust pH with formic acid.

2.4 Dimethyl Labeling of Peptides at Neutral pH (Optional)

1. 4% Formaldehyde: Light: CH_2O (stock: 37%, Sigma). Medium: CD_2O (stock: 20%, Isotec). Heavy: $^{13}CD_2O$ (stock: 20%, Isotec).
2. 0.6 M Sodium cyanoborohydride/deuteride. Light: $NaBH_3CN$ (Fluka). Heavy: $NaBD_3CN$ (Isotec).
3. 1 M Tris–HCl pH 7.3.
4. 5% Formic acid (Sigma).
5. A vacuum centrifuge.

2.5 SVP Digestion of ADP-Ribose

1. SVP as purified in Subheading 3.1.

2.6 Peptide Desalting

1. C^{18} SPE cartridges.
2. Desalting buffer A: 5% Acetonitrile, 0.1% formic acid.
3. Desalting buffer B: 80% Acetonitrile, 0.1% formic acid.

2.7 Enrichment of Phospho- and Phosphoribosylated Peptides

1. PHOS-Select™ Iron Affinity Gel (Sigma).
2. IMAC binding buffer: 40% Acetonitrile and 0.1% formic acid.
3. IMAC elution buffer: 0.5 M Potassium phosphate pH 7.
4. C^{18} SPE tips.

2.8 Separation and Analysis of Peptides by LC-MS/MS

1. High-performance liquid chromatography system hooked up to a high-mass-accuracy mass spectrometer.
2. HPLC solvent A: 0.1% Formic acid in water.
3. HPLC solvent B: 0.1% Formic acid in acetonitrile.

2.9 Data Analysis

1. Analysis program for mass spectrometry data, e.g., MaxQuant (freely available at www.maxquant.org).

3 Methods

3.1 SVP Purification

3.1.1 Affinity Purification

1. Pre-chill all buffers, tubes, and column on ice.
2. Attach a syringe to the Blue Sepharose column and push 5 column volumes (CVs) of water through, followed by 5 CVs of loading buffer.
3. Add 1 mL of loading buffer to the lyophilized SVP, and then use a blunt fill needle to pull the protein mixture into a 1 mL syringe.
4. Load the mixture onto the Blue Sepharose column, collecting flow-through in a pre-chilled microcentrifuge tube.
5. Push 5 CVs of loading buffer through the column to wash away undesired proteins.
6. Elute the sample into pre-chilled 1.5 mL tubes, collecting (5) 0.5 mL fractions.
7. Wash the column with 5 CV of water followed by 5 CV of 20% ethanol. Store at room temperature.

3.1.2 SDS-PAGE Analysis

1. Prepare 2× SDS-PAGE loading buffer, as appropriate for the SDS-PAGE gel of choice. Add 5 μL of buffer to 5 μL of either the input, flow-through, or elutions 1–5. Boil samples by incubating them in the heat block for 3 min at 95 °C (or as directed for the system used).
2. Prepare an extra input sample for use in section Protein Quantification Subheading 3.1.4; store at 4 °C.
3. Load the ladder and samples onto an SDS-PAGE gel; resolve as directed by the manufacturer.
4. Stain the gel with a protein stain as per the manufacturer's instructions (e.g., SimplyBlue™).
5. Identify fractions containing SVP, which runs slightly slower (i.e., higher) than its size of 96 kDa. Note that the protein profile for partially purified SVP will vary from one lot to another; therefore, the contaminating protein pattern may look different on your gel from that shown in Fig. 3.

3.1.3 Dialysis

1. Pre-chill 1 L of dialysis buffer in a large beaker (1.5 or 2 L) with a stir bar at the bottom.
2. Transfer fractions of interest (typically the second and third elutions) to the dialysis tubing with one end rolled and clipped. After adding sample, roll and clip the other end. Secure the clip to a piece of foam to allow the sample to float near the top of the buffer. Cover the beaker with foil or plastic wrap.
3. Place the beaker on a stir plate in the cold room (*see* **Note 1**) or, alternatively, in an ice bucket on a stir plate to ensure that the solution stays cold throughout the entire dialysis.
4. Allow the mixture to stir for 1 h at a speed just fast enough to rotate the sample.
5. Transfer the sample out of the tubing and into a pre-chilled tube; the sample volume may have increased slightly due to osmosis.

3.1.4 Protein Quantification

1. Sample 5 μL of the dialyzed mixture, combine it with 5 μL of 2× SDS-PAGE loading buffer, and incubate for 3 min at 95 °C.
2. Retrieve the extra input sample prepared in Subheading 3.1.2.
3. Prepare the BSA standard curve.
   (a) Dissolve lyophilized BSA into water to make a final concentration of 0.1 mg/mL.
   (b) Transfer 0.5 μL of stock solution to 99.5 μL of 1× SDS-PAGE loading buffer to make A (0.5 μg/mL).
   (c) Transfer 25 μL of A to 100 μL of 1× loading buffer to make a 0.1 μg/mL standard (B).
   (d) Transfer 50 μL of B to 50 μL of buffer to make a 0.05 μg/mL standard (C).
   (e) Transfer 16 μL of C to 64 μL of buffer to make a 0.01 μg/mL standard (D).
   (f) Boil the standards by incubating at 95 °C for 3 min.
   (g) Extra standard may be frozen for use later.
4. Load an SDS-PAGE gel with a protein ladder and 10 μL each of input, dialyzed sample, and protein standards A–D.
5. After resolving the gel, use a total protein stain to visualize the protein bands and an imager to digitize the gel for analysis. A scanner or camera may be used in place of a dedicated imager.
6. Measure the optical density of the protein bands using software such as ImageJ to construct a standard curve and estimate the concentration of the SVP protein in the final solution. Sample loss can be estimated based on comparing the concentrations of the protein before and after purification. This loss should be minimal (*see* **Note 2**).

7. The enzyme should be used immediately or aliquoted and stored at −80 °C for up to 1 month. Carefully avoid freeze-thaw cycles!

3.2 Cell Culture and Lysis with Optional SILAC Labeling and/or Affinity Purification

3.2.1 Cell Culture

1. As this protocol may be used with any manner of tissue culture, we leave the parameters of culture open to whatever is appropriate for the cell line and biological process of interest.
2. Note that SILAC labeling, if used, would be performed as the first step of cell culture. For a detailed protocol please see [11].

3.2.2 Affinity Purification (Optional)

If desired, a subset of the proteome may be enriched using appropriate protocols, as determined by the researcher. If this is attempted, then standard protocols may need to be adjusted in the following ways:

1. Neutral or acidic pH should be maintained, to avoid the loss of ADP-ribosylation modifications.
2. Mass spec-friendly detergents should be used to replace problematic ones—N-octylgluconate (NOG) in place of NP-40/Tween/Triton, and sodium deoxycholate in place of sodium dodecyl sulfate, for example, as these detergents are dialyzable and can therefore be removed from the purification mixture by washing the sample. Commercially available, MS-friendly detergent kits may also be used.
3. Samples should ideally be eluted from their enrichment matrices using the urea lysis buffer provided here, as it is compatible with maintaining the PTM of interest (due to pH) and enzymatic activity of SVP (which cannot tolerate guanidinium, among other common elution chemicals). If elution is not efficient, the digestion may be performed in the presence of the enrichment beads (this is referred to as on-bead digestion, as opposed to in-solution digestion following elution) (*see* **Note 3**).

3.2.3 Cell Lysis

1. Wash cells three times with ice-cold PBS. For adherent cells, this can be done directly on the culture plate. For suspension cells, this will require spinning the cells down between each wash. After washing, carefully remove all PBS—in the case of adherent cells this may require the researcher to tilt the plate 90° to drain for 30 s.
2. Add the urea lysis buffer to cells, either directly to the pellet (if cells are in suspension) or directly to the plate (if cells are adherent), followed by the scraping of the adherent cells into the lysis buffer. Optimal volume will vary based on cell size, but 200 μL of lysis buffer for 10^6 cells is fairly standard (*see* **Note 4**).

3. Transfer the lysate to a chilled microcentrifuge tube and either incubate on ice for 10 min or, if available, sonicate in a cold sonication bath with alternating rest periods between sonication cycles to prevent the sample from heating. This sonication may help with lysis and further breaking nucleic acids apart to release chromatin-associated proteins.
4. Centrifuge the samples at max speed in a chilled benchtop microcentrifuge for 10 min. Save the supernatant, discarding any pellet formed (there may not be one, as urea will break up most cellular components, including nucleic acids).
5. Quantify the protein amount using a commercial kit, which is compatible with urea (e.g., Thermo Pierce™ 660 nm Protein Assay). Adjust the protein concentration to 10 mg/mL in lysis buffer.

3.3 Reduction, Alkylation, and Digestion of Proteins at Neutral pH

1. Reduce in 1 mM TCEP: Combine 50 μL (500 μg) of the whole-cell lysate with 1 μL of 50 mM TCEP; incubate for 10 min at 37 °C.
2. Alkylate in 2 mM CAM: Add 1 μL of 100 mM CAM, and incubate for 10 min at 37 °C in the dark.
3. Add 8 μL of 50 mM $CaCl_2$.
4. Add 6 μL of 1 M $MgCl_2$.
5. Add 200 μL of 200 mM TEAB pH 7.
6. Add 10 μL of 1 mg/mL LysC/trypsin mixture.
7. Add 124 μL of water to bring the final volume to 400 μL.
8. Incubate for 16 h at 37 °C to digest proteins.
9. Proceed directly to Subheading 3.4 or 3.5, depending on whether dimethyl labeling will be performed or not.

3.4 Dimethyl Labeling of Peptides at Neutral pH (Optional)

Note that this labeling method is an alternative to SILAC or other labeling methods for relative quantitation, such as the Tandem Mass Tag (TMT™ from Thermo) or iTRAQ® (SCIEX).

1. Add 16 μL of 4% formaldehyde: Light: CH_2O. Medium: CD_2O. Heavy: $^{13}CD_2O$.
2. Vortex to mix; centrifuge to collect again.
3. Add 16 μL of 0.6 M sodium cyanoborohydride: Light and medium: $NaBH_3CN$. Heavy: $NaBD_3CN$.
4. Vortex to mix; centrifuge to collect again.
5. Label: Shake at room temperature for 1 h.
6. Quench: Add 64 μL of 1 M Tris–HCl pH 7.3; vortex.
7. Quench further: Add 32 μL of 5% formic acid; vortex and spin down.

8. Mix labeling states together; if mixing three samples the final volume will be ~1.6 mL. Concentrate on a speed-vac so urea concentration is back to ~1 M (~1.2 mL).
9. Proceed directly to SVP digestion.

3.5 SVP Digestion of ADP-Ribose

1. Add 5 μg of SVP to reaction mixture and incubate at 37 °C for 2 h.

3.6 Desalt Peptides

1. Assemble an appropriately sized C18 cartridge to bind the peptides using either gravity flow or a vacuum manifold, if available.
2. Activate the matrix with 1 CV of buffer B.
3. Equilibrate the matrix with 2 CV of buffer A.
4. Load the peptides onto the column.
5. Wash with 2 CV of buffer A.
6. Elute with 0.1 CV of buffer B, twice (combine eluates).
7. Concentrate the eluted peptides in a vacuum centrifuge, aiming to remove most, but not quite all, of the volume.
8. Resuspend peptides in 100 μL of IMAC binding buffer.

3.7 Enrichment of Phospho- and Phosphoribosylated Peptides

1. Remove 100 μL of the IMAC bead slurry/sample (25 μL of beads in a 50% slurry); wash with 1 mL of binding buffer, three times. Leave in binding buffer at a volume 10% higher than the original volume.
2. Transfer 50 μL of the washed bead slurry to each sample.
   (a) Transfer another 50 μL of bead slurry to a second tube/sample.
3. Incubate peptides on beads for 60 min with vigorous shaking at room temperature.
4. Spin beads down using a quick spin on a benchtop centrifuge.
5. Transfer the supernatant to the second set of beads for that sample; add 100 μL of fresh binding buffer back to the beads and vortex to resuspend.
6. Incubate the second peptide enrichment for 60 min at room temperature with vigorous shaking.
7. Spin beads down and discard supernatant.
8. Add 100 μL of binding buffer to beads; vortex to resuspend.
9. Spin all samples down and remove supernatant; add another 100 μL of fresh binding buffer.
10. Spin all samples down and remove supernatant, add back 25 μL of IMAC elution buffer, and resuspend.

11. Spin down, transfer eluates to a new tube, add another 25 μL of IMAC elution buffer, and resuspend.
12. Spin down, and transfer eluates to the same tube used in **step 11**.
13. Add 50 μL of desalting buffer A to eluates.
14. Desalt on a C^{18} SPE tip or cartridge with a capacity of 50 μg or more, using the same steps described in Subheading 3.6 but resuspend in desalting buffer A.
15. Samples may be briefly stored at −20 °C or analyzed immediately by LC-MS/MS.

3.8 Separation and Analysis of Peptides by LC-MS/MS

1. Equilibrate an analytical C^{18} column in 2% acetonitrile and 0.1% formic acid on a nanoflow LC equipped with a spray tip for electrospray ionization.
2. Load samples onto a column and apply a suitable acetonitrile gradient for the separation of peptides over 60–90 min. Identify peptides by CID fragmentation (*see* **Note 5**).

3.9 Data Analysis

1. Load Raw files into MaxQuant for analysis (current version is 1.6.7.0).
2. Identify samples as experiments with fractions (2/sample, 1 from each phosphoenrichment step) and mark PTM as "true."
3. Enter phosphoribose as a new modification (212.01 Da, $C_5H_9O_7P$) for D, E, K, R, and C residues.
4. Mark phosphoribosylation, phosphorylation, and carbamidomethylation as variable modifications, in addition to the default modifications. Remove carbamidomethylation from the fixed modifications list.
5. Add the appropriate FASTA file for the sample type used.
6. Set the number of threads as two less than the quantity of logical cores on the PC.
7. Run the search.
8. Results will be reported in the included viewer as well as the "txt" folder in the experiment folder.

4 Notes

1. When using equipment in the cold room it is advisable to allow the equipment to equilibrate overnight to mitigate the risk of damage to the equipment due to the formation of condensation.

2. As companies often sell enzymes based on their activity and not their concentration, and this particular enzyme is purified from natural sources and not synthesized, the researcher may notice a large range of protein concentrations between lots purchased from the same company that need to be normalized differently for every experiment.
3. It is recommended to test both methods by eluting the sample into the urea lysis buffer, saving this elution for further analysis, and then adding more urea lysis buffer to the beads to perform an on-bead digestion of the remaining sample.
4. If the solution is too thick to pipette due to nucleic acid release, the researcher may add more lysis buffer to dilute the sample.
5. While other fragmentation methods can be used for phosphoribosylated peptides, we have determined that CID is optimal for most peptides, most of the time. For a full comparison between CID, ETD, and HCD for phosphoribosylated peptides, see [5].

Acknowledgments

The authors would like to thank Kevin Johnson for his assistance at the bench, and Dr. Paul Boersema for his counsel during the optimization of the DML protocol for analysis of ADP-ribosylated peptides. This research was supported by the Intramural Research Program of NIAID, NIH.

References

1. Daniels CM, Ong SE, Leung AK (2015) The promise of proteomics for the study of ADP-ribosylation. Mol Cell 58:911–924
2. Daniels CM, Ong SE, Leung AK (2014) A phosphoproteomic approach to characterize protein mono and poly(ADP-ribosyl)ation sites from whole cell lysate. J Proteome Res 13:3510–3522
3. Zhang Y et al (2013) Site-specific characterization of the asp- and Glu-ADP-ribosylated proteome. Nat Methods 10:981–984
4. Rosenthal F et al (2011) Identification of distinct amino acids as ADP-ribose acceptor sites by mass spectrometry. Methods Mol Biol 780:57–66
5. Daniels CM, Ong SE, Leung AKL (2017) ADP-ribosylated peptide enrichment and site identification: the phosphodiesterase-based method. Methods Mol Biol 1608:79–93
6. Mendes MA et al (2003) The shielding effect of glycerol against protein ionization in electrospray mass spectrometry. Rapid Commun Mass Spectrom 17:672–677
7. Ong SE et al (2002) Stable isotope labeling by amino acids in cell culture, SILAC, as a simple and accurate approach to expression proteomics. Mol Cell Proteomics 1:376–386
8. Kruger M et al (2008) SILAC mouse for quantitative proteomics uncovers kindlin-3 as an essential factor for red blood cell function. Cell 134:353–364
9. Boersema PJ et al (2008) Triplex protein quantification based on stable isotope labeling by peptide dimethylation applied to cell and tissue lysates. Proteomics 8:4624–4632
10. Lau HT et al (2014) Comparing SILAC- and stable isotope dimethyl-labeling approaches for quantitative proteomics. J Proteome Res 13:4164–4174
11. Ong SE, Mann M (2006) A practical recipe for stable isotope labeling by amino acids in cell culture (SILAC). Nat Protoc 1:2650–2660

Chapter 12

The Analysis of *Mycobacterium tuberculosis*-Induced Bioenergetic Changes in Infected Macrophages Using an Extracellular Flux Analyzer

Bridgette M. Cumming, Vineel P. Reddy, and Adrie J. C. Steyn

Abstract

Metabolism plays an important role in the activation and effector functions of macrophages. Intracellular pathogens, such as *Mycobacterium tuberculosis*, subvert the immune functions of macrophages to establish an infection by modulating the metabolism of the macrophage. Here, we describe how the Seahorse Extracellular Flux Analyzer (XF) from Agilent Technologies can be used to study the changes in the bioenergetic metabolism of the macrophages induced by infection with mycobacteria. The XF simultaneously measures the oxygen consumption and extracellular acidification of the macrophages noninvasively in real time, and together with the addition of metabolic modulators, substrates, and inhibitors enables measurements of the rates of oxidative phosphorylation, glycolysis, and ATP production.

Key words Macrophages, *Mycobacterium tuberculosis*, Extracellular flux analyzer, Oxidative phosphorylation, Glycolysis, ATP production rate, Cellular bioenergetics

1 Introduction

Tuberculosis (TB) is the leading cause of death due to an infectious agent [1]. This is largely due to the development of multidrug-resistant and extensively drug-resistant TB, and coinfection with HIV. For these reasons, host-directed therapies are being increasingly considered as adjunctive therapy to treat TB [2–4]. Here, we describe the protocols to investigate bioenergetic, metabolic changes in macrophages infected with *Mycobacterium tuberculosis* (*Mtb*) and other mycobacterial strains, using an extracellular flux analyzer (XF) from Agilent Technologies, which is widely used for the metabolic profiling of diverse cell/tissue types. Although we have described the protocols for use in an XFe96, the same assays can be performed in an XFe24, but the number of cells per well and the volumes of assay media in the wells and in the injection ports need to be increased by a factor of 2.5. As the wells in the XFe96 and XFp have the same surface area, the seeding densities and

Suresh Mishra (ed.), *Immunometabolism: Methods and Protocols*, Methods in Molecular Biology, vol. 2184,
https://doi.org/10.1007/978-1-0716-0802-9_12,

volumes of assay media in the wells and in the injection ports for both instruments will be the same.

The XF performs accurate, real-time measurements of living cells by simultaneously quantifying the cellular oxygen consumption rate (OCR) and the extracellular acidification rate (ECAR). The OCR gives a measure of the rate of oxygen consumed by complex IV of the electron transport chain and NADPH oxidase in the cytoplasm. The ECAR gives a measure of the acidification rate due to protons extruded with lactate that are produced from pyruvate, the terminal product of glycolysis, and carbonic acid, resulting from the hydration of carbon dioxide generated by the tricarboxylic acid cycle [5]. While the assay is running, the cells can be metabolically modulated by successive injections of compounds from four ports surrounding the sensor probe of the cartridge, and real-time responses to the substrates, activators, or inhibitors of pathways can be elucidated [6]. Here, we describe protocols for the preparation of five different models of macrophages: the proliferating mouse RAW264.7 macrophage cell line; the human monocytic THP-1 cell line that is terminally differentiated into macrophages; terminally differentiated primary cells such as wild-type and genetic knock-out intraperitoneal macrophages [7] and bone marrow derived macrophages from mice [8]; and human monocyte derived macrophages [9]. We then explain the protocols for the growth, preparation and infection of these macrophage models with *Mtb*, the avirulent vaccine strain, *M. bovis* (Bacillus Calmette-Guérin), and dead *Mtb* [9] and how the modulation of the bioenergetic metabolism induced by the mycobacterial infection can be assessed in real time using extracellular flux analyses. The assays described here include the Cell Mito Stress Test, the Glycolysis Stress Test, the Glycolytic Rate Assay and the ATP Production Rate. The results of the extracellular flux analysis can be verified with ^{13}C-tracing experiments to determine the extent of incorporation of the stably labelled substrate into the intermediates of central carbon metabolism [9]. The effects of infection with clinical strains of *Mtb* could also be investigated using these protocols.

2 Materials

1. Seahorse XFe96 analyzer with Wave controller or XFp (*see* **Note 1**).
2. Seahorse XFe96 or XFp Sensor Cartridges and Cell Culture microplates.
3. Incubator, 37 °C, with 5 % CO_2 and 95% humidity.
4. Incubator, 37 °C, without CO_2.

5. Calibrated pH meter.
6. MACS® Manual separator for magnetic cell isolation.
7. Seahorse XF Calibrant solution.
8. Seahorse XF Base Medium.
9. Seahorse XF DMEM medium, pH 7.4, 5 mM HEPES (for the glycolytic rate assay).
10. Seahorse XF 1.0 M glucose solution.
11. Seahorse XF 100 mM pyruvate solution.
12. Seahorse XF 200 mM glutamine solution.
13. Ammonium-chloride-potassium (ACK) lysing buffer.
14. Antibiotic-antimycotic solution (100× ABAM).
15. DMEM medium, 4 mM L-glutamine, 4.5 g/L D-glucose, 1.5 g/L sodium bicarbonate.
16. Complete DMEM medium: DMEM medium containing 10% (v/v) fetal bovine serum (FBS) and 1x (v/v) ABAM.
17. DMEM medium for infection: Complete DMEM medium without 1x ABAM.
18. Complete RPMI-1640 medium: 10% (v/v) fetal bovine serum (FBS), 2 mM L-glutamine, 2 g/L D-glucose, 2 g/L sodium bicarbonate, 1 mM sodium pyruvate, 1× (v/v) ABAM.
19. RPMI-1640 medium for infection: Complete RPMI-1640 medium without the ABAM.
20. RPMI-1640 medium for human monocytes: 10% (v/v) human serum, 2 mM L-glutamine, 4.5 g/L D-glucose, 2 g/L sodium bicarbonate, 1 mM sodium pyruvate , 1× nonessential amino acids, 1 mM HEPES.
21. Fetal bovine serum (FBS): Heat inactivate at 56 °C for 1 h.
22. Human serum: Heat inactivate at 56 °C for 1 h and spin at 3000 *g* for 15 min to pellet fibrin.
23. Phorbol 12-myristate 13-acetate (PMA).
24. Granulocyte-macrophage colony-stimulating factor (GM-CSF).
25. Macrophage colony-stimulating factor (M-CSF).
26. 100× Nonessential amino acids: 10 mM of L-alanine, L-asparagine, L-aspartic acid, L-glutamic acid, glycine, L-proline, L-serine.
27. BBL Thioglycollate medium: 3% (v/v) solution in water.
28. Dulbecco's phosphate-buffered saline, pH 7.4 (PBS).
29. CD14^{+} human microbeads.
30. Histopaque-1077.

3 Methods

A summary of the macrophage models examined, seeding densities in the XF cell culture microplates, and differentiation conditions used is provided in Table 1 (*see* **Note 2**). The preparation of the individual macrophage models is described below.

3.1 Mouse Bone Marrow-Derived Macrophages (BMDM)

3.1.1 Isolation of BMDM from Mice

1. Sacrifice mice according to institutional animal care guidelines.
2. In a biosafety cabinet, gently peel off the skin from the mouse. Collect the tibia/fibula and femur bones from both legs and place them in a Petri plate containing PBS pH 7.4 with 1× ABAM in the biosafety cabinet to maintain sterility.
3. Carefully remove muscle from around the bones using scissors and forceps. Make sure that no hair sticks to the bones during the procedure.
4. Rinse the bones in a Petri plate containing 70% ethanol for 30 s.
5. Next, wash the bones in two sets of Petri plates containing PBS pH 7.4 and rinse the bones carefully to remove the muscle from around the bones.
6. Carefully cut the tibia and femur bones at each end using sharp scissors and forceps.
7. Draw up 10 mL of PBS, pH 7.4, with 1× ABAM solution into the 10 mL syringe with 27 G needle.
8. Hold the bone in a vertical position and inject 10 mL of PBS, 1× ABAM, into the top opening of the bone to flush bone marrow cells into the sterile Petri plate (*see* **Note 3**).

Table 1
Seeding densities and differentiation of different macrophage lineages (*see* Note 2)

Macrophage model	Seeding numbers/well in XFe96/XFp cell culture microplate	Differentiation	Period of differentiation
hMDM	80,000	10 ng/mL GM-CSF	6 days
THP-1	100,000	100 nM PMA	24 h
RAW264.7	65,000	–	–
BMDM	200,000	20 ng/mL M-CSF	6 days
Intraperitoneal macrophages	200,000	–	–

9. Pipette cells up and down several times using a sterile 5 mL pipette attached to a portable pipette aid into a Petri plate to make a single-cell suspension.
10. Pass the cells through a 70 μm cell strainer to remove bone debris.
11. Centrifuge the cells in 15 mL conical tube at 800 *g* for 5 min at 4 °C.
12. Decant the supernatant fluid and tap the tip of the conical tube to loosen the cell pellet.
13. Transfer the cells to a 50 mL conical tube.

3.1.2 ACK Lysis Procedure

1. Add 2 mL of ACK lysis buffer to cells in 50 mL conical tube to lyse the red blood cells and incubate for 2 min at room temperature (*see* **Note 4**).
2. Fill the tubes with ice-cold PBS, pH 7.4, to neutralize the ACK lysing buffer.
3. Centrifuge the cells at 800 *g* for 5 min at 4 °C.
4. Completely decant the PBS pH 7.4 and gently tap the inverted tube on paper towel to remove residual PBS.
5. Carefully thaw the cell pellet by gently tapping the bottom tip of the conical tube with your finger.
6. Resuspend the cells in 20 mL of DMEM complete medium.
7. Centrifuge the cells at 800 *g* for 5 min at 4 °C.
8. Resuspend the cells in 3–4 mL of DMEM complete medium and count the cells using a hemocytometer or an automated cell counter.

3.1.3 Seeding Cells in XF Culture Plates

1. Prepare 1×10^6 BMDMs/mL in DMEM complete medium with 20 ng/mL M-CSF.
2. Seed 2×10^5 cells per well of a cell culture microplate in 200 μL of media. In an XFe96 cell culture microplate, seed 8 wells (or one column) per condition investigated, and in an XFp cell culture microplate, seed 3 wells per condition investigated and do multiple runs if more than one condition is investigated (*see* **Note 5**).
3. Add only media (no cells) to all the wells of columns 1 and 12 of the XFe96 cell culture microplate and to wells A and H of the XFp cell culture microplate as background control required by the Wave controller (*see* **Note 6**).
4. Incubate the cells at 37 °C with 5% CO_2, 95% humidity, for 3 days.

5. Change the media after 3 days with DMEM complete medium containing 20 ng/mL M-CSF and incubate for another 3 days.
6. After a total of 6 days of differentiation, infect the BMDMs as described in Subheading 3.6.

3.2 Mouse Intraperitoneal Macrophages

3.2.1 Injection of 3% BBL Thioglycollate Medium into the Mouse Peritoneum

1. Carefully inject 1 mL of 3% BBL Thioglycollate medium into the peritoneum of the mouse.
2. Wait for 4 days for macrophages to be recruited to the mouse peritoneum.

3.2.2 Isolation of Intraperitoneal Macrophages from Mice

1. After 4 days, sacrifice mice according to the institutional animal care guidelines.
2. Make a small cut in the skin of the mouse abdomen and peel off the skin carefully without puncturing the peritoneum (*see* **Note 7**).
3. Prepare PBS pH 7.4 containing 5% FBS and keep on ice.
4. Take up 5 mL of PBS pH 7.4 with 5% FBS into the 5 mL syringe and attach 27 G needle.
5. Inject 5 mL of this solution carefully into the peritoneal cavity.
6. Massage the mouse abdomen for 30 s.
7. Carefully remove the intraperitoneal macrophages from the peritoneum by drawing the cell suspension into the syringe (3–4 mL).
8. Cut the peritoneum and carefully remove the remaining intraperitoneal macrophages using a 1 mL pipette.
9. Pass the cell suspension through a 70 μm cell strainer and collect the intraperitoneal macrophages in a 15 mL conical tube.
10. Centrifuge the intraperitoneal macrophages at 800 *g* for 5 min at 4 °C.
11. Decant the PBS and tap the tip of the conical tube to loosen the cell pellet.

3.2.3 ACK Lysis Procedure

1. Add 1 mL of ACK lysis buffer to the intraperitoneal macrophages and incubate them at room temperature for 2 min (*see* **Note 4**).
2. Fill the tubes with ice-cold PBS, pH 7.4, to neutralize the ACK lysing buffer.
3. Centrifuge the cells at 800 *g* for 5 min at 4 °C.
4. Completely decant the PBS and gently tap the inverted tube on paper towel to remove residual PBS.

5. Carefully thaw the cell pellet by gently tapping the bottom tip of the conical tube with your finger.
6. Resuspend the cells in 12 mL of RPMI complete medium.
7. Centrifuge the cells at 800 *g* for 5 min at 4 °C.
8. Resuspend the cells in 3–4 mL of RPMI complete medium and count the cells using a hemocytometer or an automated cell counter.

3.2.4 Seeding Cells in XF Cell Culture Plates

1. Prepare 1×10^6 intraperitoneal macrophages/mL in the RPMI complete medium.
2. Seed 2×10^5 cells per well in a XFe96 or XFp cell culture microplate in 200 μL of media (*see* **Note 5**).
3. Add 200 μL RPMI complete medium only (no cells) to the wells in columns 1 and 12 of the XFe96 cell culture microplate and to wells A and H of the XFp cell culture microplate as background control (*see* **Note 6**).
4. Incubate the cells at 37 °C, 5% CO_2, 95% humidity, overnight.
5. The following day, infect the intraperitoneal macrophages as described in Subheading 3.6.

3.3 RAW264.7 Macrophages (ATCC® TIB-71™)

1. Culture RAW264.7 cells as per the ATCC protocols (https://www.atcc.org/products/all/TIB-71.aspx).
2. After checking the health of the macrophages under an inverted microscope, remove the supernatant fluid from the adherent RAW264.7 cells in a T75 flask, wash the cells with 2 mL DMEM complete medium, add 5–10 mL fresh DMEM complete medium depending on the cell confluency, and scrape the macrophages from the bottom of the culture flask using a cell scraper.
3. Dilute 10 μL of the suspension of the RAW264.7 macrophages 1:2 in trypan blue to exclude the dead cells and count the number of viable RAW264.7 macrophages using a hemocytometer or an automated cell counter. Calculate the volume of cells needed to prepare 10 mL of 8.125×10^5 cells/mL suspension in DMEM complete medium.
4. Seed the RAW264.7 cells into the XF cell culture microplate at 65,000 cells per well (80 μL) and incubate in a 37 °C, 5% CO_2, 95% humidity, incubator.
5. Allow the cells to adhere to the base of the well for 4 h prior to mycobacterial infection.
6. Do not seed cells into columns 1 and 12 of the XFe96 cell culture microplate or wells A and H of the XFp cell culture microplate. Fill these wells with media only (*see* **Note 6**).

3.4 THP-1 Monocytes (ATCC® TIB-202™)

1. Culture THP-1 cells as per the ATCC protocols (https://www.atcc.org/products/all/TIB-202.aspx).
2. Count the cultured THP-1 cells and calculate the volume of suspension culture needed to prepare 10 mL of 1.25×10^6 cells/mL suspension.
3. Spin down the calculated volume of cultured THP-1 cells at 800 *g* to pellet the cells and resuspend the cells in 10 mL fresh RPMI-1640 complete medium.
4. Add phorbol 12-myristate 13-acetate to a final concentration of 100 nM in the 10 mL cell suspension.
5. Seed the THP-1 cells into the XFe96 cell culture microplates at 100,000 cells per well (80 μL) and incubate overnight in a 37 °C, 5% CO_2, 95% humidity, incubator to allow for the differentiation of the monocytes into macrophages.
6. Infect the differentiated THP-1 cells within 24 h of addition of 12-myristate 13-acetate to the THP-1 cells and perform the XF assay after 18–24 h of infection. Longer infection periods can be investigated at low multiplicities of infection of 0.1 or lower.

3.5 Human Monocyte-Derived Macrophages (hMDM)

1. Isolate peripheral blood mononuclear cells (PBMCs) from buffy coats using density gradient centrifugation. Dilute 8 mL of the buffy coat in 27 mL PBS, pH 7.4, and layer the diluted buffy coat over 15 mL Histopaque 1077 in a 50 mL conical tube.
2. Centrifuge (swinging bucket rotor) the layers at 400 *g* for 35 min at room temperature without acceleration or braking.
3. Remove the plasma carefully within 0.5 cm of the PBMC layer using a 10 mL serological pipette, collect the PBMC layer using a 5 mL serological pipette into a conical tube, and wash the cells twice with PBS pH 7.4 with centrifugation at 400 *g* for 10 min to pellet the cells in between washes, to remove the platelets.
4. Isolate $CD14^+$ monocytes from the PBMC using magnetic activated cell sorting (MACS, Miltenyi Biotec) and $CD14^+$ microbeads (human).
5. Count the isolated monocytes using trypan blue and a hemocytometer or a cell counter to prepare 10 mL solution of 1×10^6 cells/mL in the RPMI-1640 medium.
6. Add GM-CSF to this cell suspension to a final concentration of 10 ng/mL and seed the human monocytes into the XFe96 or XFp cell culture microplate at 80,000 cells per well (80 μL).
7. Incubate the monocytes in a 37 °C, 5% CO_2, 95% humidity, incubator for 6 days to allow for the differentiation of the monocytes into macrophages.

8. Change the media on the third day of the differentiation ensuring that the media still contains 10 ng/mL GM-CSF.
9. Infect the hMDMs with mycobacteria on the sixth day of differentiation and run the XF assay after 18–24 h of infection.

3.6 Culturing of Mycobacteria and Infection of Macrophages

1. Thaw a fresh *Mtb* stock frozen at −80 °C in 7H9 media containing 0.2% (v/v) glycerol), 10% (v/v) OADC, and 0.01% (v/v) tyloxapol and incubate at 37 °C until an OD_{600} of 1.0.
2. Subculture the *Mtb* and use this subculture to infect the macrophages when it reaches an OD_{600} of 0.6–0.8.
3. Pellet the mycobacteria by centrifugation at 3000 *g* for 5 min, discard the supernatant, and resuspend the mycobacteria in PBS using pipetting and sonication (3 × 30 s).
4. Measure the OD_{600} to determine the concentration of mycobacteria (1.0 $OD_{600} = 1 \times 10^8$ *Mtb* per mL) and calculate the volume of mycobacteria required for each multiplicity of infection (MOI) based on the number of monocytes/macrophages seeded into the well.
5. Culture *M. bovis* BCG under the same conditions as the *Mtb* and treat the mycobacteria in an identical manner as the *Mtb* to infect the macrophages.
6. Prepare the dead *Mtb* by heating a known concentration of *Mtb* at 80 °C for 20 min. Dilute the heat-killed *Mtb* in PBS to infect the macrophages at different MOI based on its concentration (OD_{600}) prior to heat killing.

3.7 Workflow for the XF Assays

3.7.1 Day Prior to the Run

1. Start up the XF analyzer instrument and Wave controller (*see* **Note 8**).
2. Hydrate a XF sensor cartridge by removing the sensor cartridge from the utility plate, placing it upside down on the bench and filling each well of the utility plate with 200 μL of the calibrant solution.
3. In the case of the XFp cell culture microplate, add 400 μL of sterile water to the moats surrounding the wells of the utility plate.
4. Place the cartridge back onto the utility plate lowering the sensors into the calibrant solution in each well and incubate the cartridge in a 37 °C non-CO_2 incubator overnight.
5. Prepare the Assay template on Wave software. If using the Wave software on your desktop, select an instrument, XFe96 or XFp, and double-click on the appropriate assay template file from the "Template View," or create your own assay template by selecting the "Blank" template.

6. If you chose a "Blank" template, click on the "Protocol" tab. Add measurement periods or injections by clicking on the "Measure" or "Injection" tabs. Modify the length of mixing or measuring by clicking on "Edit Measurement Details." Typically, a measurement consists of 3 min of "mixing" and 3 min of "measuring," repeated for 3 cycles (*see* **Note 9**). Under the "Injection" tab, select the port to be injected.
7. Longer measurement times can be achieved by increasing the number of cycles. Define the different groups included in your experiment by clicking on the "Group definitions" tab and edit the names of the groups present by clicking on the current name of the group. Add extra groups by clicking on the "Add groups" tab.
8. If desired, add any further details of the experiment by clicking on the "Add" buttons next to "Injection strategies," "Pretreatments," "Assay media," and "Cell Type" descriptions.
9. In the glycolytic rate assay, define the Assay media as "Glycolytic Rate Assay medium (DMEM-based)" to set the Buffer factor to 2.6 mmol/L/pH.
10. Double-click on "Plate Map" and indicate where the defined groups have been plated on the plate map by selecting the group and then clicking on the wells of the plate map.
11. Save the changes to the assay template. If the assay template was created on your desktop software, export the assay template to a USB drive or a network drive linked to the XF Wave controller.

    A user guide on how to set up an assay template using the XF Wave software can be found on the Agilent website: https://www.agilent.com/cs/library/usermanuals/public/S7894-10000_Rev_B_Wave_2_4_User_Guide.pdf

3.7.2 Day of the Assay

1. Warm the appropriate XF assay medium to 37 °C and adjust pH to 7.4.
2. Using the appropriate XF assay media, dilute stock solutions of the required compounds for the XF assay to a concentration that is 10× the desired final concentration in the microplate well, unless otherwise indicated (*see* **Note 10**).
3. Warm the prepared compounds to 37 °C, readjust the pH to 7.4, and keep the compounds at 37 °C.
4. As the final volume in the wells of the XF cell culture microplate prior to the start of the XF run is 180 μL, the volumes of the compounds to be loaded into the ports to ensure a 10× dilution of the compounds once in the well are as follows: port A: 20 μL; port B: 22 μL; port C: 25 μL; and port D: 27 μL.

5. Using the loading guides supplied with the XF cartridge, load the cartridge with the appropriate volumes of the prepared compounds in the allocated ports as indicated in each Assay protocol (Subheading 3.8).
6. Place the cartridge back in the 37 °C non-CO_2 incubator for a minimum of 30 min prior to the assay (*see* **Note 11**).
7. Remove the XF cell culture plate from the 37 °C, 5% CO_2, 95% humidity, incubator and examine the cells under an inverted microscope to confirm consistent plating and proper cell morphology.
8. Carefully remove the culture medium from the cells to leave a minimum of 20 μL of medium in the base of the well.
9. Wash the cells gently twice with 200 μL of pre-warmed (37 °C) XF assay medium (relevant to the assay) to remove the fetal calf serum and sodium bicarbonate present in the cell culture media (*see* **Note 12**).
10. After the second wash, add 160 μL of medium to each well to give a final volume of 180 μL (depending on the assay). In the case of the XF Glycolytic Rate Assay and the XF Real-Time ATP Rate assay, the second wash is delayed until cells have been incubated for 1 h at 37 °C, in a non-CO_2 incubator and immediately prior to starting the experiment.
11. View the cells under the microscope to ensure that cells are not washed away.
12. Incubate the plate for 30–60 min at 37 °C in a non-CO_2 incubator prior to the assay to remove the CO_2, which is dissolved in the assay medium (*see* **Note 13**).
13. Once the XF cell culture plate is ready for the run, open the designed assay template on the Wave software and click on the "Run Assay" table.
14. Add any extra details of the assay in the "Assay Summary" view that you wish to save and then click "Start run."
15. The Wave controller will then give instructions to load the cartridge with the utility plate into the XF Flux Analyzer. Ensure that the lid is removed from the cartridge and the cartridge and plate are orientated correctly such that the blue-marked corner faces you. Then click on the "Continue" tab or the "I'm ready" tab shown in a dialogue box of the software to close the tray door of the analyzer.
16. After the sensors of the cartridge are calibrated, the Wave controller will display a message to load the cell plate. Click on "Open tray" to eject the utility plate and load the cell plate. Once again, be sure to remove the lid of the cell plate and orientate the cell plate correctly with the blue triangle

orientated toward you before clicking on “Load cell plate” or “I’m ready” to close the tray door and begin the assay.

17. Upon assay completion, remove the cell plate and the cartridge from the XF analyzer as prompted by the dialogue boxes of the software.
18. Save the completed assay results as an XFD file on the Wave software.
19. To normalize the OCR and ECAR readings using protein concentration, carefully remove medium from the cell plate to leave behind 20 μL (the volume removed will depend on the assay that was run and the number of injections during the run).
20. Add 10 μL of formalin to each well to fix any pathogens so the plate can be removed from the BSL3 to measure absorbance of the Bradford reagent using a plate reader.
21. Add 10 μL of 25 mM NaOH to all the wells and perform a Bradford assay [10], using the wells in lanes 1 and 12 of the XFe96 cell culture microplate that were used for the background control in the XF assay for the standards of the Bradford assay or BCA assay.
22. Open the saved assay XFD file by clicking on the file name under the “Results” view of the Wave software and click on the “Normalize” tab. Insert the calculated protein concentrations for the wells used in the XF assay into the normalization plate map. Insert the normalization units and click on “Apply.”
23. Export the assay results as desired using the “Export” Tab. The raw data of the profiles of the assay can be exported into “Excel” or “GraphPad Prism.”
24. The BioTeK Cytation series of instruments can also be used to determine the exact number of cells in each well to normalize the OCR and ECAR readings to cell number.

3.8 Assays

The concentrations of the compounds used in the XF assays need to be optimized for each macrophage model or cell line. The concentrations recommended in the Agilent User Guides for each assay did not always generate the optimal response required for the assay in different macrophage models. Hence, we titrated the concentrations of the compounds on the macrophage models during an XF run to determine effective concentrations of the compounds. Table 2 provides optimized concentrations of the compounds for the indicated macrophage models.

3.8.1 XF Cell Mito Stress Test

The Cell Mito Stress Test (CMST) measures parameters of mitochondrial function by directly measuring the OCR of the cells in response to mitochondrial modulators that are injected

Table 2
Final concentrations of the compounds in the wells of the XFe96 used for the indicated macrophage models

Compound	hMDM	THP-1	RAW264.7	BMDMs	Intraperitoneal macrophages
Oligomycin	1.5 μM	1.5 μM	1.5 μM	1.5 μM	1.5 μM
FCCP	1 μM	1 μM	1.5 μM	1.5 μM	1.5 μM
Antimycin A	2.5 μM	0.5 μM	0.5 μM	0.5 μM	0.5 μM
Rotenone	2.5 μM	0.5 μM	0.5 μM	0.5 μM	0.5 μM
2-Deoxyglucose	2.5 mM	2.5 mM		50 mM	50 mM

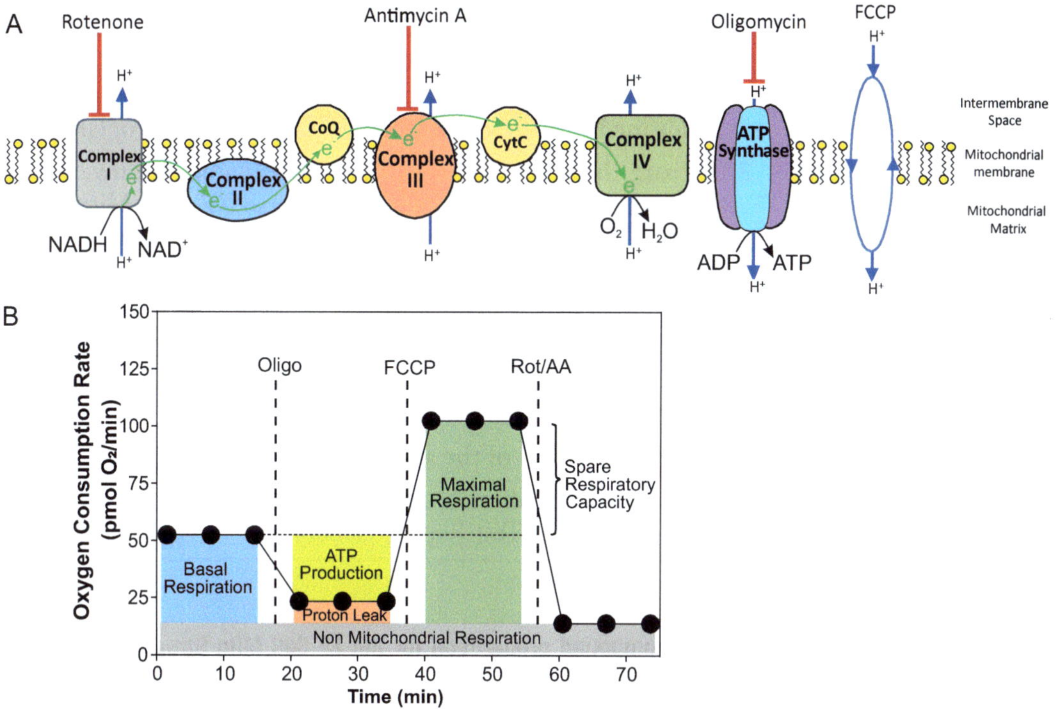

Fig. 1 The XF Cell Mito Stress Test. (**a**) Representation of the electron transport chain (ETC) demonstrating where the compounds used in the Cell Mito Stress Test target the ETC and the mitochondrial membrane. (**b**) The CMST profile demonstrating how the respiratory parameters are calculated

sequentially. After measuring basal respiration, oligomycin, an inhibitor of ATP synthase (complex V), is injected, and the OCR response observed (a decrease in OCR in cells utilizing OXPHOS to produce ATP) gives a measure of the oxygen consumed to produce mitochondrial ATP (ATP production) (Fig. 1a, b). A second injection of FCCP, which is an ionophore, collapses the proton gradient and depolarizes the mitochondrial membrane. This increases the electron flow through the electron transport chain (ETC) inducing maximal oxygen consumption at complex

IV. The FCCP-stimulated OCR is used to calculate the spare respiratory capacity, which is the difference between the maximal OCR and basal OCR. Spare respiratory capacity indicates the ability of the cell to respond to increased ATP demand or under conditions of stress. The third injection is a mixture of rotenone and antimycin A, which inhibits complexes I and III of the ETC, respectively. This shuts down the ETC and consequently mitochondrial respiration, and enables the calculation of the rate of oxygen consumed by processes outside the mitochondria (Fig. 1a, b).

1. Prepare the CMST XF media by supplementing the XF Base Medium with 10 mM glucose, 1 mM sodium pyruvate, and 2 mM glutamine.
2. After warming the CMST XF media to 37 °C, adjust the pH of the media to 7.4 with 1 M NaOH and keep at 37 °C for the preparation of the CMST compounds and for exchange of the spent media of the cells in the XF cell culture microplate.
3. Prepare the oligomycin, FCCP, and rotenone-antimycin A as indicated in Table 3 using the Seahorse XF Cell Mito Stress Test kit, and load in ports A, B, and C, respectively, as shown in Table 3. Load the CMST media in the ports of the background wells.
4. Prepare the cell culture plate as described previously and after uploading the assay template designed for the assay onto the Wave controller, run the assay.
5. After completion of the assay, normalize the OCR readings.
6. Following normalization, export the results to the Seahorse XF Cell Mito Stress Test Report Generator.

Table 3
Agilent Seahorse XF-recommended compound preparation for the XF Cell Mito Stress Test on an XFe96 or XFp

Compound	XF	CMST medium (μL)	Stock conc. (μM)	10× (conc. in port) (μM)	Final conc. in well (μM)	Injection volume (μL)	Port
Oligomycin	XF96	630	100	5	0.5	20	A
FCCP		720	100	15	1.5	22	B
Rotenone and antimycin A		540	50	5	0.5	25	C
Oligomycin	XFp	252	50	15	1.5	20	A
FCCP		288	50	15	1.5	22	B
Rotenone and antimycin A		216	25	5	0.5	25	C

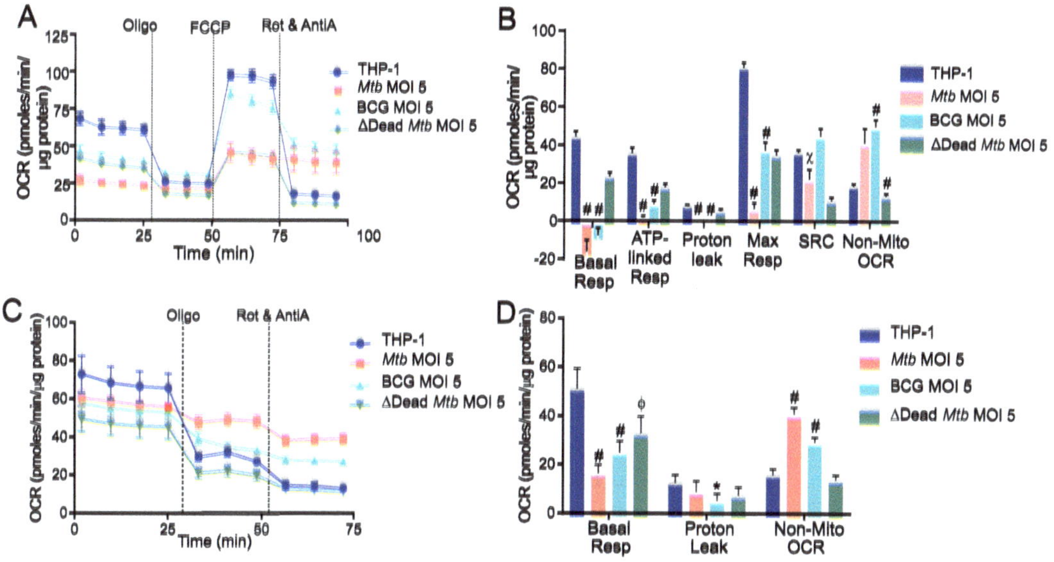

Fig. 2 (**a**) XF Cell Mito Stress Test profiles and (**b**) respiratory parameters of differentiated THP-1 macrophages infected with *Mtb*, *M. bovis* BCG, and dead *Mtb* at a MOI of 5. (**c**) Modified Cell Mito Stress Test (without the FCCP) and (**d**) respiratory parameters to determine the non-mitochondrial respiration, which is needed to calculate the basal respiration and proton leak

7. When the file is opened in Excel, enable the Macro by clicking on "Enable editing". Click on "Edit current group selection," and select your samples by clicking on the tick boxes of the samples or conditions that you want to analyze, followed by clicking "Update summary." The values of the respiratory parameters are given under the "Measures sheet" tab.
8. If the values for the non-mitochondrial OCR exceed the initial OCR values resulting in a negative basal respiration (Fig. 2a, b), repeat the assay *without* the FCCP injection (Fig. 2c). Manually calculate the basal respiration by subtracting the last OCR measurement after the addition of rotenone and antimycin A (non-mitochondrial OCR) from the third initial OCR reading prior to the first injection of oligomycin (Fig. 2c, d). Calculate the proton leak by subtracting the non-mitochondrial OCR from the ATP-linked respiration (after addition of oligomycin).

3.8.2 XF Glycolysis Stress Test

The glycolysis stress test (GST) measures the ECAR induced by glucose metabolism. Glycolysis converts glucose in the cell to pyruvate, which can be converted to either lactate in the cytoplasm or used in the TCA cycle that generates CO_2 in the mitochondria. Both processes result in the production and extrusion of protons into the extracellular medium, which results in the acidification of the medium surrounding the cell. Initially, the cells are incubated in the GST medium without glucose or pyruvate and the ECAR is measured. This gives a measure of the non-glycolytic acidification. The first injection is a saturating concentration of glucose that is utilized in glycolysis and the resulting pyruvate can be converted into lactate and/or used in the TCA cycle. Both processes extrude protons into the surrounding medium inducing a rapid increase in the ECAR. This glucose-induced response is often referred to as the rate of glycolysis under basal conditions; however, it may be referred to as the rate of glucose metabolism under basal conditions (taking both protons generated in glycolysis and protons associated with carbonic acid generated from the carbon dioxide produced by the TCA cycle into account). Oligomycin, the inhibitor of ATP synthase in the mitochondria, is injected secondly, thereby inhibiting mitochondrial ATP production. This shifts the ATP requirements of the cell onto glycolysis, thereby increasing the rate of glycolysis and increasing ECAR to give the value of the maximum glycolytic capacity. The final injection is 2-deoxyglucose (2-DG) that binds to and inhibits hexokinase, the first enzyme in glycolysis, resulting in the inhibition of glycolysis. The resulting decrease in ECAR confirms that the increased ECAR observed after the injection of oligomycin is indeed due to glycolysis. The difference between the rate of glucose metabolism and the maximal glycolytic capacity is called the glycolytic reserve (Fig. 3). Figure 3b

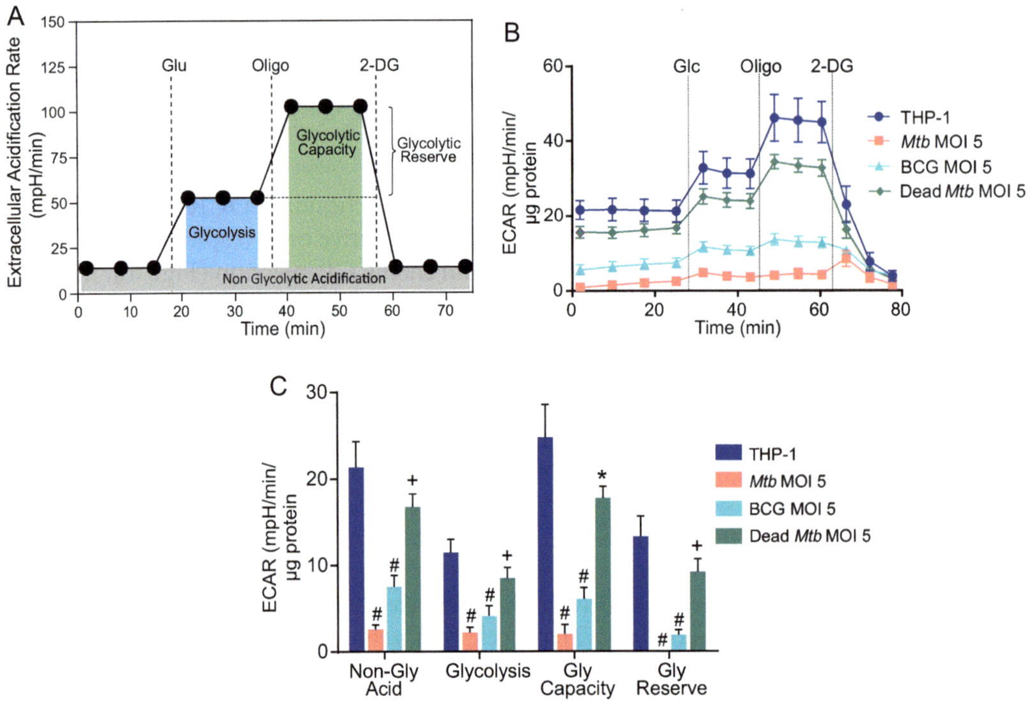

Fig. 3 Profile of the XF Glycolysis Stress Test demonstrating how the parameters of glucose metabolism are determined from the assay

Table 4
Agilent Seahorse XF-recommended compound preparation for the XF Glycolysis Stress Test on an XFe96 or XFp

Compound	XF	GST medium (μL)	Stock conc.	10× (conc. in port)	Final conc. in well	Injection volume (μL)	Port
Glucose	XF96	3000	100 mM	–	10 mM	20	A
Oligomycin		300	50 μM	15 μM	1.5 μM	22	B
2-DG		3000	500 mM	–	50 mM	25	C
Glucose	XFp	300	100 mM	–	10 mM	20	A
Oligomycin		288	50 μM	15 μM	1.5 μM	22	B
2-DG		300	500 mM	–	50 mM	25	C

demonstrates the typical ECAR profile of the Glycolysis Stress Test of uninfected THP-1 macrophages and how infection with *Mtb* significantly reduces the ECAR due to glycolysis and the glycolytic reserve of the THP-1 cells in comparison to infection with dead *Mtb* (Fig. 3c). Infection with the vaccine strain, *M. bovis* BCG also reduced the glycolytic parameters of the macrophages but to a lesser extent than *Mtb*.

1. Prepare the GST assay medium by supplementing Seahorse XF Base Medium with 1 mM glutamine.
2. Warm the assay medium to 37 °C and adjust the pH to 7.4 with 1 M NaOH. Use it to prepare the substrate and compounds for the GST and keep the media at 37 °C to exchange the medium in the cell culture plate.
3. Prepare glucose, oligomycin, and 2-DG from the Seahorse XF GST kit as indicted in Table 4 and load the compounds in ports A, B, and C, respectively, as shown in Table 4. Load the GST medium in ports A, B, and C of the background control wells (wells without cells).
4. Prepare the cell culture plate, run the assay, and normalize the results as described previously (Subheading 3.7.2).
5. After completion of the assay and normalization, "Export" the assay results to the Seahorse XF Glycolysis Stress Test Report Generator, which calculates the glycolytic parameters and has the values of these parameters in the "Measures Sheet" tab of the Excel file.

3.8.3 Glycolytic Rate Assay (GRA)

The Glycolytic Rate Assay utilizes both the OCR and ECAR measurements to calculate the total proton efflux rate (PER) from both glycolysis and carbonic acid (Fig. 4a). The amount of oxygen consumed by the cell is used to calculate the contribution of CO_2

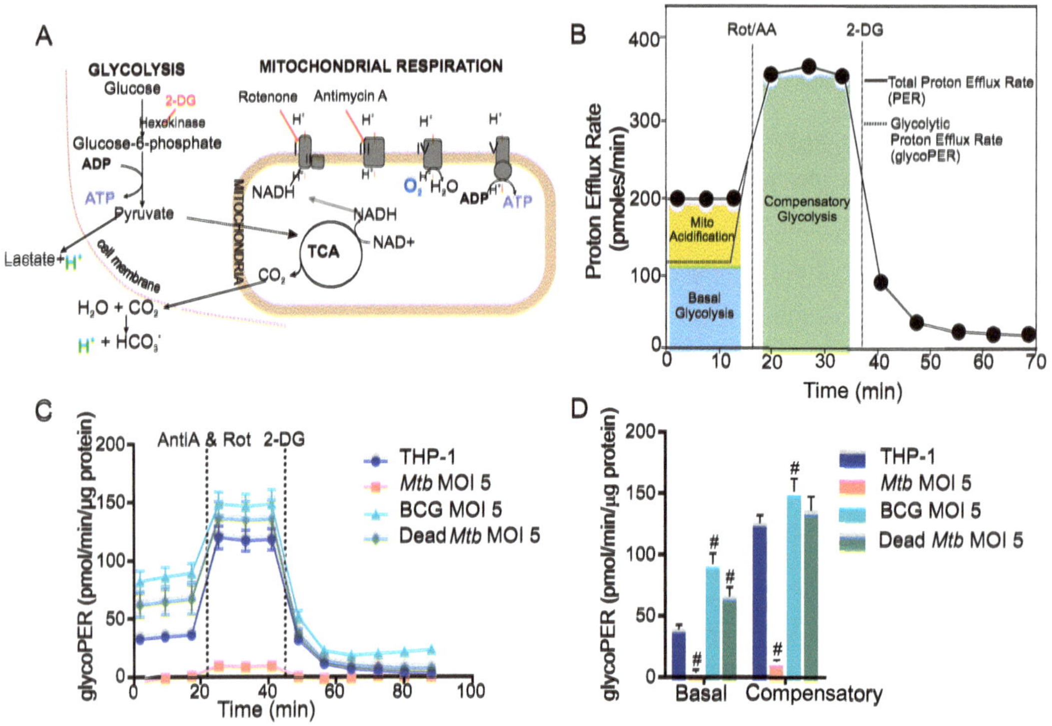

Fig. 4 (**a**) Representation of the proton efflux due to glycolysis and mitochondrial respiration that are measured by the XF Glycolytic Rate Assay (GRA). (**b**) The GRA profile generated by the Seahorse XF Glycolytic Rate Assay Report Generator demonstrating total PER and glycoPER and how to determine mitoPER

generated by the mitochondria to the total acidification. This is then subtracted from the total proton efflux rate to determine the glycolytic proton efflux rate (glycoPER). The advantage of this assay is that it allows for the real-time lactate measurements of changes in glycolysis rates that go undetected in long-term lactate accumulation assays.

In the assay, three basal rate readings are measured in XF assay medium containing glucose, glutamine, pyruvate, and HEPES buffer to calculate the total PER. Then rotenone and antimycin A are injected to inhibit mitochondrial respiration and thus the production of the carbonic acid-derived protons, followed by three further readings (Fig. 4b). These readings are used to calculate the rate of proton efflux from respiration. Inhibition of the ETC by rotenone and antimycin A will induce compensatory glycolysis to meet the cell's ATP demands, thus enabling calculation of the compensatory glycoPER. Finally, 2-DG is injected to inhibit hexokinase, the first enzyme in the glycolytic pathway, and thus glycolysis. The decrease in PER serves as confirmation that the PER prior to the addition of 2-DG was indeed due to glycolysis. Perform the calculations to determine glycoPER and mitoPER post-data acquisition using the Seahorse XF Glycolytic Rate Assay Report Generator (Fig. 4b). Figure 4c demonstrates the typical glycoPER profile

Table 5
Agilent Seahorse XF-recommended compound preparation for the XF Glycolytic Rate Assay on an XFe96

Compound	GRA medium	Stock conc.	10× (conc. in port)	Final conc. in well	Injection volume (μL)	Port
Rotenone and antimycin A	540 μL	50 μM	5 μM	0.5 μM	20	A
2-Deoxyglucose		500 mM	500 mM	50 mM	22	B

of the uninfected THP-1 macrophages and how the infections with the dead *Mtb* or *M. bovis* (BCG) vaccine strain significantly increase the glycoPER of the macrophages in contrast to the significant reduction in the glycoPER of *Mtb* infected THP-1 cells (Fig. 4d).

1. Prepare the assay medium for the GRA by supplementing 50 mL of XF DMEM medium pH 7.4 (Agilent Technologies), which contains 5 mM HEPES (*see* **Note 14**), with 500 μL of 1 M glucose (final concentration 10 mM), 500 μL 100 mM pyruvate (final concentration 1 mM), and 500 μL of 200 mM glutamine (final concentration 2 mM glutamine).
2. Warm the assay medium to 37 °C and check the pH is 7.4.
3. Use the warm medium to prepare the GRA compounds (a mixture of rotenone and antimycin A, and 2-deoxyglucose) for the glycolytic rate assay (GRA) and load the compounds into the drug ports as indicated in Table 5.
4. Prepare the cell culture plate, run the assay, and normalize the XF values with the protein concentration.
5. Prior to exporting the results to the Glycolytic Rate Assay Report Generator, define the "Assay medium" and the "Buffer factor."
6. To define the Assay medium, click on "Modify" and under the "Group definitions" tab, click on the adjacent "Add" button/tab. In the "Edit Assay Medium" dialogue box that appears to the left, click on the pull-down menu of "Media" and select the "Glycolytic Rate Assay Medium (DMEM-based)." Scroll down in this dialogue box and check that the "Buffer Factor" at the bottom of this box is 2.6 mmol/L/pH.
7. To define the buffer factor, click on "Assay medium" and click on "Configure" beside "Buffer Capacity" and the "Buffer Factor" in the adjacent dialogue box. A table entitled "Configure Background Buffer Factor" will open. Click on the tick boxes in the "Default Buffer Factor" column to select the default buffer factor for the indicated rows and columns. Then click on "Save" at the bottom of the dialogue box and

in the "Modify Assay Mode" view, click on "Apply" to apply the buffer factor to the calculations needed to calculate the proton efflux rate.

8. Export the assay results to the Seahorse XF Glycolytic Rate Assay Report Generator and save the file in the desired location. Open the saved file in Excel and enable the Macro by clicking on "Enable editing."
9. Click on "Edit current group selection" and select your samples by clicking on the tick boxes beside the names of the samples/conditions that you wish to analyze.
10. The average assay parameters (basal glycoPER, compensatory glycoPER, mitoPER) calculated from the GRA are given in the "Measures Sheet" tab below the profiles.

3.9 XF Real-Time ATP Rate Assay

The XF Real-Time ATP rate assay measures the total ATP production rates in the cells and distinguishes between the fractions of ATP produced from the mitochondrial oxidative phosphorylation (OXPHOS) and that produced from glycolysis. These are the two main pathways responsible for ATP production in mammalian cells. While OXPHOS consumes oxygen, thereby inducing OCR, both pathways contribute to the ECAR. In the Real-Time ATP Rate assay, both ECAR and OCR are measured under basal conditions, followed by the addition of oligomycin that inhibits ATP synthase

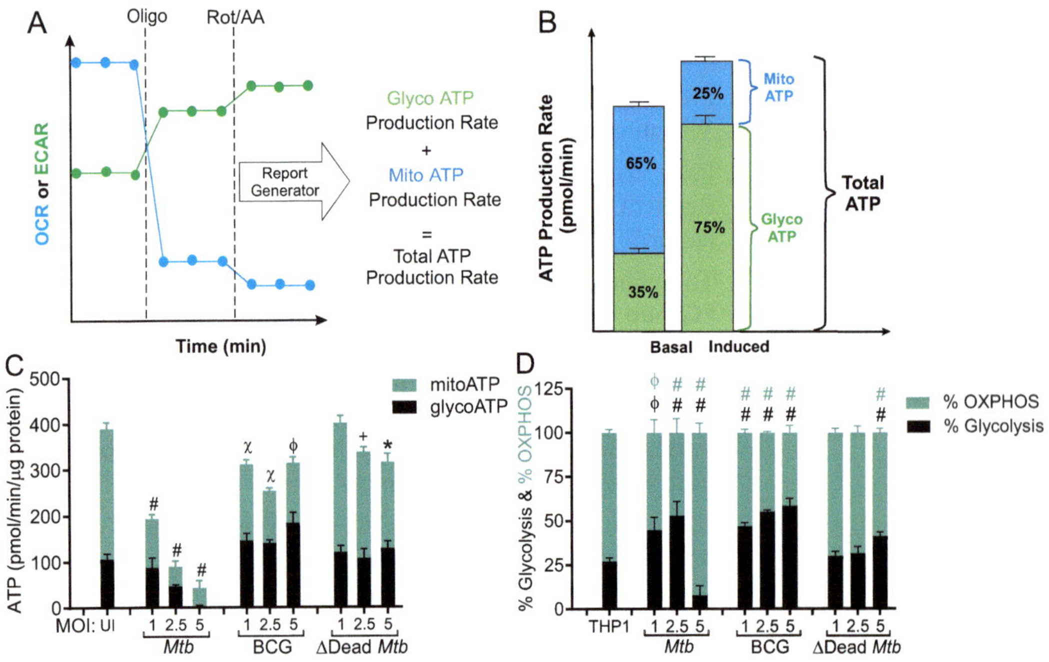

Fig. 5 (**a**) Typical OCR and ECAR profiles obtained in the Real-Time ATP Rate assay and (**b**) the output from the XF Real-Time ATP Rate Assay Report Generator

(Fig. 5a). The OCR that is inhibited by oligomycin is equivalent to the OCR coupled to the mitochondrial ATP production and is used to calculate the contribution of OXPHOS to ATP production rate using a P/O ratio of 2.79. The second injection is a mixture of rotenone and antimycin A that will inhibit the ETC and OXPHOS enabling the measurement of the ECAR, and calculation of PER, due to glycolysis. This is used to calculate the contribution of glycolysis to the ATP production rate. These calculations are performed post-data acquisition with the Seahorse XF Real-Time ATP Rate Assay Report Generator (Fig. 5b). Figure 5c depicts the typical output of the ATP Rate Assay demonstrating that although infection of the macrophages with *Mtb* or *M. bovis* BCG decrease the ATP production rate, glycolysis contributes to the total ATP production to a greater extent than in uninfected THP-1 macrophages (Fig. 5d). In dead-Mtb infected macrophages, the contribtution from glycolysis to the total ATP production only increases at high MOIs.

1. Prepare the assay media for the Real-Time ATP Rate assay by supplementing 100 mL of XF DMEM medium, pH 7.4 (this medium contains 5 mM HEPES), with 10 mM glucose, 1 mM pyruvate, and 2 mM glutamine.
2. Warm the assay medium and check the pH is 7.4; otherwise adjust the pH with 1 M NaOH.
3. Use the warm media to prepare the compounds for the assay (oligomycin and a mixture of rotenone and antimycin A) from the XF Real-Time ATP Rate Assay Kit and load the compounds into the drug ports as indicated in Table 6.
4. Upload the designed assay template on the Wave controller, prepare the cell culture plate, run the assay, and normalize as described in Subheading 3.7.
5. In the assay template on the Wave software, ensure that the assay medium is described, and the buffer capacity and buffer factor are configured as described in the XF Glycolytic Rate assay (Subheading 3.8.3, **steps 5–7**) before exporting the results to the Seahorse XF Real-Time ATP Rate Report Generator.

Table 6
Agilent Seahorse XF-recommended compound preparation for running the XF Real-Time ATP Rate Assay on an XFe96

Compound	Assay medium (μL)	Stock conc. (μM)	10× (conc. in port) (μM)	Final conc. in well (μM)	Injection volume (μL)	Port
Oligomycin	420	150	15	1.5	20	A
Rotenone and antimycin A	540	50	5	0.5	22	B

4 Notes

1. To study the infection with live *M. tuberculosis*, the extracellular flux analyzer needs to be placed in a biosafety level (BSL) 2 cabinet in a BSL3 laboratory.
2. These are the seeding densities that are optimum for the macrophage models described here. It is advisable to determine the optimum seeding conditions for any macrophage model or cell line by titrating the cell numbers in an XF cell culture microplate and determining the OCR and ECAR over 5–8 readings. Plot the constant OCR values against the cell number in order to determine the linear range of OCR versus seeding density. It is preferable to obtain a confluent cell layer at the base of the microplate well; however, this seeding density must fall within the linear response of OCR to seeding densities.
3. If all the cells have been removed from the bone, the bone should be a clear white color. If not, repeat injection of 10 mL PBS with 1× ABAM into the top of the bone and collect the flow through into the sterile Petri plate.
4. It is crucial to restrict the exposure of the cells to the ACK lysing buffer to 2 min. Longer exposure times reduce the yield of BMDMs.
5. Seed 8 replicates (wells) per condition investigated on the XFe96, e.g., 8 wells of uninfected macrophages from wild-type mice, 8 wells of uninfected macrophages from knockout mice, 8 wells of *Mtb*-infected macrophages from wild-type mice, and 8 wells of *Mtb*-infected macrophages from knockout mice. In an XFp, seed 3 wells of the XFp cell culture microplate with the cells from the wild-type mice and 3 wells with the cells from the knockout mice. When the infection of these cells is being investigated, the same plate layout would be used in another XFp cell culture microplate and the cells would be infected as described. To increase the number of observations on the XFp, two uninfected macrophage XFp microplates are analyzed and two infected macrophage XFp microplates are analyzed on the XFp.
6. The wells that contain only media *without cells* in the XF cell culture microplates serve as background controls during the XF assay run and should be indicated on the plate map of the assay template. They are necessary for the Wave software algorithms to calculate the OCR and ECAR. Furthermore, after the XF assay run, use these wells in the XFe96 cell culture microplate for the protein standards needed in the Bradford assay following the XF run to determine the protein concentration of each well for normalization of the XF data.

7. It is crucial not to puncture the peritoneum of the mouse at this point, as this will result in spillage of the intraperitoneal macrophages out of the peritoneum.
8. The XF requires a minimum of 5 h for the temperature to reach and stabilize at 37 °C. Thus, it is advisable to turn on the XF analyzer the day before the assay is run.
9. A minimum of three measurements and mixing steps between every injection is recommended to allow for the compound to have maximal effect on the cells.
10. The compounds for each assay can be prepared from the Seahorse XF kits for the respective assays designed for the XFe96, XFe24, or XFp. The kits provide six foil pouches with each pouch containing all the reagents for a complete assay in the XF of choice—there are separate kits for each XF instrument. If the Seahorse XF kits are purchased, prepare the compound stock solutions and working solutions as described in the Agilent Seahorse XF User Guides for each assay. Alternatively, purchase oligomycin, FCCP, rotenone, antimycin A, 2-deoxyglucose, and glucose from Sigma. Stock solutions (2.5 mM) of oligomycin, FCCP, rotenone, and antimycin A stock solutions are prepared in DMSO, and 2-deoxyglucose (500 mM) and glucose (100 mM) are prepared in the relevant assay medium.
11. Once the drugs have been loaded in the cartridge, the cartridge should not be incubated at 37 °C for longer than 30 min as this will result in a change in the pH of the loaded compounds.
12. Fetal calf serum and sodium bicarbonate will diminish the ECAR signal due to their buffering capacity.
13. The CO_2 must be removed from the assay medium as the CO_2 reacts with H_2O to form carbonic aid, which acidifies the medium and skews the ECAR readings.
14. A low concentration of HEPES buffer (5 mM) provides consistent buffer capacity values across the time frame of the assay. Although it may lower the raw ECAR signal slightly, the HEPES buffer will keep the ECAR consistent, improving the accuracy of the transformation to the PER.

References

1. World Health Organisation (2018) Global tuberculosis report 2018. World Health Organization, Geneva
2. Palucci I, Delogu G (2018) Host directed therapies for tuberculosis: futures strategies for an ancient disease. Chemotherapy 63 (3):172–180
3. Tobin DM (2015) Host-directed therapies for tuberculosis. Cold Spring Harbor Perspect Med 5(10):a021196
4. Zumla A, Maeurer M, Chakaya J et al (2015) Towards host-directed therapies for tuberculosis. Nat Rev Drug Discov 14(8):511–512
5. Ferrick DA, Neilson A, Beeson C (2008) Advances in measuring cellular bioenergetics

using extracellular flux. Drug Discov Today 13 (5–6):268–274
6. Nicholls DG, Darley-Usmar VM, Wu M et al (2010) Bioenergetic profile experiment using C2C12 myoblast cells. J Vis Exp 46:e2511
7. M. Rahman A, Cumming BM, Addicott KW et al. (2020) Hydrogen sulfide dysregulates the immune response by suppressing central carbon metabolism to promote tuberculosis. PNAS 117(12):6663–6674
8. Reddy VP, Chinta KC, Saini V et al. (2018) Ferritin H deficiency in myeloid compartments dysregulates host energy metabolism and increases susceptibility to *Mycobacterium tuberculosis* infection. Front Immunol 9:860
9. Cumming BM, Addicott KW, Adamson JH et al (2018) *Mycobacterium tuberculosis* induces decelerated bioenergetic metabolism in human macrophages. eLIFE 7:e39169
10. Kruger NJ (2009) The Bradford method for protein quantification. In: Walker JM (ed) The protein protocols handbook. Humana Press, Totowa

Chapter 13

Analyzing the Metabolic Phenotype of Bone Marrow-Derived Dendritic Cells by Assessing Their Oxygen Consumption and Extracellular Acidification

Hsi-Ju Wei, John J. Letterio, and Tej K. Pareek

Abstract

Dendritic cells (DCs) are the bridge between innate and T cell-dependent adaptive immunity, and are promising therapeutic targets for cancer and immune-mediated disorders. In the recent past, DCs have gained significant interest to manipulate them for the treatment of cancer and immune-mediated disorders. This can be achieved by differentiating them into either immunogenic or tolerogenic DCs (TolDCs), by modulating their metabolic pathways, including glycolysis, oxidative phosphorylation, and fatty acid metabolism, to orchestrate their desired function. For immunogenic DCs, this maturation shifts the metabolic profile to a glycolytic metabolic state and leads to the use of glucose as a carbon source, whereas TolDCs prefer oxidative phosphorylation (OXPHOS) and fatty acid oxidation for their energy resource.

Understanding the metabolic regulation of DC subsets and functions at large not only will improve our understanding of DC biology and immune regulation, but can also open up opportunities for treating immune-mediated ailments and cancers by tweaking endogenous T-cell responses through DC-based immunotherapies. Here we describe a method to analyze this dichotomous metabolic reprogramming of the DCs for generating reliable and effective DC cell therapy products. We, hereby, report how to measure the OXPHOS and glycolysis level of DCs. We focus on the metabolic reprogramming of TolDCs using a pharmacological nuclear factor (erythroid-derived 2)-like-2 factor (Nrf2) activator as an example to illustrate the metabolic profile of TolDCs.

Key words Dendritic cells, Oxidative phosphorylation, Glycolysis, Immunometabolism

1 Introduction

Dendritic cells (DCs) are an innate immune cell population with the capacity to process and present antigenic peptides on major histocompatibility complex (MHC) molecules to antigen-specific T cells. Therefore, the hallmark function of DCs is to uphold immune homeostasis by inducing T cell-mediated immunity to foreign antigens, and tolerance to self-antigens, and maintaining the proper balance between immunogenicity and tolerance. DCs exert this immune homeostasis management role in both the innate and

Suresh Mishra (ed.), *Immunometabolism: Methods and Protocols*, Methods in Molecular Biology, vol. 2184,
https://doi.org/10.1007/978-1-0716-0802-9_13,

adaptive immune response by way of antigen presentation, cytokine secretion, and polarization of T cells [1]. Upon the activation from immature DCs (iDCs), the mature DCs (mDCs) protect against pathogenic invasion [2]; however, the tolerance is also critical by being the least nonresponsive to the self-antigen. Among this, one specific type of DCs (the tolerogenic dendritic cells or TolDCs) are critical in maintaining tolerance [3]. Unlike the mDCs, the dominant anti-inflammatory signatures in TolDCs are characterized by the reduction of stimulatory ligands [4], the secretion of immunosuppressive cytokines [5], the regulation of T-cell polarization [6], and a distinct cellular metabolic profile [7]. The failure of tolerance results in the development of autoimmune or inflammatory diseases [8].

Metabolism is the process by which cells acquire and process nutrients to fulfill energy and biosynthetic requirements for biological functions. Cellular metabolism has been identified as a key component in immune cell functions, and has been recently recognized as an essential regulator of DC development and functional responses. Recent studies revealed that genes related to metabolic pathways show fundamentally different expressions between iDCs, mDCs, and TolDCs [9]. High mitochondrial activity and induction of genes related to oxidative phosphorylation (OXPHOS) provide the catabolic profile and high-energy demand that associate with the DC tolerogenic capacity [9, 10]. However, the maturation of DCs through Toll-like receptors (TLRs) by utilizing LPS stimulation shifts the metabolic profile from OXPHOS to glycolysis, due to the upregulation of inducible nitric oxide synthase (iNOS) and nitric oxide (NO) [11].

Previously, the metabolic changes in DCs used to be measured by the traditional cellular metabolic assays, such as radioactive assays or alternative colorimetric tracer assays. Both measurements rely on the process of cell lysis and require a large number of cells [12]. The alternative colorimetric tracer assays even lack the sensitivity and robustness. Recently, the Seahorse Extracellular Flux (XF) Analyzer has been developed to provide the real-time measurements of cellular metabolism. The OXPHOS from mitochondrial respiration and glycolysis in cells are simultaneously measured as the quantified levels of oxygen consumption (OCR) and extracellular acidification (ECAR), respectively. This assay provides researchers with a real-time measurement and manipulation of metabolic pathways in cells with small numbers of cells and no radioactive materials or lysis of cells' processes. Here, we describe a detailed protocol of how to study the cellular metabolic profile of DCs by using 8-well miniplate Seahorse XFp Analyzer. This apparatus is ideal for a quick and easy setup to perform routine measurements. However, the protocol that we describe here can also be utilized as a high-throughput method in the 96-well format XFe96 Analyzer. We will provide an example for measuring metabolic profile, including mitochondrial respiration and glycolysis of mDCs by TLR stimulation. As another example, we will describe changes in metabolism of TolDCs, induced by a nuclear factor (erythroid-derived 2)-like-2 factor (Nrf2) regulator, 2-cyano-3,12-dioxooleana-1,9(11)-dien-28-oic acid-difluoro-propyl-amide (CDDO-DFPA). Bone marrow-derived dendritic cells (BMDCs) isolated from C57BL/6 or Nrf2$^{-/-}$ mice are used as the cell source in this chapter. This protocol allows us to achieve new insights into DC metabolism [13]. Furthermore, the metabolic reprogramming of DCs by pharmacological agents can be investigated as the screening method for any potential therapeutics in the clinic.

2 Materials

2.1 BMDC Preparation and Treatment

1. DC culture medium: RPMI-1640 Plus L-glutamine, 10% FBS, 1% nonessential amino acid (100×), 10 mM HEPES, 50 nM β-mercaptoethanol, and 5% penicillin/streptomycin (*see* **Note 1**).
2. GM-CSF.
3. IL-4.
4. Anti-CD11c antibody for FACS.
5. LPS.
6. CDDO-DFPA.

2.2 Metabolism Assay

1. XFp extracellular flux analyzer (Agilent Technologies).
2. Poly-D-lysine.
3. OCR assay medium: XF Base Medium, 10 mM glucose, 1 mM pyruvate, and 2 mM L-glutamine, pH 7.4.
4. ECAR assay medium: XF Base Medium, 2 mM L-glutamine, pH 7.4.
5. Oligomycin.
6. Carbonyl cyanide-4-(trifluoromethoxy) phenylhydrazone (FCCP).
7. Rotenone/antimycin A.
8. Glucose.
9. 2-Deoxy-D-glucose (2-DG).
10. Report generator.
11. Wave software.
12. Seahorse XFp Carrier Tray.
13. Seahorse XFp FluxPak: The Seahorse XFp sensor cartridge and XFp miniplate.
14. Seahorse XF Calibrant.

3 Methods

3.1 BMDC Preparation

1. Isolate cells from the BM of C57BL/6 or Nrf2$^{-/-}$ mice (or any other models or source of interest) and differentiate into BMDCs as described earlier [14].
2. Isolate bone marrow cells and culture in DC culture medium with GM-CSF (15 ng/mL) and IL-4 (10 ng/mL).
3. At day 7, harvest the BMDCs and confirm the CD11c expression by FACS analysis (*see* **Note 2**).

3.2 BMDC Treatment

1. One day before the harvesting of the BMDCs, coat poly-D-lysine (50 μg/mL) in XFp Cell Culture Miniplates overnight at 37 °C, with 5% CO_2 and in a 95% humidity cell incubator.
2. Add 100 μL of Milli-Q H_2O or PBS to the moat chambers.
3. Treat 100 μL of 6×10^4 BMDCs in each well with or without CDDO-DFPA 1 h followed by LPS (10 ng/mL) for 24 h in wells B–G of poly-D-lysine-coated XFp Cell Culture Miniplates (proceed to Subheading 3.4, **step 2** or Subheading 3.5, **step 2** as applicable) (*see* **Note 3**).

3.3 Sensor Cartridge Hydrating for the XFp Analyzer

1. Fill 200 μL of the Seahorse XF Calibrant in the XFp Cell Culture Miniplates well and 400 μL in the moat chambers.
2. Put the XFp Sensor Cartridge back to the miniplate containing calibrant (*see* **Note 4**).
3. Place the cartridge assembly at 37 °C in a dry incubator without CO_2 overnight (*see* **Note 5**).

3.4 Mitochondrial Stress Test

1. Warm the OCR assay medium to 37 °C.
2. Centrifuge the XFp Cell Culture Miniplates at 300 *g* for 5 min.
3. Gently remove all but 20 μL of the DC culture medium from each well.
4. Add 160 μL of OCR assay medium to the wells.
5. Repeat Subheading 3.4, **steps 3** and **4**.
6. Add 180 μL of OCR assay medium to wells A and H for the background correction wells.
7. Place the cell in a non-CO_2 37 °C incubator for 45 min before analysis.
8. Prepare control compounds of mitochondrial stress test for the sensor cartridge: 10 μM oligomycin, 5 μM FCCP, and 5 μM rotenone/antimycin A (*see* **Note 6**).
9. Load 20 μL of oligomycin into port A (final concentration: 1 μM), 22 μL of FCCP into port B (final concentration: 0.5 μM), and 25 μL of rotenone/antimycin A into port C (final concentration: 0.5 μM) (*see* **Notes** 7 and **8**).

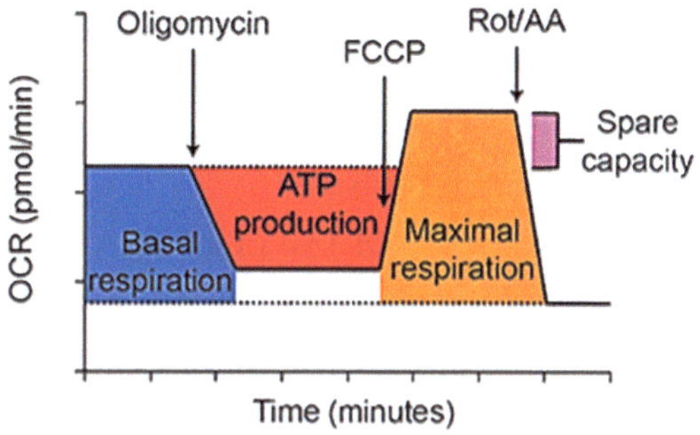

Fig. 1 Schematic representation of real-time mitochondrial respiration. Reproduced from Wei et al. (2018) with permission from the *Journal of Autoimmunity* [16]

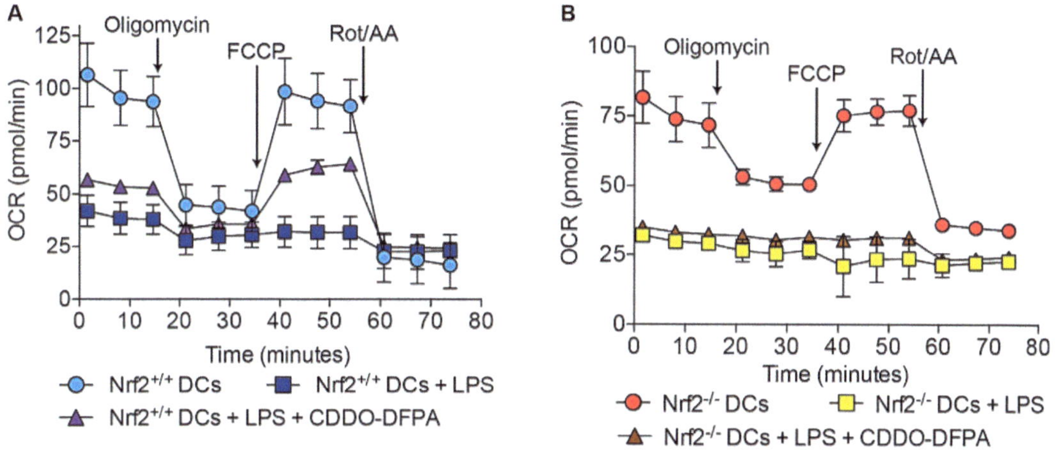

Fig. 2 Characterization of mitochondrial function of Nrf2$^{+/+}$ and Nrf2$^{-/-}$ DCs. Nrf2$^{+/+}$ or Nrf2$^{-/-}$ BMDCs were pretreated in the presence or absence of CDDO-DFPA (400 nM) for 1 h before exposure to LPS (10 ng/mL) for 24 h. (**a**) Representative kinetic study of mitochondrial OCR (pmol/min) in Nrf2$^{+/+}$ DCs (light blue), Nrf2$^{+/+}$ DCs + LPS (blue), and Nrf2$^{+/+}$ DCs + LPS + CDDO-DFPA (purple) with sequential addition of oligomycin, FCCP, and rotenone/antimycin A. (**b**) Representative kinetic study of mitochondrial OCR (pmol/min) in Nrf2$^{-/-}$ DCs (red), Nrf2$^{-/-}$ DCs + LPS (yellow), and Nrf2$^{-/-}$ DCs + LPS + CDDO-DFPA (brown) with sequential addition of oligomycin, FCCP, and rotenone/antimycin A. Reproduced from Wei et al. (2018) with permission from the *Journal of Autoimmunity* [16]

10. Run the default template of Agilent Seahorse Cell Mito Stress Test in XFp extracellular flux analyzer (*see* Subheading 3.6).
11. Perform a complete OCR study in four consecutive stages: basal respiration, mitochondrial complex V inhibition (oligomycin), maximal respiration induction (FCCP), and electron transportation chain (ETC) inhibition (rotenone/antimycin A). A schematic figure has been shown to represent the stages during the real-time mitochondrial respiration (Fig. 1).
12. An example of a mitochondrial stress test performed for CDDO-DFPA treatment on LPS-stimulated BMDCs of Nrf2$^{+/+}$ or Nrf2$^{-/-}$ mice is shown in Fig. 2.

3.5 Glycolysis Stress Test

1. Warm the ECAR assay medium to 37 °C.
2. Centrifuge the XFp Cell Culture Miniplates at 300 *g* for 5 min.
3. Gently remove all but 20 μL of the DC culture medium from each well.
4. Add 160 μL of ECAR assay medium to the wells.
5. Repeat Subheading 3.5, **steps 3** and **4**.
6. Add 180 μL of ECAR assay medium to wells A and H for the background correction wells.
7. Place the cell in a non-CO_2 37 °C incubator for 45 min before analysis.
8. Prepare control compounds of glycolysis stress test for the sensor cartridge: 100 mM glucose, 10 μM oligomycin, and 500 mM 2-DG (*see* **Note 6**).
9. Load 20 μL of glucose into port A (final concentration: 10 mM), 22 μL of oligomycin into port B (final concentration: 1 μM), and 25 μL of 2-DG into port C (final concentration: 50 mM) (*see* **Notes 7** and **8**).
10. Run the default template of Seahorse Glycolysis Stress Test in the XFp extracellular flux analyzer (*see* Subheading 3.6).
11. A complete ECAR assay consists of four stages: basal, glycolysis induction (glucose), maximal glycolysis induction (oligomycin), and glycolysis inhibition (2-DG). A schematic figure has been shown to represent the stages during real-time glycolysis (Fig. 3).
12. An example of a glycolysis stress test performed for CDDO-DFPA treatment on the LPS-stimulated BMDCs of Nrf2$^{+/+}$ or Nrf2$^{-/-}$ mice is shown in Fig. 4.

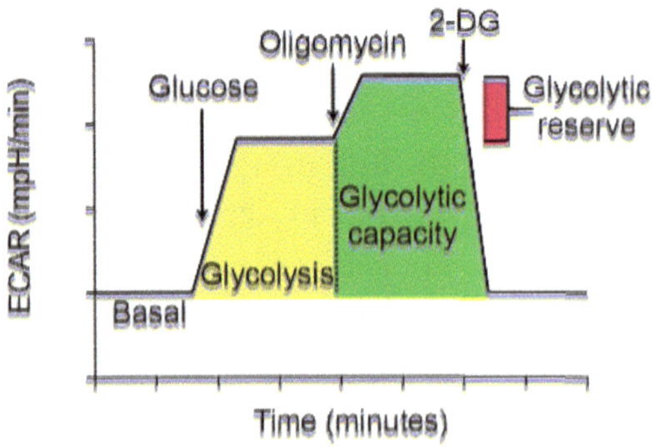

Fig. 3 Schematic representation of a real-time glycolysis. Reproduced from Wei et al. (2018) with permission from the *Journal of Autoimmunity* [16]

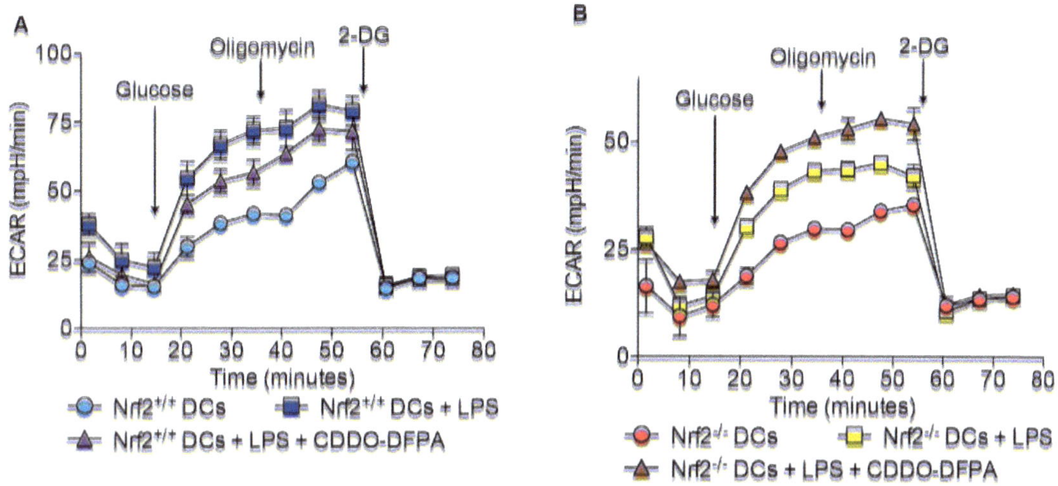

Fig. 4 Characterization of the glycolytic function of $Nrf2^{+/+}$ and $Nrf2^{-/-}$ DCs. $Nrf2^{+/+}$ or $Nrf2^{-/-}$ BMDCs were pretreated in the presence or absence of CDDO-DFPA (400 nM) for 1 h before exposure to LPS (10 ng/mL) for 24 h. (**a**) Representative kinetic study of glycolytic ECAR (mpH/min) in $Nrf2^{+/+}$ DCs (light blue), $Nrf2^{+/+}$ DCs + LPS (blue), and $Nrf2^{+/+}$ DCs + LPS + CDDO-DFPA (purple) with sequential addition of glucose, oligomycin, and 2-DG. (**b**) Representative kinetic study of glycolytic ECAR (mpH/min) in $Nrf2^{-/-}$ DCs (red), $Nrf2^{-/-}$ DCs + LPS (yellow), and $Nrf2^{-/-}$ DCs + LPS + CDDO-DFPA (brown) with sequential addition of glucose, oligomycin, and 2-DG. Reproduced from Wei et al. (2018) with permission from the *Journal of Autoimmunity* [16]

3.6 Running the XFp Analyzer Assay

1. Power on the XFp Analyzer by turning on the power switch on the back of the instrument.
2. Wait for at least 20 min for the XFp Analyzer to warm and equilibrate to the designed temperature.
3. Press "Start" from the XFp Analyzer.
4. Select the template ("Cell Mito Stress Test" or "Glycolysis Stress Test") to perform a specific assay (*see* **Note 9**).
5. Verify the assay condition and make any modification to the groups.
6. Check all desired steps, including "Equilibration."
7. Increase or decrease the number of measurements for each step if it is necessary.
8. Review whether all the information and setting are as desired.
9. Press "Start Assay."
10. Remove the lid, place the loaded and hydrated assay cartridge on the analyzer tray, and click "Continue." Make sure that the orientation of the cartridge on the tray of the analyzer is correct. The calibration process will take approximately 20 min.
11. After the calibration process, remove the utility plate from the tray and place the XFp Cell Culture Miniplates with BMDCs on the tray to start the assay. Ensure to remove the lid from the cell plate.

12. Press "Continue" to initiate the process from the equilibration step.
13. During the measuring, the tabs of "Overview" and "OCR vs. ECAR" are the two real-time views as data acquired from the XFp Analyzer.

3.7 Data Analysis

1. The data from the XFp analyzer provides the OCR and ECAR results, as well as the oxygen consumption and the pH changing value, respectively.
2. Import the raw data while designating each injection port and the control compounds to the Wave software.
3. The Report Generator and Wave software (*see* **Note 10**) can now generate the quantified results for OXPHOS and glycolysis.
4. The Report Generator allows for automatic calculation and generates the key parameters of mitochondrial functions and glycolytic functions by importing the information of sequential compound injections in the mitochondrial stress test and the glycolysis stress test, respectively.
5. The results of the characterization of the mitochondrial and glycolytic function of Nrf2$^{+/+}$ and Nrf2$^{-/-}$ DCs have been generated from Report Generator as an example in Figs. 5 and 6.
6. The results can now be exported to Excel and GraphPad Prism for future use. An extensive guide for the Wave software and Report Generators, which covers data analysis and operation, can be found on the website of Agilent Technologies.

4 Notes

1. The endotoxin level must be less than 0.1 EU/mL in the FBS.
2. The average of CD11c^{+} BMDCs from day 7 will be approximately 80–85% of the expanded cell population.
3. It is critical to seed the number of cells to obtain a confluent monolayer. Therefore, it is essential to characterize a specific cell type when it is first applied to this protocol. We have investigated the appropriated seeding number for the BMDCs. The seeding number of the DCs from other sources can be referred from other literature [15].
4. Gently put the cartridge back to the miniplate to avoid any air bubbles. Use the XFp Carrier Tray (Agilent Technologies) for handling and incubation of plates.

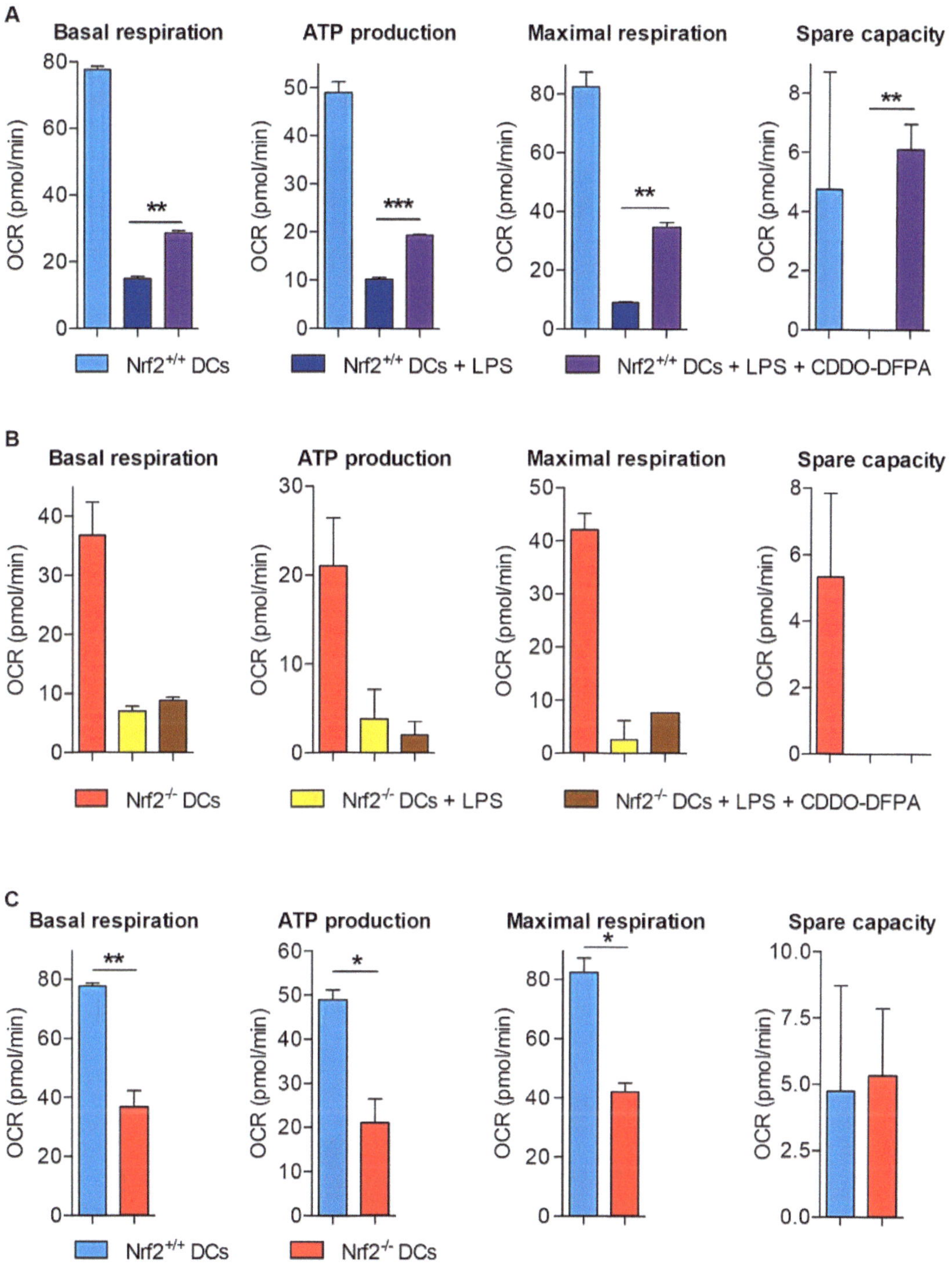

Fig. 5 (**a**) OCR quantification of basal respiration, ATP production, maximal respiration, and spare capacity of Nrf2$^{+/+}$ DCs (light blue), Nrf2$^{+/+}$ DCs + LPS (blue), and Nrf2$^{+/+}$ DCs + LPS + CDDO-DFPA (purple). (**b**) OCR quantification of basal respiration, ATP production, maximal respiration, and spare capacity of Nrf2$^{-/-}$ DCs (red), Nrf2$^{-/-}$ DCs + LPS (yellow), and Nrf2$^{-/-}$DCs + LPS + CDDO-DFPA (brown). (C) OCR quantification of basal respiration, ATP production, maximal respiration, and spare capacity of Nrf2$^{+/+}$ (light blue) and Nrf2$^{-/-}$ DCs (red). The results are expressed as mean $\pm$ SD of three experiments. $*P < 0.05$, $**P < 0.01$, $***P < 0.001$ compared with the LPS-treated groups. Unpaired student t-test. Reproduced from Wei et al. (2018) with permission from the *Journal of Autoimmunity* [16]

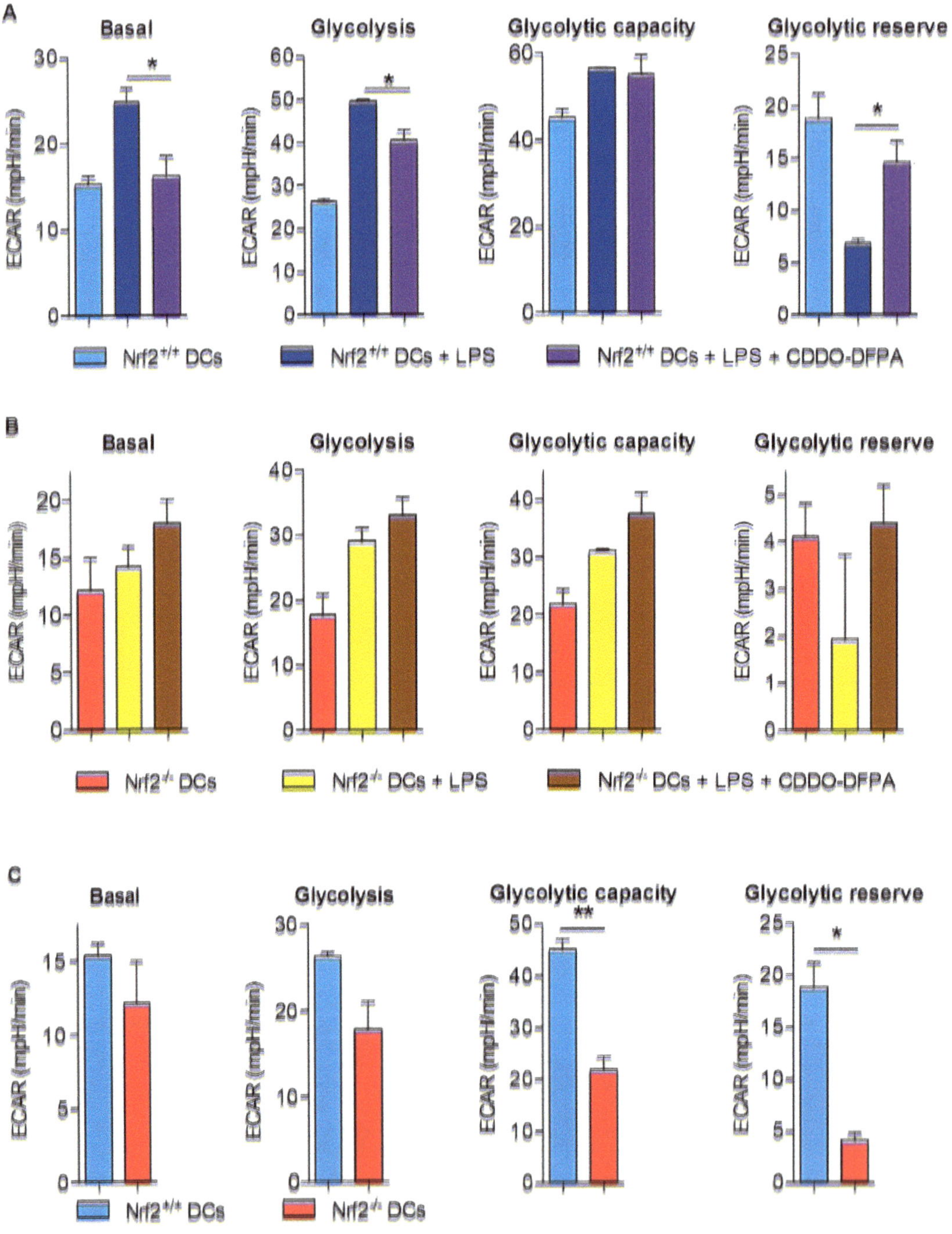

Fig. 6 (**a**) ECAR quantification of basal, glycolysis, glycolytic capacity, and glycolytic reserve of Nrf2$^{+/+}$ DCs (light blue), Nrf2$^{+/+}$DCs + LPS (blue), and Nrf2$^{+/+}$ DCs + LPS + CDDO-DFPA (purple). (**b**) ECAR quantification of basal, glycolysis, glycolytic capacity, and glycolytic reserve of Nrf2$^{-/-}$ DCs (red), Nrf2$^{-/-}$ DCs + LPS (yellow), and Nrf2$^{-/-}$ DCs + LPS + CDDO-DFPA (brown). (**c**) ECAR quantification of basal, glycolysis, glycolytic capacity, and glycolytic reserve of Nrf2$^{+/+}$ (light blue) and Nrf2$^{-/-}$ DCs (red). The results are expressed as mean ± SD of three experiments. $^{*}P < 0.05$, $^{**}P < 0.01$, $^{***}P < 0.001$ compared with the LPS-treated groups. Unpaired student t-test. Reproduced from Wei et al. (2018) with permission from the *Journal of Autoimmunity* [16]

5. To prevent the evaporation of the calibrant, a Milli-Q H_2O reservoir should be put into the incubator to maintain the humidity. In addition, CO_2 reacts with H_2O that acidifies the medium. Because the glycolysis is determined by any changes in extracellular pH, ECAR value will be incorrect reading.
6. The compounds should be reconstituted and freshly made on the same day. Do not refreeze and reuse.
7. Loading of the control compound into the port should be gentle and careful. Lower the tip into the port as far as it goes without any resistance. Forcing the tip into the port will cause the leakage of solution from the injection port. Gently dispense the control compound into the port, and carefully withdraw the tip from the port to avoid creating any air bubbles. Taping the cartridge to remove the air bubbles will also cause the leakage of the solution from the injection port.
8. In this protocol, we are generating either mDCs or TolDCs by TLR stimulation or Nrf2 activation, respectively. DCs were treated with LPS or Nrf2 prior to the metabolic assay. However, the remaining one port (port D) is empty and designed for researchers to directly inject during the measuring of the metabolic assay. Since the analyzer uses compressed air to inject compounds from the ports into the wells sequentially at different stages, It is important to carefully design the strategy for selecting the stage where the interested agent is added to the cells.
9. If the experiment is designed for directly adding the interested agent during the measuring of the metabolic assay, select the template that contains "Acute Injection" for the assay.
10. The obtained metabolic value may not be the real effect from the treatment, but rather due to the cell proliferation or the cell density during the analysis. It is an option to normalize the readouts based on the cell number after the analysis. Protein quantification methods, such as the bicinchoninic acid (BCA) assay, can be used in this protocol. Entering the ratio between each sample into the Wave software can enable it to automatically perform normalization.

Acknowledgments

We would like to acknowledge the support of the Jane and Lee Seidman Chair in Pediatric Cancer Innovation (John Letterio). This work was supported by the Angie Fowler Adolescent and

Young Adult Cancer Research Initiative at the Case Comprehensive Cancer Center, and the Callahan Graduate Scholar Award to Hsi-Ju Wei from the F.J. Callahan Foundation.

References

1. Steinman RM (2007) Lasker basic medical research award. Dendritic cells: versatile controllers of the immune system. Nat Med 13 (10):1155–1159. https://doi.org/10.1038/nm1643
2. Geginat J, Nizzoli G, Paroni M, Maglie S, Larghi P, Pascolo S, Abrignani S (2015) Immunity to pathogens taught by specialized human dendritic cell subsets. Front Immunol 6:527. https://doi.org/10.3389/fimmu.2015.00527
3. Morelli AE, Thomson AW (2007) Tolerogenic dendritic cells and the quest for transplant tolerance. Nat Rev Immunol 7(8):610–621. https://doi.org/10.1038/nri2132
4. Hubo M, Trinschek B, Kryczanowsky F, Tuettenberg A, Steinbrink K, Jonuleit H (2013) Costimulatory molecules on immunogenic versus tolerogenic human dendritic cells. Front Immunol 4:82. https://doi.org/10.3389/fimmu.2013.00082
5. Rutella S, Danese S, Leone G (2006) Tolerogenic dendritic cells: cytokine modulation comes of age. Blood 108(5):1435–1440. https://doi.org/10.1182/blood-2006-03-006403
6. Garcia-Gonzalez P, Ubilla-Olguin G, Catalan D, Schinnerling K, Aguillon JC (2016) Tolerogenic dendritic cells for reprogramming of lymphocyte responses in autoimmune diseases. Autoimmun Rev 15 (11):1071–1080. https://doi.org/10.1016/j.autrev.2016.07.032
7. Sim WJ, Ahl PJ, Connolly JE (2016) Metabolism is central to tolerogenic dendritic cell function. Mediat Inflamm 2016:10. https://doi.org/10.1155/2016/2636701
8. Kamradt T, Mitchison NA (2001) Tolerance and autoimmunity. N Engl J Med 344 (9):655–664. https://doi.org/10.1056/NEJM200103013440907
9. Malinarich F, Duan K, Hamid RA, Bijin A, Lin WX, Poidinger M, Fairhurst AM, Connolly JE (2015) High mitochondrial respiration and glycolytic capacity represent a metabolic phenotype of human tolerogenic dendritic cells. J Immunol 194(11):5174–5186. https://doi.org/10.4049/jimmunol.1303316
10. Nikolic T, Roep BO (2013) Regulatory multitasking of tolerogenic dendritic cells—lessons taken from vitamin d3-treated tolerogenic dendritic cells. Front Immunol 4:113. https://doi.org/10.3389/fimmu.2013.00113
11. Everts B, Amiel E, van der Windt GJ, Freitas TC, Chott R, Yarasheski KE, Pearce EL, Pearce EJ (2012) Commitment to glycolysis sustains survival of NO-producing inflammatory dendritic cells. Blood 120(7):1422–1431. https://doi.org/10.1182/blood-2012-03-419747
12. Ferrick DA, Neilson A, Beeson C (2008) Advances in measuring cellular bioenergetics using extracellular flux. Drug Discov Today 13(5–6):268–274. https://doi.org/10.1016/j.drudis.2007.12.008
13. Pearce EJ, Everts B (2015) Dendritic cell metabolism. Nat Rev Immunol 15(1):18–29. https://doi.org/10.1038/nri3771
14. Wei HJ, Letterio JJ, Pareek TK (2018) Development and functional characterization of murine tolerogenic dendritic cells. J Vis Exp 135. https://doi.org/10.3791/57637
15. Pelgrom LR, van der Ham AJ, Everts B (2016) Analysis of TLR-induced metabolic changes in dendritic cells using the seahorse XF(e)96 extracellular flux analyzer. Methods Mol Biol 1390:273–285. https://doi.org/10.1007/978-1-4939-3335-8_17
16. Wei HJ, Gupta A, Kao WM, Almudallal O, Letterio JJ, Pareek TK (2018) Nrf2-mediated metabolic reprogramming of tolerogenic dendritic cells is protective against aplastic anemia. J Autoimmun 94:33. https://doi.org/10.1016/j.jaut.2018.07.005

Chapter 14

The Evaluation of Mitochondrial Membrane Potential Using Fluorescent Dyes or a Membrane-Permeable Cation (TPP^+) Electrode in Isolated Mitochondria and Intact Cells

João S. Teodoro, Ivo F. Machado, Ana C. Castela, Anabela P. Rolo, and Carlos M. Palmeira

Abstract

The proton electrochemical gradient generated by respiratory chain activity accounts for over 90% of all available ATP and, as such, its evaluation and accurate measurements regarding its total values and fluctuations is an invaluable component in the understanding of mitochondrial functions. Consequently, alterations in electric potential across the inner mitochondrial membrane generated by differential protonic accumulations and transport are known as the mitochondrial membrane potential, or $\Delta\psi$, and are reflective of the functional metabolic status of mitochondria. There are several experimental approaches to measure $\Delta\psi$, ranging from fluorometric evaluations to electrochemical probes. Here we discuss the advantages and disadvantages of several of these methods, ranging from one that is dependent on the movement of a particular ion (tetraphenylphosphonium (TPP^+) with a selective electrode) to the selection of a fluorescent dye from various types to achieve the same goal. The evaluation of the accumulation and movements of TPP^+ across the inner mitochondrial membrane, or the fluorescence of accumulated dye particles, is a sensitive and accurate method of evaluating the $\Delta\psi$ in respiring mitochondria (either isolated or still inside the cell).

Key words TPP^+-selective electrode, Membrane potential, Mitochondria, Metabolic states, TMRM, TMRE, Rh123; JC-1, $DiOC_6(3)$

1 Introduction

Mitochondria possess two membranes, of which the inner is virtually impermeable. Within this membrane lies the respiratory chain and the phosphorylative system, which actions lead to the generation of a transmembrane electrochemical potential of ejected protons. This gradient is then used by several proteins for various effects; the most well known is the phosphorylation of ADP to ATP by the F1-F0 ATP synthase. The vast majority of ATP present within eukaryotic cells is produced this way [1]. The entire sequence of biochemical reactions that take place leading to ATP

Suresh Mishra (ed.), *Immunometabolism: Methods and Protocols*, Methods in Molecular Biology, vol. 2184, https://doi.org/10.1007/978-1-0716-0802-9_14,

synthesis was discovered and elegantly described by Peter Mitchell, and is known as the chemiosmotic theory [2]. Despite the knowledge generated since, it is still not fully understood how oxidative phosphorylation is regulated, since there are numerous modulators of this process that can act either directly or indirectly, resulting in the alteration of the activity of this process to better suit the cell's needs.

The electrochemical proton-motive gradient is created by the charges and osmotic imbalance of protons across the inner mitochondrial membrane. The respiratory chain oxidizes substrates (most notably NADH and succinate), receiving electrons in the process. These electrons are then transported across the respiratory chain via redox carriers toward molecular oxygen (O_2) in an energetically favorable fashion. However, the said transport is impossible without being accompanied by the vectorial ejection of protons from the mitochondrial matrix toward the intramembrane space (with some notable exceptions) [3]. From all of the four respiratory chain complexes, all but the second (succinate dehydrogenase) are capable of ejecting protons when electrons flow through them. As such, protons are ejected by complex I (NADH:ubiquinone reductase), III (ubiquinol:cytochrome *c* reductase), and IV (cytochrome *c* oxidase). Upon reaching complex IV, four electrons are donated to O_2, generating H_2O (Fig. 1).

Complex I oxidizes NADH generated during the activity of the Krebs cycle (or transported to the mitochondrial matrix by specific shuttles), resulting in NAD^+, electrons, and accompanying ejection of protons. Similarly, complex II (which is also part of the Krebs cycle) oxidizes succinate to fumarate, resulting in the reduction of FAD^+ to $FADH_2$ and, as previously mentioned, without protonic ejection. $FADH_2$ is bound to complex II and is the source of the electrons this complex transports. From either complex I or II, electrons are then shuttled toward ubiquinone (an inner membrane-soluble protein), reducing it to ubiquinol. This protein then transports the electrons toward complex III, which places

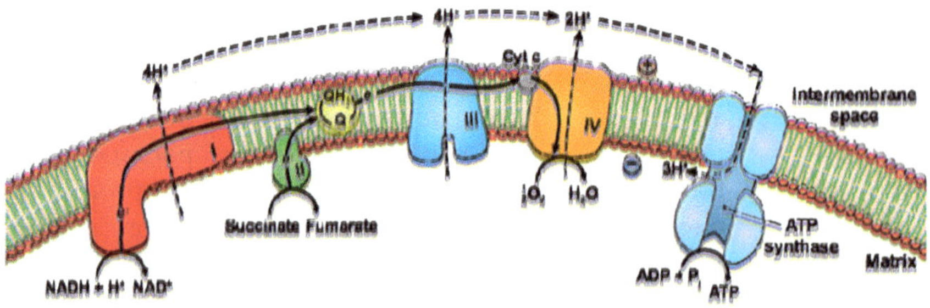

Fig. 1 The electron transport chain and the F1-FO ATP synthase. Electrons are transported from the donors (NADH or succinate) toward molecular oxygen, resulting in the formation of water. Simultaneously, protons are ejected toward the intermembrane space, creating a gradient across the proton-impermeable inner mitochondrial membrane. This highly energetic gradient can be used by the ATP synthase to generate ATP

them in another soluble protein (but this time only bound to the cytosolic leaflet, outside of the membrane), cytochrome *c*. Reduced cytochrome *c* finally leads the electrons toward complex IV where they are finally joined to O_2 [4].

Complex I is composed by more than 40 proteins, of which only 7 are encoded in the mitochondrial genome. Of noticeable interest are a prosthetic flavin mononucleotide and six Fe-S centers, as well as the binding site for ubiquinone, where the transient semiquinone radical is formed, before achieving full reduction in the form of ubiquinol. A widely known and used modulator of complex I activity is rotenone, a lipophilic inhibitor of electron transport toward ubiquinone [5]. It is also noteworthy that complex I is the major entry point of electrons into the respiratory chain, and as such it is no surprise that many mitochondrial based illnesses and injuries target complex I activity.

Complex III is formed by 11 proteins, of which only 1 is encoded in the mitochondrial genome. Noteworthy members of this complex include cytochrome *b*, a Rieske Fe-S protein, and cytochrome *c1*. Notable modulators of its activity include myxothiazol, which prevents electron transport from ubiquinol into this complex, more specifically toward a Rieske Fe-S protein. Similarly, but this time in the exit side of the complex, where cytochrome *c* binds, antimycin A is capable of preventing further electronic transport.

Complex IV is formed by 14 proteins, of which 3 are mitochondrial DNA encoded. It is here that over 90% of the cell's O_2 consumption takes place, which occurs in two steps. First, in this enzymatic complex's active site, there is the formation of transient oxide anions ($O_2^{\cdot -}$), which then react with matrix protons, resulting in H_2O. This way, the generation of the highly reactive superoxide radical anion ($O_2^{\cdot -}$) is severely reduced, since partially reduced oxygen is not easily released due to the high affinity of complex IV toward these species. However, despite all this, up to 5% of all O_2 consumed still results in superoxide formation. The inhibition of complex IV's activity with, for example, carbon monoxide, nitric oxide, or cyanide, greatly increases superoxide generation. Regardless, the reduction of O_2 to H_2O leads to matrix alkalization, which in turn promotes transmembrane electrochemical proton gradient stability.

Thus, an electrochemical gradient of protons is generated across the inner mitochondrial membrane, which is also known as the proton-motive force (Δp). It is thus comprised of two indivisible components, an electrical membrane potential ($\Delta\psi$) and a pH gradient (ΔpH). Under normal physiological conditions, most of this gradient is in the form of $\Delta\psi$, and can have a magnitude of over −220 mV [6]. As mentioned, since the matrix side of the membrane is more alkaline and negatively charged, mitochondria can build up enormous quantities of positively charged, lipophilic compounds and even some acids.

This gradient is most notably used by complex V, also known as the F1-FO ATP synthase [2, 7]. This protein complex is comprised of two major units, both with a wide array of various proteins (of which only two are encoded by the mitochondrial genome). The extrinsic, matrix-side F1 subunit is the catalytic subsection of the complex, where bound ADP and inorganic phosphate are covalently bound in the form of ATP. The FO transmembrane component of this complex is a simple protonic channel across the inner mitochondrial membrane, which is directly inhibited by oligomycin A (hence the O in FO). The favorable transport of protons across the FO section drives the F1 activity. As expected, both the NADH/NAD+ (redox potential) and the ATP/ADP (phosphorylation potential) ratios are powerful regulators of the rate of ATP synthesis [8].

As such, the understanding of the status of mitochondrial functions associated with the membrane potential is crucial to the study of mitochondrial functions in health and disease. We now will discuss some methods to assess the mitochondrial membrane potential in various common laboratory models, isolated mitochondria in suspension, and an in vitro model such as a cell line culture.

1.1 Δψ Estimation with a TPP⁺ Electrode

The membrane potential of mitochondria, due to their small size, cannot be directly measured by microelectrodes. A study by Bakeeva et al. [9] reports that fat-soluble ions can passively spread across the inner membrane. When the membrane (of cells or organelles) is hyperpolarized, a lipid-soluble cation added to the medium is electrophoretically transferred to the cells or organelles, and thus the concentration of liposoluble cations in the medium decreases. Examples of said ions include rhodamine 123, tetramethylrhodamine methyl ester ($TMRM^+$), tetraphenylphosphonium (TPP^+), or methyl triphenylphosphonium ($TPMP^+$) [10]. Taking the example of TPP^+, when present in the medium, it can cross the inner mitochondrial membrane and accumulate until the electrochemical equilibrium is reached, which is distributed according to the Nernst equation:

$$\Delta\psi\ (\text{mV}) = -59 \log \frac{[\text{TPP}^+]_{\text{in}}}{[\text{TPP}^+]_{\text{out}}},\ \text{at}\ \ 25^{\circ}\text{C}$$

where $[\text{TPP}^+]_{\text{in}}$ represents the ion concentration in the matrix and $[\text{TPP}^+]_{\text{out}}$ represents the ion concentration in the medium. Possible differences in activity coefficients are generally neglected. The volume of the mitochondrial matrix is usually assumed to be 1.1 μL/mg of protein (Fig. 2). Another, more decomposed alternative to the equation that can also be used to estimate membrane potential is

$$\Delta\psi\ (\text{mV}) = 59 \log\ (v/V) - 59\left(10^{\Delta E/59} - 1\right),\ \text{at}\ \ 25^{\circ}\text{C}$$

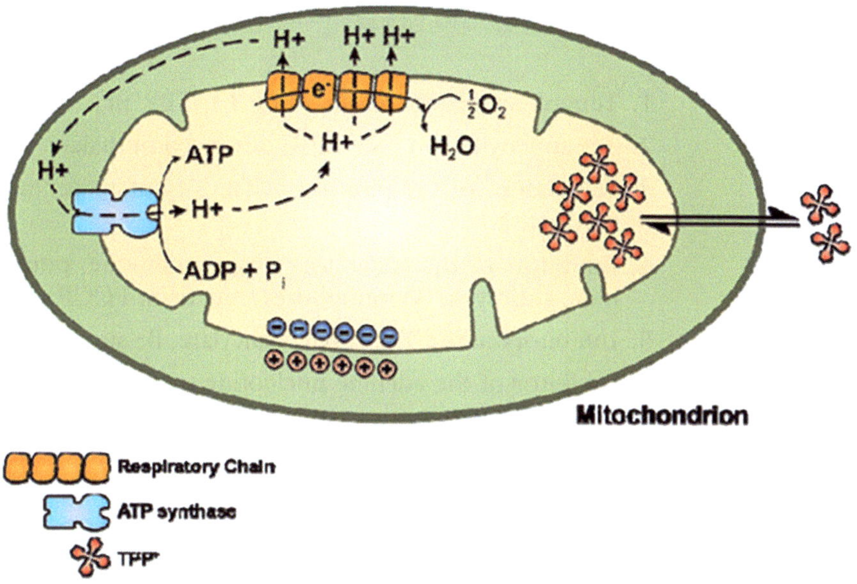

Fig. 2 The interconnection of TPP^+ movements and the mitochondrial membrane potential. When there is a membrane potential across the inner mitochondrial membrane, a net negative charge accumulation inside the mitochondrial matrix exists, attracting the permeable TPP^+ ions, reducing their concentration in the medium which is detectable with a TPP^+-sensitive electrode. Generation of ATP by the ATP synthase diminishes this gradient, "pushing" TPP^+ electrons out of the mitochondrion

where v is the mitochondrial volume, V is the reaction medium volume, and ΔE is the deflection of the electrode potential from the baseline. Thus, the passive TPP^+ membrane binding is ignored, which may lead to an overestimation of $\Delta\psi$. However, this effect is very small and could be a problem if the total $\Delta\psi$ is low, around 90 mV; since typical respiring, coupled mitochondria have a $\Delta\psi$ of approximately 200 mV/mg of protein, this effect is insignificant.

Any changes in membrane potential, when mitochondria go from "state a" to "state b" are given by

$$\Delta\psi/59 = \log(10\Delta Ea/59 - 1) - \log(10\Delta Eb/59 - 1), \text{ at } 25^\circ C$$

This shows that it is essential to measure ΔEa and ΔEb, and not just the difference between Ea and Eb.

Although there are several methods for measuring $\Delta\psi$, the use of the tetraphenylphosphonium selective electrode (TPP^+) is still a method of choice because of its sensitivity. This method is based on the accumulation of TPP^+ by energized mitochondria, which present a negative charge on the matrix due to proton ejection.

2 Materials

2.1 Reagents

1. Tetraphenylphosphonium (TPP^+Cl, TPP^+Br).
2. Tetraphenylboron (Na^+TPB) diisooctyl phthalate.
3. Substrates of respiratory chain (glutamate/malate and succinate).
4. Inhibitors of the respiratory chain (rotenone, potassium cyanide, antimycin A) ionophores (nigericin, FCCP).
5. Inhibitors of the TCA cycle (salicylate, Br-succinimide).
6. Inhibitor of the adenine nucleotide translocase (carboxyatractyloside, atractyloside).
7. Inhibitor of calcium uniporter (ruthenium red).
8. Dimethyl sulfoxide (DMSO).
9. Cell culture medium (for example, Dulbecco's modified Eagle medium, DMEM).
10. Fetal bovine serum (FBS).
11. Proton ionophore (dinitrophenol, DNP).
12. A fluorescent probe (for example, 6.6 μM TMRM, prepared from a 100 μM stock solution in DMSO).

2.2 Preparation of a TPP^+ Electrode

1. Membrane potentials ($\Delta\psi$) are evaluated using a TPP^+-sensitive electrode. This electrode is composed of a polyvinyl chloride (PVC) membrane containing tetraphenylboron as an ion exchanger, prepared according to Kamo et al. [11] and using a calomel electrode as a reference.
2. The PVC membrane (Fig. 3) is prepared by preparing a solution of 0.34 mg of tetraphenylboron sodium salt (Na^+ TPB), 16 mg of polyvinyl chloride (high molecular weight), and 57 μL of diisooctyl phthalate and tetrahydrofuran (for a final volume of 500 μL), and allowing it to evaporate on a glass plate bounded by a 1.9 cm diameter glass ring (*see* **Note 1**). This leaves a transparent membrane roughly 0.2 mm thick.
3. A portion of the membrane is placed in tetrahydrofuran in a PVC tube (2 mm internal diameter). Care should be taken to avoid tetrahydrofuran causing extensive dissolution of the central part of the solid membrane through which the TPP^+ concentration is sensed.
4. In order for the tetrahydrofuran to be rapidly evaporated, gentle suction and blowing in the tube should be performed. Any membrane material that overlaps the tubing should be cut with scissors or a surgical blade.
5. The complete electrode is immersed in a reference solution containing 0.1–0.2 mL TPP^+ 10 mM (*see* **Note 3**). The electrode contains a silver wire with a membrane-free AgCl coating

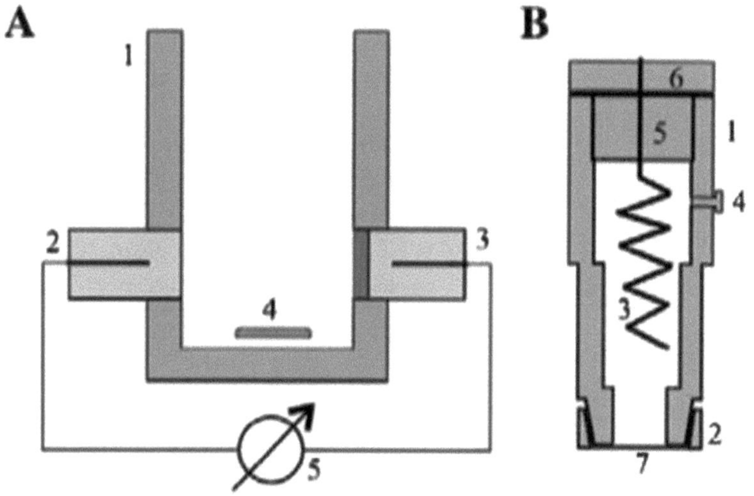

Fig. 3 (**a**) Schematic drawing of the membrane potential measuring circuit: (*1*) the measuring chamber; (*2*) the Ag/AgCl reference electrode; (*3*) the TPP^+-selective electrode; (*4*) the magnetic stirrer; (*5*) PC with measuring board. (**b**) Scheme of TPP^+-selective electrode: (*1*) the electrode body; (*2*) the acrylic ring; (*3*) the Ag/AgCl wire; (*4*) the opening for internal electrolyte filling; (*5*) the Plexiglass block where the wire Ag is sealed; (*6*) the Plexiglass ring with the floor to secure the acrylic block inside the electrode body; (*7*) selective PVC membrane for TPP^+

that is connected to a suitable electrometer. For electrode usage, it is necessary to immerse it overnight in a 10 mM TPP^+ solution (*see* **Note 4**). The electromotive force is measured between the TPP^+ electrode and a calomel electrode in the sample solution.

A good TPP^+ electrode should have a linear voltage response to log [TPP^+] with an inclination of 59 at 25 °C according to the Nernst equation

$$\Delta\psi = \frac{RT}{zF} \ln \frac{a_{\text{out}}}{a_{\text{in}}}$$

where R, T, z, and F are the universal gas constant, absolute temperature, valence, and Faraday constant, respectively, and a_{out} and a_{in} are the activities of the fat-soluble ions within the cell/organelle and the medium. An easily observed characteristic with successive additions of a 1 mM TPP^+ solution and doubling of the previous concentration with each addition is the repetition of the electrode's response. As

$$\Delta E = 2 : 3RT/nF \log[C1/C2]$$

$$\text{If } C1/C2 \text{ is } 2, \text{and } 2.3RT/nF \text{ is } 59 \text{ mV, then}$$

$$\Delta E = 17.8\text{mV}$$

If the system-coupled recorder has a 20 mV scale, each TPP^+ concentration doubling pulse should produce a similar response of approximately 89% of the scale (*see* **Note 4**).

2.3 Buffers

1. Homogenization medium (medium A): 225 mM Mannitol, 75 mM sucrose, 0.5 mM EGTA, 0.5 mM EDTA, 0.1% BSA (fatty free), 10 mM HEPES, pH 7, 4.
2. Washing medium (medium B): 225 mM Mannitol, 75 mM sucrose, 10 mM HEPES, pH 7.4.
3. Respiration (reaction) medium (medium C): 130 mM Sucrose, 50 mM KCl, 5 mM $MgCl_2$, 5 mM KH_2PO_4, and 10 mM HEPES, pH 7.4.

3 Methods

3.1 The Isolation of Hepatic Mitochondria

All materials should be kept on ice during the procedure to ensure the maintenance of a low temperature throughout the process.

1. Take a rat (about 250 g), fasted overnight, and sacrifice it by cervical dislocation under anesthesia.
2. Cut the rat abdomen under the rib cage using scissors and tweezers. Remove the liver as quickly as possible and place it in a beaker containing cold medium A.
3. Remove as much fibrous tissue and fat adherent to the liver as possible.
4. Chop the liver with scissors into small pieces. Wash it at least twice by changing the cold medium A so that as much blood and fats as possible are removed.
5. Add approximately 6 mL of ice-cold medium A per gram of minced liver. Transfer everything to a previously chilled (glass) Potter-Elvehjem homogenizer with a PTFE pestle.
6. Homogenize the tissue with the help of the pestle spinning at 300 rpm with 3–4 up-and-down movements (making sure that the pestle reaches the bottom of the homogenizer in the first or second movement).
7. Pre-cool two centrifuge tubes and transfer the previous homogenate to them. In a refrigerated centrifuge, equilibrate the tubes and centrifuge at 4 °C, 800 *g*, for 10 min to remove the denser components from the homogenate (nuclei, red blood cells, fragmented cells).
8. Carefully decant the supernatant (wasting a small amount of supernatant to avoid contamination with the pelleted particles) into new, also cooled, centrifuge tubes and centrifuge at 4 °C, 10,000 *g*, for 10 min.

9. Remove as much of the supernatant as possible. Submerge the pellet with medium A. Gently but quickly resuspend the pellet with the help of a brush. Mitochondria form a soft brown sediment. If a red spot in the center is noted, it should be discarded for it is formed by pelleted red blood cells. Often, a moving surface layer is observed, which must be removed during decantation for it is formed by injured mitochondria.
10. Add roughly 15–20 mL of medium A and centrifuge the suspension at 10,000 *g* for 10 min.
11. Repeat **steps 9** and **10** twice, replacing medium A with medium B.
12. Finally, carefully resuspend the purified mitochondrial fraction (pellet) with the aid of a small brush in about 1–2 mL of medium B. Place the now ultrapure mitochondrial fraction in 1.5 mL tubes.
13. Quantify the mitochondrial protein content with a standard protein assay (for example, use the biuret method).

3.2 Δψ Fluctuations Associated with the Phosphorylation-Dephosphorylation Cycle

1. Membrane potential oscillations (Δψ) (Fig. 4) are measured in an open, thermostated reaction chamber (25 °C) under constant magnetic stirring (*see* **Note 5**).
2. TPP^+ should be added to the reaction medium at a recommended concentration of 3 μM to obtain high measurement sensitivity, and to avoid any toxic effects to the mitochondria [12].
3. An ion-selective electrode should be used to determine the concentration of the chemical probe (i.e., TPP^+) in the medium. This allows one to determine changes in probe accumulation, and therefore continuously monitor Δψ and absolute millivolt scale [13].
4. $TPMP^+$ and TPP^+ ions have the advantage of being used at submicromolar concentrations, avoiding interference with mitochondrial metabolism and the in and out diffusion of the probe.
5. In a toxicological approach, the real-time measurement of Δψ in isolated mitochondria can have multiple uses. Different chemicals can be incubated with isolated mitochondrial fractions, and their Δψ effects can be recorded, allowing a view of the possible mitochondrial targets of the tested compound.
6. In addition, mitochondrial fractions isolated from animal mod els, treated with a chemical entity or possessing a pathological condition, may also be used [14].

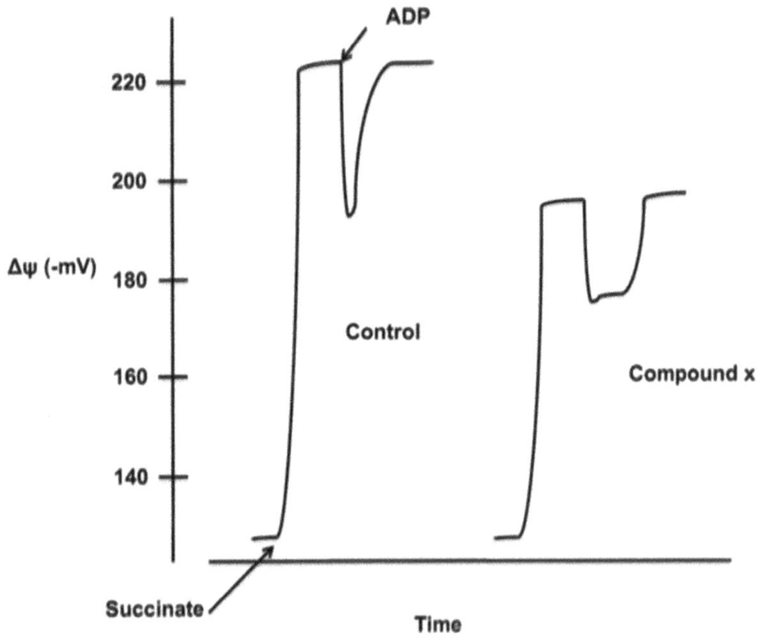

Fig. 4 Typical markings obtained with a TPP^+ electrode. After the addition of compound *x*, the initial mitochondrial membrane potential decrease is observed, simultaneously with an increase in the retardation phase, which indicates the time required for $\Delta\psi$ stabilization to occur after the addition of ADP

3.3 Measuring Mitochondrial Membrane Potential ($\Delta\psi_m$) with Cationic Fluorescent Probes

1. The mitochondrial membrane potential ($\Delta\psi_m$) can also be assessed by referring to lipophilic cationic fluorescent probes, by tracking their movement across the inner mitochondrial membrane and their accumulation in the mitochondrial matrix, according to the mitochondrial membrane potential. This is particularly helpful in situations where mitochondrial isolation is not useful or feasible, such as in cases of low quantities of material available, or the interest of other, non-mitochondrial players in mitochondrial membrane potential evaluation.
2. Currently, the most commonly used fluorescent compounds are tetramethylrhodamine methyl (TMRM) and ethyl (TMRE) esters, rhodamine 123 (Rh123), 5,5′,6,6′-tetrachloro-1,1′,3,3-′-tetraethylbenzamidazolocarbocyanine (JC-1), and 3,3-′-dihexyloxacarbocyanine iodide ($DiOC_6$) [3]. These fluorescent probes can be divided into two main categories: the rhodamine and rhodamine derivatives, which include TMRM, TMRE, and Rh123, and the carbocyanines, which encompass JC-1 and $DiOC_6$ [3].
3. The fluorescent probes are more advantageous than TPP^+ for the monitoring of $\Delta\psi_m$ on intact cells (Table 2), since it is more difficult to isolate an appropriate number of viable mitochondria from cells in culture than from tissues. Moreover, with the use of fluorescent probes, and taking advantage of fluorescence

Table 2
A summary of the advantages and disadvantages of each mitochondrial potentiometric probe

Probe	Advantages	Disadvantages
TPP^+	Best to measure $\Delta\psi_m$ of isolated mitochondria	Requires mitochondrial isolation Toxic at elevated concentrations
TMRM, TMRE	TMRM is the least toxic fluorescent probe to mitochondria	TMRE binds non-specifically to mitochondria membrane and inhibits oxidative phosphorylation at low concentrations
Rh123	Less sensitive to $\Delta\psi_p$ changes	Nonspecific binding to mitochondria
JC-1	Easy to monitor $\Delta\psi_m$ due to monomeric (green fluorescent) and aggregate (red fluorescent) forms	Red/green fluorescence ratio is very prone to the incorrect determination of $\Delta\psi_m$ Different redistribution times for the monomeric and aggregate forms lead to misleading $\Delta\psi_m$ interpretations
$DiOC_6(3)$		Stains the endoplasmic reticulum Very powerful inhibitor of oxidative phosphorylation (for concentrations >1 nM)

microscopy or a scintillation counter, it is also possible to accurately evaluate the mitochondrial membrane potential, with other potential advantages (for example, if fluorescent microscopy is available, it is possible to achieve an insight into the distribution and morphology of the polarized mitochondria network inside cells).

Before conducting the experiment to monitor $\Delta\psi_m$ with cationic fluorescent probes, it is essential that certain aspects are clearly understood in order to acquire reliable data. The fluorescent probes mentioned here can undergo a process called quenching, which is briefly characterized by a decrease in the fluorescence intensity. This process is influenced by the concentration of the probes, such that when it reaches a threshold in the mitochondrial matrix, the quenching process occurs (*see* **Note 6**). Besides this factor, it is also of importance that the probes be carefully chosen in accordance with the experiment that is going to be performed.

3.3.1 Quench Versus Non-quench Modes

The concentration of the probes is a critical factor for the interpretation of these experiments. Higher probe concentrations (quench mode, *see* **Note 7**) are used for the detection of rapid changes in $\Delta\psi_m$, which are because of the fact that as the probe concentration increases, the relation between the concentration and the fluorescence intensity is nonlinear [15]. On the other hand, lower probe concentrations (non-quench mode, *see* **Note 7**) can be used to avoid quenching in the mitochondrial matrix, and are normally employed to monitor slower $\Delta\psi_m$ changes or even to compare different mitochondrial populations.

Table 1
Spectral properties of fluorescent cationic probes for the measurement of $\Delta\psi_m$

Probe	Excitation maximum (nm)	Emission maximum (nm)
TMRM, TMRE	553	576
Rh123	507	529
JC-1	498	525 (monomer) 595 (aggregate)
$DiOC_6(3)$	489	506

Based on [16, 17]

3.3.2 Choosing the Most Adequate Fluorescent Probe

Currently, there are several fluorescent probes (Table 1) that can be used for the monitoring of $\Delta\psi_m$ [15, 16]; however some are more advantageous than others, depending on the experiment that is being conducted.

For the cationic fluorescent probes to distribute in the cell, and in mitochondria, they have to cross their respective membranes, which is an action dependent on the probes' permeability capacity. This means that besides $\Delta\psi_m$, one must also take into consideration the plasma membrane potential ($\Delta\psi_p$), and whether it can affect the experiment. Regardless, the $\Delta\psi_p$ has less of an impact when using slow permeant probes.

Another factor that may need to be considered is the presence of multidrug resistance transporters (MDR) in the cells. These transporters have as common substrates the rhodamine and rhodamine derivate probes, which will lead to their efflux, thus affecting their redistribution in the cell [17]. Therefore, if it is suspected that the cells have active MDR, they should be inhibited to appropriately measure the $\Delta\psi_m$. Cyclosporin H or verapamil is a known inhibitor of MDR that can be used.

TMRM and TMRE

TMRM and TMRE (Fig. 5) are structurally similar to Rh123 (Fig. 6). Although TMRM and TMRE are used in several studies, they have some limitations that need to be clearly understood so that they can be overcome. For instance, they were discovered to bind to mitochondria membranes and to inhibit cellular respiration, with TMRE having a more powerful effect than TMRM [15, 18]. However, at lower concentrations, these effects can be disregarded [15]. Both TMRM and TMRE are highly cell permeant, meaning that they easily cross the cell membrane, requiring less time to reach equilibrium in the mitochondrial matrix. These probes can be used to perform experiments in both quench (>50 nM) and non-quench (<30 nM) modes.

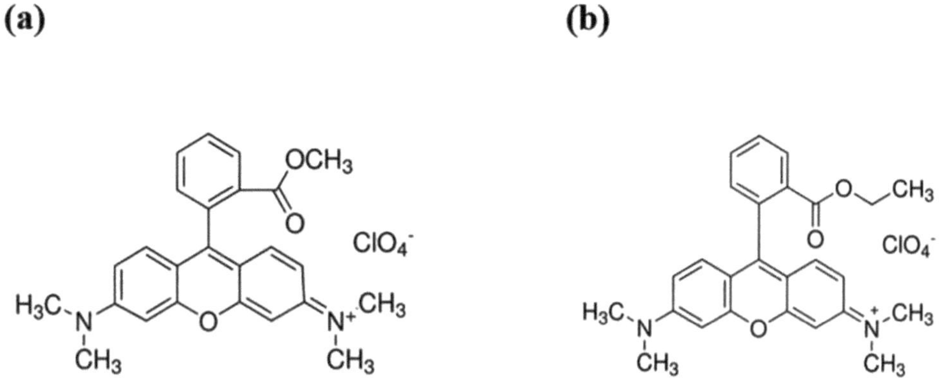

Fig. 5 Chemical structure of the cationic fluorescent probes: (**a**) tetramethylrhodamine methyl ester (TMRM) and (**b**) tetramethylrhodamine ethyl ester (TMRE)

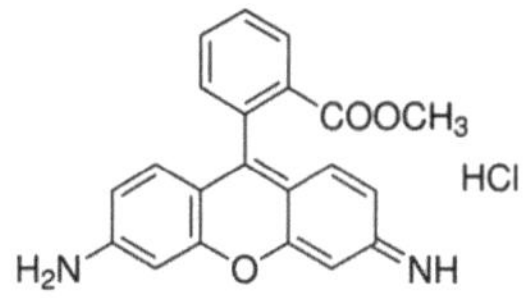

Fig. 6 Chemical structure of the cationic fluorescent probe rhodamine 123 (Rh123)

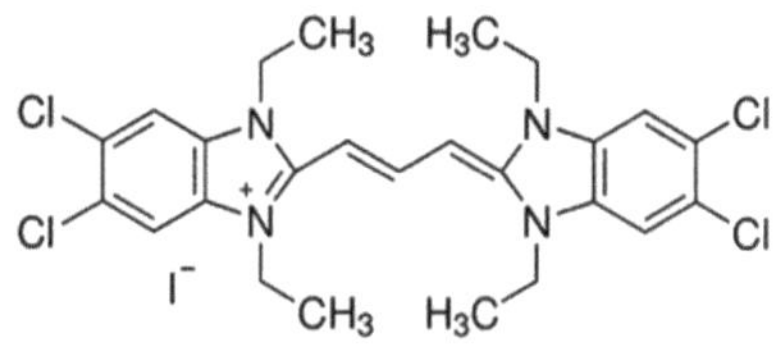

Fig. 7 Chemical structure of the cationic fluorescent probe 5,5′,6,6′-tetrachloro-1,1′,3,3′-tetraethylbenzamidazolocarbocyanine (JC-1)

Rh123

Rh123 (Fig. 6) non-specifically binds to mitochondria and inhibits mitochondrial respiration less than TMRE but more than TMRM [18]. It is more hydrophilic and therefore less cell membrane permeant than TMRM or TMRE; thus, changes in the plasma membrane potential do not affect Rh123 redistribution as much as the other two rhodamine derivate probes. Therefore, due to these characteristics, it is mostly utilized in quench mode (1–10 μM).

JC-1

The carbocyanine probe JC-1 (Fig. 7) can also be used to monitor $\Delta\psi_m$, although it behaves differently than the other aforementioned fluorescent probes. When the concentration of JC-1 reaches a threshold, its emission spectrum shifts from green to red, due to the formation of aggregates (J-aggregates), which, unlike the aggregate forms of TMRM and Rh123, are fluorescent. Because of this shifting, this probe appears to be very appealing, since not only it allows one to monitor its redistribution, but it is also possible to visualize its accumulation and aggregation in the mitochondrial matrix. Simultaneously, the hyperpolarization of mitochondria increases the formation of J-aggregates in the mitochondrial matrix. The $\Delta\psi_m$ is typically determined from the ratio between the red fluorescence from the J-aggregates of the mitochondrial matrix and the green fluorescence from the monomeric JC-1 of the cytoplasm. However, this probe has also drawbacks that interfere with the correct assessment of $\Delta\psi_m$. For instance, in a cell population that has a heterogeneous surface-to-volume ratio, as in neurons, where their soma and neurites have significantly different ratios, the measurement of $\Delta\psi_m$ may be incorrect. This is due to the fact that to measure $\Delta\psi_m$, the probe relies on the formation of aggregates, and since it is slowly permeant, the threshold concentration required for the formation of J-aggregates is not reached at the same period in mitochondria of the soma and of neurites [15, 16].

DiOC$_6$(3)

DiOC$_6$(3) (Fig. 8) is also commonly employed to measure $\Delta\psi_m$; however, it must be used with extreme care, since it has been reported not to be specific for mitochondria, being also used as a staining agent for the endoplasmic reticulum [19]. In addition, it also has toxic effects for mitochondria, being a potent inhibitor of the respiratory chain's complex I. Therefore, to avoid these nonspecific and perilous effects, it must be employed at very low concentrations (<1 nM), being typically used for analysis by flow cytometry [16, 20], which can be seen as a limitation of this probe, since it cannot be used with other instruments, which would require higher probe concentrations to accurately and reliably measure $\Delta\psi_m$.

In summary, considering the advantages and disadvantages of the abovementioned fluorescent probes, choosing the best approach to be used must not be taken lightly, and the probes' characteristics, such as their permeance, toxicity, and concentration, must be carefully evaluated.

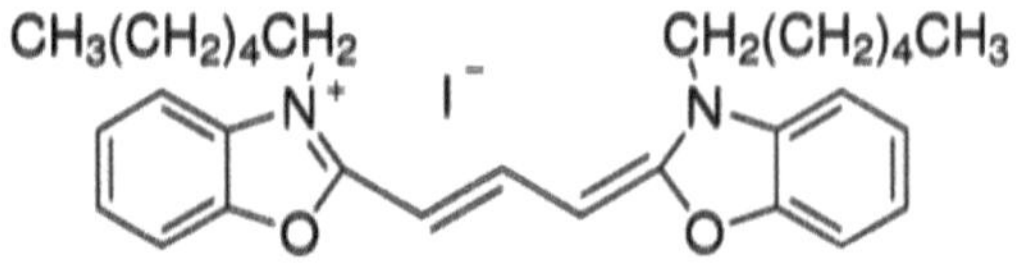

Fig. 8 Chemical structure of the cationic fluorescent probe 3,3′-dihexyloxacarbocyanine iodide (DiOC6(3))

3.4 The Preparation of Mammalian Monolayer Cell Culture

1. Grow the cells in 75 cm^2 (T75) flasks in culture medium (DMEM) with the appropriate supplements (e.g., FBS and 1% antibiotic-antimycotic) at 37 °C in a humidified 5% CO_2 atmosphere.
2. To seed the cells in the appropriate container to conduct the experiments, detach the cells by removing the culture medium followed by trypsinization.
3. Count and seed cells in multi-well plates in order to achieve about 70% confluence on the day of the experiment.

3.5 The Measurement of $\Delta\psi_m$ in Mammalian Adherent Cells Using TMRM in Quench Mode on a Microplate Reader

1. Aspirate the cell culture medium.
2. Add 6.6 μM of TMRM in 1 mL of culture medium, without FBS (*see* **Notes 8** and **9**), and incubate the cells for 15–45 min at 37 °C in the dark.
3. Remove the medium and replace it with a warm culture medium, without FBS (*see* **Note 10**).
4. Analyze the basal green fluorescence with excitation and emission wavelengths of 485 nm and 590 nm, respectively.
5. At the end of the 5-min period, incubate the cells with 1 μM DNP and measure fluorescence using these same emission and excitation wavelengths (*see* **Note 11**).
6. Calculate the membrane potential analyzing the delta fluorescence between basal and DNP-stimulated fluorescence.

4 Notes

1. To prevent tetrahydrofuran from evaporating overnight at room temperature, cover the glass ring with a glass beaker.
2. This solution must first be degassed.
3. Electrode shelf life: The TPP^+ selective electrode has an average useful life of 2 months.
4. Membranes prepared as described have a shelf life of many years if stored dry by letting the solutions evaporate. They can be easily rehydrated overnight by placing them in a TPP^+ 10 mM solution (both inside and outside of the electrode) overnight and testing the electrode for Nernstian linearity.
5. Rules for the use of ion-specific electrodes:
   (a) In order to achieve stable and consistent electrode signals with low background noise, the following details should be noted:
   (b) Protect all electrode connections with nonconductive material.
   (c) Protect the reaction chamber in a Faraday cage.

(d) Maintain a constant temperature and magnetic stirring.
(e) Use a medium with reasonable ionic strength (e.g., 0.1 M KCl).
(f) Maintain a sufficient pH buffering capacity.

6. Problems may arise from hydrophobic substances and reagents adhering to the PVC membrane. These problems may be overcome by:
   (a) Using these substances in as few quantities as possible.
   (b) Mixing hydrophobic substances with the biological material before contacting them with the electrode.
   (c) Washing the electrode or biological membrane material with 70% ethanol after each experiment: Ethanol washing shortens the life of the electrode by extracting some ionophore from the PVC membrane. However, this does not affect the potential or inclination of the electrode since the membrane contains about a thousand times more ionophore than is needed for optimal function.
7. If the probes are above a critical concentration inside the mitochondrial matrix, they begin to aggregate. The aggregates are nonfluorescent, and thus lead to fluorescence quenching.
8. The probe concentration should be optimized for each experiment (due to the cell type). Nevertheless, quenching is typically achieved with a probe concentration higher than 50 nM.
9. TMRM is light sensitive; thus, it must not be directly exposed to intense light.
10. The culture medium should not contain either FBS or phenol red, since they are likeable causes of background interference with the analysis of the probes.
11. DNP is a proton ionophore used to completely depolarize mitochondria and to give an estimate of $\Delta\psi_m$.

References

1. Wallace DC (2009) Mitochondria, bioenergetics, and the epigenome in eukaryotic and human evolution. Cold Spring Harb Symp Quant Biol 74:383–393
2. Mitchell P (1966) Chemiosmotic coupling in oxidative and photosynthetic phosphorylation. Biol Rev 41:445
3. Jastroch M, Divakaruni AS, Mookerjee, Treberg SJR, Brand MD (2010) Mitochondrial proton and electron leaks. Essays Biochem 47:53–67
4. Crofts AR (2004) The Q-cycle—a personal perspective. Photosynth Res 80:223–243
5. Murai M, Miyoshi H (2016) Current topics on inhibitors of respiratory complex I. Biochim Biophys Acta 1857:884–891
6. Azzone GF, Petronilli V, Zoratti M (1984) Cross-talk between redox- and ATP-driven H + pumps. Biochem Soc Trans 12:414–416
7. Gerle C (2016) On the structural possibility of pore-forming mitochondrial FoF1 ATP synthase. Biochim Biophys Acta 1857:1191–1196
8. Brown GC (1992) Control of respiration and ATP synthesis in mammalian mitochondria and cells. Biochem J 284:1–13

9. Bakeeva LE, Grinius LL, Jasaitis AA, Kuliene VV, Levitsky D, Liberman EA, Severina II, Skulachev VP (1970) Conversion of biomembrane-produced energy into electric form. II. Intact mitochondria. Biochim Biophys Acta 216:13–21
10. Ross MF, Kelso GF, Blaikie FH, James AM, Cocheme HM, Filipovska A, Da Ros T, Hurd TR, Smith RA, Murphy MP (2005) Lipophilic triphenylphosphonium cations as tools in mitochondrial bioenergetics and free radical biology. Biochem Mosc 70:222–230
11. Kamo N, Muratsugu M, Hongoh R, Kobotake Y (1979) Membrane potential of mitochondria measured with an electrode sensitive to tetraphenyl phosphonium and relationship between proton electrochemical potential and phosphorylation potential in steady state. J Membr Biol 49:105–121
12. Jensen BD, Gunther TE (1984) The use of tetraphenylphosphonium (TPP+) to measure membrane potentials in mitochondria: membrane binding and respiratory effects. Biophys J 49:105–121
13. Plásek J, Vojtísková A, Houstek J (2005) Flow-cytometric monitoring of mitochondrial depolarisation: from fluorescence intensities to millivolts. J Photochem Photobiol 78:99–108
14. Moreno AJ, Santos DL, Magalhães-Novais S, Oliveira PJ (2015) Measuring mitochondrial membrane potential with a tetraphenylphosphonium-selective electrode. Curr Protoc Toxicol 65:1–16
15. Perry SW, Norman JP, Barbieri J, Brown EB, Gelbard HA (2011) Mitochondrial membrane potential probes and the proton gradient: a practical usage guide. BioTechniques 50:98–115
16. Nicholls DG, Ward MW (2000) Mitochondrial membrane potential and neuronal glutamate excitotoxicity: mortality and millivolts. Trends Neurosci 23:166–174
17. Dykens JA, Stout AK (2001) Assessment of mitochondrial membrane potential in situ using single potentiometric dyes and a novel fluorescence resonance energy transfer technique. Methods Cell Biol 65:285–309
18. Scaduto RC Jr, Grotyohann LW (1999) Measurement of mitochondrial membrane potential using fluorescent rhodamine derivatives. Biophys J 76:469–477
19. Sabnis RW, Deligeorgiev TG, Jachak MN, Dalvi TS (1997) DiOC6(3): a useful dye for staining the endoplasmic reticulum. Biotech Histochem 72:253–258
20. Rottenberg H, Wu S (1998) Quantitative assay by flow cytometry of the mitochondrial membrane potential in intact cells. Biochim Biophys Acta 1404:393–404

Chapter 15

The Multiparametric Analysis of Mitochondrial Dynamics in T Cells from Cryopreserved Peripheral Blood Mononuclear Cells (PBMCs)

Jesse J. R. Masson, Matias Ostrowski, Gabriel Duette, Man K. S. Lee, Andrew J. Murphy, Suzanne M. Crowe, and Clovis S. Palmer

Abstract

The analysis of mitochondrial dynamics within immune cells allows us to understand how fundamental metabolism influences immune cell functions, and how dysregulated immunometabolic processes impact biology and disease pathogenesis. For example, during infections, mitochondrial fission and fusion coincide with effector and memory T-cell differentiation, respectively, resulting in metabolic reprogramming. As frozen cells are generally not optimal for immunometabolic analyses, and given the logistic difficulties of analysis on cells within a few hours of blood collection, we have optimized and validated a simple cryopreservation protocol for peripheral blood mononuclear cells, yielding >95% cellular viability, as well as preserved metabolic and immunologic properties. Combining fluorescent dyes with cell surface antibodies, we demonstrate how to analyze mitochondrial density, membrane potential, and reactive oxygen species production in CD4 and CD8 T cells from cryopreserved clinical samples.

Key words Mitochondrial density, Mitochondrial membrane potential, ROS, CD4 and CD8 T cells, Immunometabolism

1 Introduction

The field of immunometabolism provides novel insights into the role of metabolic reprogramming in regulating immune cell differentiation and functions [1, 2]. Mitochondria play a critical role in metabolic transitions and differentiation of immune cells. For example, the development of effector T cells coincides with high mitochondrial fission rates and a metabolic shift away from oxidative phosphorylation toward glycolysis to increase biomass production and cell division [3]. Alternatively, the differentiation into memory T cells is accompanied by high mitochondrial fusion and a shift toward mitochondrial oxidative phosphorylation for maximal ATP production [3, 4]. Mitochondrial ATP production is

Suresh Mishra (ed.), *Immunometabolism: Methods and Protocols*, Methods in Molecular Biology, vol. 2184,
https://doi.org/10.1007/978-1-0716-0802-9_15, © Springer Science+Business Media, LLC, part of Springer Nature 2020

achieved by maintaining a mitochondrial membrane potential generated by protein complexes that create a hydrogen gradient across the mitochondria [5]. Upon dysfunction or a loss of viability, mitochondria lose their ability to maintain a proton membrane potential, resulting in an increase in apoptotic signaling via Fas/Fas ligand binding, a mechanism that is increased by reactive oxygen species (ROS) which signals cellular stress and cell death cascades [6]. Exploring immune cell metabolism has been important in understanding the differentiation and the survival and functions of these cells while also providing novel insights into the immunopathogenesis of important health conditions such as obesity [7], aging [8], autoimmune disease [9, 10], cancer [11, 12], and various infections [13–15].

Herein, a limitation to exploring immunometabolic readouts is the requirement of live cells analyzed directly ex vivo. Commercially available (and commonly used) cryoprotective agents often result in an average of 13.3% reduction in peripheral blood mononuclear cell (PBMC) viability, with compromised T-cell functionalities [16]. Several researchers have documented PBMC viability below 75% [17], and inconsistent and variable results regarding the immunological effects of cryopreservation [18]. Here, we provide a step-by-step validated method [19, 20] which allows immunometabolic analysis to be applied to frozen stored cells, allowing adaptation to these readouts in clinical studies [21, 22]. To demonstrate this, we isolated and cryopreserved PBMCs obtaining >95% viable cells and assessed the mitochondrial dynamics within CD4 and CD8 T cells employing Mitotracker Green (mitochondrial mass), Mitotracker Red, DIOC6 (mitochondrial membrane potential), and hydroethidine (intracellular superoxide). The assessment of the immunologic and metabolic functionalities of T cells employing the protocols described herein has been previously documented [19, 20, 22].

2 Materials

Prepare all reagents and solutions using deionized water passed through a Millipore filtration system, autoclaved and stored at room temperature.

2.1 Peripheral Blood Mononuclear Cell Isolation

1. Lymphoprep™ (Axis-Shield) is a ready-made, sterile, and endotoxin-free solution for the isolation of PBMCs via a density gradient.
2. Dulbecco's phosphate-buffered saline (PBS).
3. Freezing solution: Roswell Park Memorial Institute medium-1640 (RPMI) and dimethyl sulfoxide (DMSO) at a 4:1 ratio (v/v). Store at 4 °C for up to 48 h.

4. Thawing solution: RPMI medium freshly supplemented with 10% human serum, 200 mM penicillin/streptomycin, and 2 mmol/L L-glutamine. Store at 4 °C for up to 2 weeks.
5. Trypan blue.
6. Anticoagulant (ethylenediaminetetraacetic acid (EDTA) or citrate recommended) 10 mL blood collection tubes.
7. Cryogenic vials (2 mL).

2.2 Fluorochromatic Antibody Preparation and Staining

1. Wash buffer: 1× PBS and albumin from bovine serum (BSA) at 0.5% (w/v).
2. Compensation beads.
3. Antibody master mix: Combine CD3 (PE), CD4 (PE-Cy7), and CD8 (APC-Cy7) antibodies provided by BD Bioscience at a 1:1:1 ratio (2 μL of each antibody per 100 μL cell suspension) in sterile FACS tubes.
4. Store on ice, and in the absence of light, to prevent photobleaching.

2.3 Mitochondrial Staining Stock (See Note 1)

1. DIOC6: Make an original stock of 100 mM DIOC6 by dissolving product in 1.74 mL of DMSO. Prepare 1 μM working solution from original 100 mM stock using serial dilution with 1× phosphate-buffered saline (1× PBS). Cells should be incubated in 20 nM final concentration (*see* **Note 2**).
2. Hydroethidine: Prepare an original stock of 79 mM hydroethidine by dissolving product in 1 mL of DMSO. Prepare 100 μM working solution from 79 mM original stock using serial dilution with 1× PBS. Cells should be incubated in 2 μM final concentration (*see* **Note 3**).
3. Mitotracker Green: Make an original stock of 1 mM Mitotracker Green as per instructions by the supplier by dissolving in DMSO supplied by the manufacturer. Prepare a 0.5 μM working solution from original stock using serial dilution with 1× PBS. Cells should be incubated at 20 nM final concentration.
4. Mitotracker Red: Make an original stock of 1 mM Mitotracker Red as per instructions by the supplier by dissolving in 91.98 μL DMSO supplied by the manufacturer. Then, prepare a 0.25 μM working solution from original stock using serial dilution with 1× PBS. Cells should be incubated at 5 nM final concentration.

3 Methods

Conduct all procedures at room temperature in a sterile working hood, unless otherwise specified.

3.1 The Isolation and Cryopreservation of Peripheral Blood Mononuclear Cells

1. Obtain 30 mL of whole human blood using venepuncture by filling three individual 10 mL anticoagulant tubes. Transfer all blood into a sterile 50 mL Falcon tube (*see* **Note 4**).
2. Spin blood at 400 *g* for 10 min, with brakes off. Gently remove tube to avoid disruption.
3. Transfer all plasma to a sterile 10 mL Falcon tube and store on ice for later use. Add 1× PBS to the remaining blood sediment in the 50 mL tube until the volume reaches 30 mL. Gently swirl to mix.
4. Using a sterile pipette, transfer 15 mL of Lymphoprep™ to a new sterile 50 mL Falcon tube. Using a 50 mL pipette very gently transfer the 30 mL solution of blood/1× PBS mixture (from **step 3**) to the surface of the Lymphoprep™ by gently slanting the tube, and without disturbing the underlying Lymphoprep. Ensure that the blood and PBS mixture creates a distinct and separate layer above the Lymphoprep™, which must remain undisturbed as shown (Fig. 1a).
5. Then, centrifuge the blood/PBS/Lymphoprep™ without brakes at 400 *g* at a temperature of 4 °C for 30 min, before gently removing the tube without disturbing the density gradient. During this period, label the 2 mL Corning cryogenic vials tubes (ID, date, cell concentration, etc.).
6. Carefully harvest the PBMC layer from the gradient using a sterile Pasteur pipette and transfer the cells to a new sterile 50 mL Falcon tube containing 30 mL of 1× PBS. This layer is distinguishable as a thin, white, cloudy layer as shown in Fig. 1b. Add 1× PBS to the Falcon tube to make cell solution to 50 mL.
7. Centrifuge the PBMC solution at 400 *g* at 4 °C for 10 min without brakes. Gently remove the tube without disturbing the PBMC pellet. Gently remove and discard the liquid from the tube without disturbing the PBMCs using a sterile pipette.
8. Resuspend the pelleted PBMCs in 2 mL of 1× PBS making the solution up to 50 mL with 1× PBS and centrifuge again at 400 *g* for 10 min without brakes. Wash cells in approximately 30 mL of RPMI medium by centrifugation.
9. Gently remove and discard the RPMI medium, ensuring not to disturb the sedimented PBMCs. Gently resuspend PBMCs in autologous plasma (approximately 3 mL) isolated in

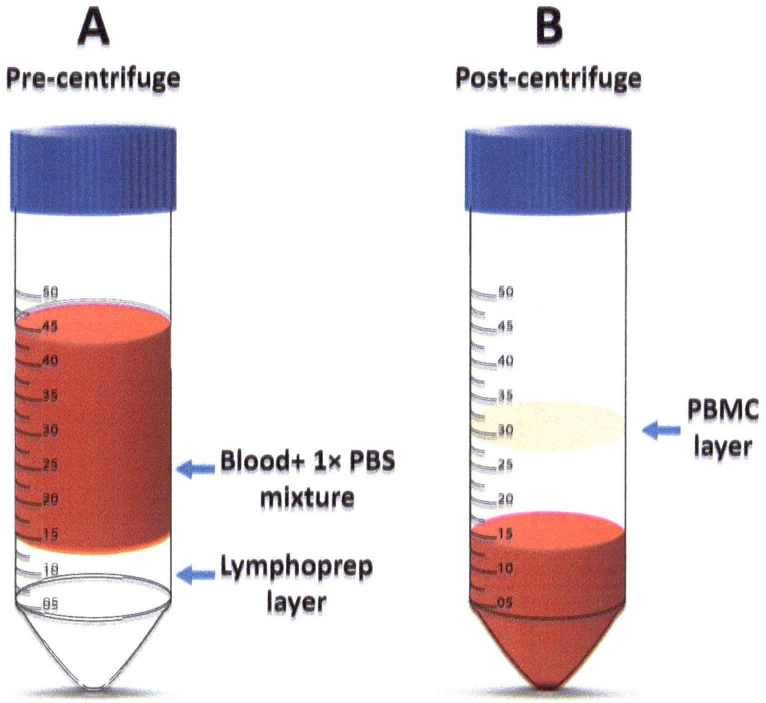

Fig. 1 (**a**) An example of how blood is layered above the Lymphoprep™ before centrifugation. (**b**) The PBMC layer can be extracted when the blood has been centrifuged by recovering the thin cloudy layer using a sterile Pasteur pipette

Subheading 3.1, and place on ice (remove 5–10 μL of the PBMC/plasma mix and assess viability and cell concentration by trypan blue stain).

10. Add an equal volume of the ice-cold cryopreservation solution to the PBMCs resuspended in autologous plasma using a sterile Pasteur pipette or 1000 μL pipette. This is a critical step (*see* **Note 5**). It is recommended to store cells at a density of 2–10 × 10^6 PBMCs/mL.
11. Aliquot 1 mL of PBMC/cryopreservation mixture into 2 mL cryopreservation vials (cryovials). Place vials in Mr. Frosty and immediately transfer aliquoted PBMCs to a −80 °C freezer for 4–12 h (leave cells in Mr. Frosty). Using dry ice, transfer the frozen PBMCs to liquid nitrogen for long-term storage (*see* **Note 6**).
12. If a larger quantity of blood or a larger collection of individual samples is being processed, it is recommended to add the cryopreservation mixture to one sample at a time to avoid delayed transfer to the −80 °C freezer.

3.2 Thawing PBMCs

1. Remove the frozen PBMC aliquots from liquid nitrogen and immediately place on dry ice for transportation.
2. To thaw and revitalize PBMCs, place the cryovials containing PBMCs into a water bath set at 40 °C until the frozen solution is dislodged from the tube while ice is still visible.
3. Immediately transfer the dislodged PBMCs to a sterile 10 mL Falcon tube containing 9 mL of thawing solution and gently invert to completely thaw PBMCs.
4. Centrifuge the thawed PBMCs at 400 *g* for 10 min at 4 °C without brakes.
5. Discard the supernatant and resuspend the pelleted PBMCs in 10 mL of 1× PBS before washing and centrifuging again.
6. Discard the supernatant and resuspend in approximately 600 μL (this can be varied depending on cell density, optimal 1–2 × 10^6/mL) of wash buffer.
7. Combine 5 μL of the PBMC mixture with 5 μL of trypan blue and assess cell viability and concentration using a Countess™ automated cell counter or using hemocytometer. This should typically yield viability >95% (*see* **Note 7**).

3.3 Cell Surface Stain and Mitochondrial Analysis

1. Aliquot 100 μL of PBMCs into separate 5 mL polypropylene round-bottom FACS tubes for analysis.
2. Prepare compensation beads by adding a single drop of both positive and negative beads into 5 mL polystyrene round-bottom FACS tubes and adding the required antibodies accordingly.
3. Transfer the desired volume of the antibody master mix to the 100 μL PBMC sample in each of the 5 mL polystyrene round-bottom FACS tubes and gently vortex. Incubate all tubes in a nontransparent ice bucket (tubes must be in direct contact with ice) in the dark (e.g., in cupboard) for 30 min.
4. Resuspend PBMCs in 4 mL of wash buffer and centrifuge at 400 *g* for 5 min.
5. Resuspend pellet and centrifuge again. Discard the supernatant and resuspend PBMCs in 100 μL of 1× PBS.
6. To the PBMC sample, add 2 μL of 100 μM hydroethidine, 0.25 μM Mitotracker Red, and 1 μM DIOC6 or 4 μL 0.5 μM Mitotracker Green.
7. Incubate for 37 °C in the dark for 30 min.
8. Add 4 mL of ice-cold 1× PBS to samples and centrifuge at 400 *g* for 5 min.
9. Discard the supernatant and resuspend PBMCs in 300 μL of ice-cold 1× PBS for flow cytometric analysis.

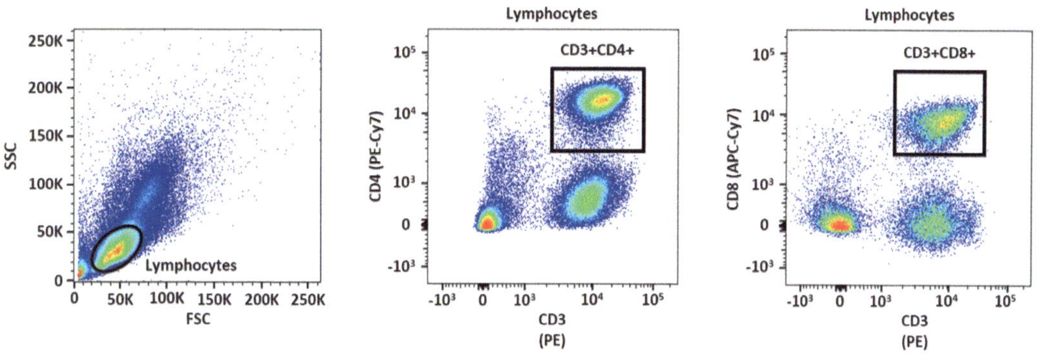

Fig. 2 The gating strategy used to define CD3 + CD4+ and CD3 + CD8+ T-cell populations

3.4 Gating and Analysis of Mitochondrial Dynamics Using Flow Cytometry

Using the gating strategy provided in Fig. 2, define the lymphocyte population using side scatter and forward scatter, and then gate the CD4+ and CD8+ T-cell populations as demonstrated. Analyze the mean fluorescent intensity (MFI) of hydroethidine (excitation/emission = 518 nm/606 nm), Mitotracker Red (excitation/emission = 579 nm/599 nm), DIOC6 (excitation/emission = 484 nm/501 nm), and Mitotracker Green (excitation/emission = 490 nm/516 nm) within these populations. ROS is detected by hydroethidine, mitochondrial density by Mitotracker Green, and mitochondrial membrane potential measured by Mitotracker Red and DIOC6. Representative results as fluorescence intensity from a healthy control are shown in Fig. 3.

4 Notes

1. Mitochondrial stain working solutions must be made fresh at the beginning of each experiment and must avoid exposure to light during the experimental process to prevent photobleaching.
2. DIOC6 is a positively charged molecule that binds to the negatively charged mitochondrial membrane. When cells are stressed, the mitochondrial membrane potential is reduced, resulting in a reduced negative charge available for DIOC6 binding. T-cell activation is associated with decreased mitochondrial membrane potential and increased ROS production [23].
3. Hydroethidine detects superoxides in the mitochondria producing a highly specific red fluorescent product, 2-hydroxyethidium upon oxidation.
4. It is important that blood samples are processed within 2 h of venepuncture to avoid ex vivo activation or loss of surface proteins. Furthermore, poor isolation, cryopreservation, and

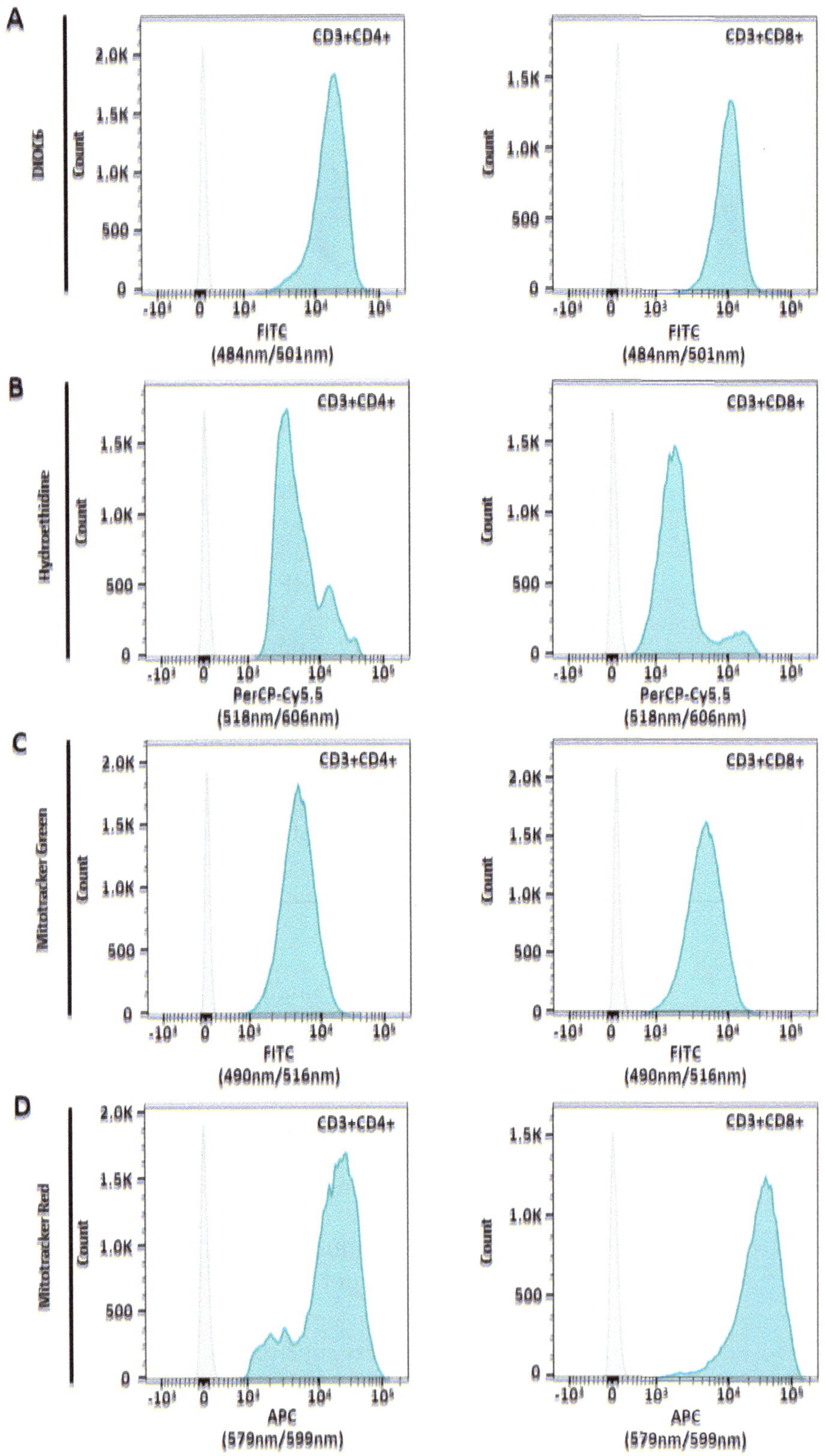

Fig. 3 The fluorescence of (**a**) DIOC6, (**b**) hydroethidine, (**c**) Mitotracker Green, and (**d**) Mitotracker Red expressed in CD4 and CD8 T cells. Gray and blue histograms represent unstained cells and those stained with mitochondrial dyes, respectively. The emission and excitation spectra of the dyes, plus the approximate flow cytometer acquisition channel, are shown on the *x*-axis

thaw techniques will induce stress and affect the activation status of cells. This protocol includes the PBMC isolation assay that, when performed as described, it typically yields a viability of >95% with preserved metabolic and immunologic functionalities [22].

5. The ice-cold cryopreservation solution must be added drop by drop, while gently swirling the PBMC solution while occasionally resting the tube on ice. Ensure that this step is performed with all tubes on ice so that the cryopreservation solution and cells are kept ice cold to avoid cellular toxicity, and thus improve cell viability.

6. Cryovials containing PBMC/cryopreservation solution mixture can be placed on ice or in a pre-chilled container containing isopropyl (Mr. Frosty) for no greater than 3 min, before transferring to −80 °C freezer.

7. We caution using preparations with <80% viability for immunometabolic analysis.

References

1. Hotamisligil GS (2017) Foundations of immunometabolism and implications for metabolic health and disease. Immunity 47(3):406–420
2. Palmer CS, Ostrowski M, Balderson B, Christian N, Crowe SM (2015) Glucose metabolism regulates T cell activation, differentiation, and functions. Front Immunol 6:1
3. Buck MD, O'Sullivan D, Geltink RIK, Curtis JD, Chang C, Sanin DE et al (2016) Mitochondrial dynamics controls T cell fate through metabolic programming. Cell 166(1):63–76
4. Shehata HM, Murphy AJ, Lee MKS, Gardiner CM, Crowe SM, Sanjabi S et al (2017) Sugar or fat? - metabolic requirements for immunity to viral infections. Front Immunol 8:1311
5. Bulthuis EP, Adjobo-Hermans MJW, Willems PHGM, Koopman WJH (2019) Mitochondrial morphofunction in mammalian cells. Antioxid Redox Signal 30(18):2066–2109
6. Devadas S, Zaritskaya L, Rhee SG, Oberley L, Williams MS (2002) Discrete generation of superoxide and hydrogen peroxide by T cell receptor stimulation: selective regulation of mitogen-activated protein kinase activation and fas ligand expression. J Exp Med 195:59–70
7. Godfrey C, Bremer A, Alba D, Apovian C, Koethe JR, Koliwad S et al (2019) Obesity and fat metabolism in HIV-infected individuals: Immunopathogenic mechanisms and clinical implications. J Infect Dis 220 (3):420–431
8. McGuire PJ (2019) Mitochondrial dysfunction and the aging immune system. Biology (Basel) 8(2):26
9. Luu M, Pautz S, Kohl V, Singh R, Romero R, Lucas S et al (2019) The short-chain fatty acid pentanoate suppresses autoimmunity by modulating the metabolic-epigenetic crosstalk in lymphocytes. Nat Commun 10(1):760
10. Piper DW, Pym BM, Toy S, Gellatly R, Byth K, Seville P (1986) The effect of maintenance cimetidine therapy on the medical, social and economic aspects of patients with chronic gastric ulcers. A placebo-controlled prospective study. Med J Aust 145(8):400–403
11. Schurich A, Magalhaes I, Mattsson J (2019) Metabolic regulation of CAR T cell function by the hypoxic microenvironment in solid tumors. Immunotherapy 11(4):335–345
12. O'Sullivan D, Sanin DE, Pearce EJ, Pearce EL (2019) Metabolic interventions in the immune response to cancer. Nat Rev Immunol 19 (5):324–335
13. Shi L, Jiang Q, Bushkin Y, Subbian S, Tyagi S (2019) Biphasic dynamics of macrophage immunometabolism during *Mycobacterium tuberculosis* infection. MBio 10(2):e02550-18
14. Palmer CS, Cherry CL, Sada-Ovalle I, Singh A, Crowe SM (2016) Glucose metabolism in T cells and monocytes: new perspectives in HIV pathogenesis. EBioMedicine 6:31–41

15. Pallett LJ, Schmidt N, Schurich A (2019) T cell metabolism in chronic viral infection. Clin Exp Immunol 197(2):143–152
16. Liang X, Hu X, Hu Y, Zeng W, Zeng G, Ren Y et al (2019) Recovery and functionality of cryopreserved peripheral blood mononuclear cells using five different xeno-free cryoprotective solutions. Cryobiology 86:25–32
17. Sun J, Gao M, Li K, Qin L, Sun H, Qiao G (2018) Quality of cryopreserved peripheral blood mononuclear cells recovered from the hepatitis/AIDS biobank. Biopreserv Biobank 16:397. https://doi.org/10.1089/bio.2018.0050
18. Anderson J, Toh ZQ, Reitsma A, Do LAH, Nathanielsz J, Licciardi PV (2019) Effect of peripheral blood mononuclear cell cryopreservation on innate and adaptive immune responses. J Immunol Methods 465:61–66
19. Palmer C, Hampartzoumian T, Lloyd A, Zekry A (2008) A novel role for adiponectin in regulating the immune responses in chronic hepatitis C virus infection. Hepatology 48 (2):374–384
20. Palmer CS, Ostrowski M, Gouillou M, Tsai L, Yu D, Zhou J et al (2014) Increased glucose metabolic activity is associated with CD4+ T-cell activation and depletion during chronic HIV infection. AIDS 28(3):297–309
21. Yeoh HL, Cheng AC, Cherry CL, Weir JM, Meikle PJ, Hoy JF et al (2017) Immunometabolic and lipidomic markers associated with the frailty index and quality of life in aging HIV+ men on antiretroviral therapy. EBioMedicine 22:112–121
22. Masson JJR, Murphy AJ, Lee MKS, Ostrowski M, Crowe SM, Palmer CS (2017) Assessment of metabolic and mitochondrial dynamics in CD4+ and CD8+ T cells in virologically suppressed HIV-positive individuals on combination antiretroviral therapy. PLoS One 12(8): e0183931
23. Ledderose C, Bao Y, Lidicky M, Zipperle J, Li L, Strasser K et al (2014) Mitochondria are gate-keepers of T cell function by producing the ATP that drives purinergic signaling. J Biol Chem 289(37):25936–25945

Chapter 16

The Measurement of Whole-Body Glucose Homeostasis in Mice

Yang Xin Zi Xu, Sudharsana R. Ande, and Suresh Mishra

Abstract

The resident immune cells (e.g., macrophages) present in major metabolic tissues, such as adipose and liver tissues, play crucial roles in respective tissue homeostasis through cross talk with metabolic tissues, and consequently in metabolic homeostasis at the systemic level, and their dysregulation contributes to metabolic dysregulation at large, as well as many associated diseases. Moreover, the cross talk between different resident immune cells and metabolic tissues is not limited to an intra-organ level but also includes interorgan cross talk, as they work in a coordinated manner throughout the body, such as in adipose tissue, skeletal muscle, and liver. Thus, it is important to determine the impact of altered immune functions on metabolic homeostasis and vice versa, to enhance our knowledge of immunometabolic biology. Glucose and insulin tolerance tests are simple methods that enable the measurement and analysis of the overall glucose homeostasis at the systemic level. Here we describe the process of performing metabolic tests for glucose homeostasis in mice, as mouse models are often used for defining the mechanistic underpinnings of physiology and pathophysiology related to immunometabolism, and in preclinical studies.

Key words Glucose metabolism, Glucose tolerance test, Insulin tolerance test, Immunometabolism, Meta-inflammation

1 Introduction

A growing understanding of immunometabolism over the last few decades has provided substantial evidence suggesting a constant cross talk between the metabolic and immune systems in normal physiology and pathophysiology. This includes different cell types within a metabolic tissue (e.g., adipose tissue), and different metabolic tissues (e.g., adipose tissue and liver) within an organism that support this extensive cross talk, such as association between chronic inflammation and metabolic dysregulation [1]. As the use of genetic mouse models is crucial to advancing our knowledge of immunometabolism at the systemic level in physiology and pathophysiology, here we describe simple methods for assessing glucose homeostasis in mice, as it relates to the effects of chronic low-grade

Suresh Mishra (ed.), *Immunometabolism: Methods and Protocols*, Methods in Molecular Biology, vol. 2184, https://doi.org/10.1007/978-1-0716-0802-9_16, © Springer Science+Business Media, LLC, part of Springer Nature 2020 inflammation on metabolic events. Particularly, we will focus on the glucose tolerance test (GTT) and the insulin tolerance test (ITT), as they are often overlooked in protocol books on immunological methods. These tests measure the changes in blood glucose levels over 60–120-min intervals, following the administration of a bolus glucose (for the GTT) or insulin (for the ITT) [2]. Of note, the GTT is the only means of finding impaired glucose tolerance [3]. Over the years, variations of the GTT have emerged with differences in the dosage of glucose, the route of glucose administration, the fasting period before the test, the duration of blood sampling during tests, and the state of consciousness (i.e., anesthetized or without anesthesia) [3]. These variations will not be included in the methods described here; however, they can be found in a study performed by Andrikopoulos et al. [3], where they have systematically validated these variables.

2 Materials

2.1 GTT and ITT

Prepare all solutions using ultrapure water and analytical grade reagents. Prepare all injectable reagents prior to use. Follow all waste disposal regulations for blood products and sharps.

1. Six or more mice for each experimental group (*see* **Notes 1** and **2**).
2. 1× Phosphate-buffered saline (PBS).
3. 10 and 20% (w/v) β-D(+)-glucose.
4. 1 cc Syringe and 27-gauge needle.
5. 70% Alcohol.
6. Timer.
7. Sterile surgical scissors and scalpel.
8. Glucometer and glucose strips (*see* **Note 3**).
9. Mouse insulin ELISA kit.
10. Weighing scale.
11. Human recombinant insulin.
12. Anesthesia (*see* **Note 4**).

3 Methods

The following experimental procedures involving animals (mice) should receive ethical approval according to institutional regulations. A humane treatment of animals should be practiced at all times.

3.1 Glucose Tolerance Test

1. Fast the experimental mice 14–18 h (overnight) in fresh cages with water supply (*see* **Note 5**).
2. The next morning prior to experiment, prepare 20% glucose in 1× PBS and sterilize the solution by 0.2 μm filtration.
3. Weigh each mouse and calculate the glucose (20% wt/vol) injection volume for 1 g glucose/1 kg body weight using the following formula: 20% β-D(+)-glucose (μL) = body weight (g) × 5 (*see* **Note 6**).
4. Make a 1–2 mm cut from the tail tip with sharp scissors or a scalpel. Collect saphenous vein blood (50 μL) by direct flow or by gentle massage of the tail into a capillary action tube or Eppendorf tube for glucose and insulin determinations (*see* **Note 7**). This is the fasting baseline glucose level and designated as time 0 (*see* **Note 8**).
5. Restrain the mouse by the scruff of the neck. Sterilize the abdominal region with 70% ethanol and inject glucose intraperitoneally with a 27-gauge sterile needle. Start injection of the next mouse after 3–5 min (*see* **Notes 9** and **10**).
6. Release the mouse back into the cage (*see* **Note 11**) and start the timer. Collect blood at 15, 30, 60, 90, and 120 min for glucose and insulin determinations (Table 1). Collect blood (50 μL) by gently removing the scab and repeating the massaging procedure described above (*see* **Notes** 7 and **12**).
7. Allow blood to clot at room temperature for about 30 min. Collect serum by centrifugation at 400 *g* for 10 min and transfer the supernatant to a clean tube.
8. Label tubes and store samples at −80 °C for further analysis.
9. Measure serum insulin concentration with commercially available ELISA kit according to the manufacturer's protocol.

3.2 Insulin Tolerance Test

Similar to GTT, all solutions should be prepared prior to use and proper disposal of biohazardous materials and sharps should be followed. The following experimental procedures involving animals (mice) should receive ethical approval according to institutional regulation. The humane treatment of animals should be practiced at all times.

1. Fast the experimental mice in fresh cages with water supply for 5–6 h.
2. Prepare 10% glucose in 1× PBS and sterilize the solution by 0.2 μm filtration for potentially hypoglycemic mice.
3. Dilute commercially available human recombinant insulin to 0.1 IU/mL (add 10 μL of 100 IU/mL in 10 mL of 1× PBS).
4. Weigh each mouse and calculate the insulin injection volume for 1 IU/kg using the following formula: insulin (μL) = body weight (g) × 10 (*see* **Note 13**).

Table 1
A sample GTT/ITT recording sheet

Animal Information				Blood glucose (mmol/L or mg/dl) at x minutes				
Treatment Group	DOB	Sex	Weight (g)	0	15	30	60	120
Notes:								

5. Make a 1–2 mm cut from the tail tip distal to the bone with sharp scissors or a scalpel. Collect saphenous vein blood (50 μL) by direct flow or by gentle massage of the tail into a capillary tube or Eppendorf tube (*see* **Note 7**). This is the fasting baseline insulin level and designated as time 0 (*see* **Note 8**).
6. Sterilize the abdominal region of the mouse with 70% ethanol. Hold the back of the animal firmly, and inject insulin intraperitoneally with a 27-gauge sterile needle.
7. Release the mouse back into the cage (*see* **Note 11**) and start the timer. Collect blood at 15, 30, 60, 90, and 120 min and measure glucose level with the glucometer (Table 1). Collect blood (50 μL) by gently removing the scab and repeating the massaging procedure described above (*see* **Note 7**).
8. Closely monitor the mice for signs of hypoglycemia including shakiness and seizure-like events/myoclonic jerking or when blood glucose drops below 2.2 mmol/L or 40 mg/dL (*see* **Note 14**).
9. Allow blood to clot at room temperature for about 30 min. Collect serum by centrifugation at 400 *g* for 10 min and transfer the supernatant to a clean tube.
10. Label tubes and store samples at −80 °C for further analysis.

11. Measure serum insulin concentration with commercially available mouse insulin ELISA kit according to the manufacturer's protocol.
12. Calculate the rate constant for blood glucose disappearance with the Lundbaek formula: $0.693/t^{1/2} \times 100$, where $t^{1/2}$ to be calculated from the slope of the least square analysis of the blood glucose concentrations during the linear phase of glucose decline [4].
13. Both GTT and ITT are stress inducing and can alter blood glucose levels and short-term physiology of the animals. Therefore, investigators should allow ample time between the two tests (*see* **Note 15**).

3.3 Data Interpretation

The results from the GTT and ITT should be presented as a time course of glucose levels (Fig. 1), and use the trapezoidal rule to determine the area under the curve above baseline glucose for the GTT [2, 4]. In addition, the ITT results can be expressed as the inverse area under the curve below baseline glucose. Data may also be presented as a percentage of basal glucose; however, caution is required in data interpretation if basal fasting glucose levels differ in different experimental groups (*see* **Note 16**). Glucose and insulin profiles during GTT and ITT should be analyzed by repeated-measures ANOVA with time as the within-subject variable and age and genotype as between-subject factors using a statistical software [5].

4 Notes

1. Mice in different experimental groups should be studied in parallel, including control mice, because they display variations in hormone levels and associated metabolism at different times during a 24-h cycle.
2. Glucose tolerance and insulin sensitivity decline with age in mice. Therefore, metabolic measurements should be carried out in age-matched mice. In addition, sex can also influence the phenotype in a genetic model or response to interventions. Thus, mice in different experimental groups should also be matched for sex.
3. Commercially available handheld whole-blood glucose monitors are very easy to use and reliable; however, they are developed for humans.
4. Investigators may choose to anesthetize mice to minimize stress. However, anesthesia affects heart rate and blood flow and induces hyperglycemia in mice [2, 3], which may not yield physiological results. Therefore, it is recommended to perform the described procedures in conscious mice.

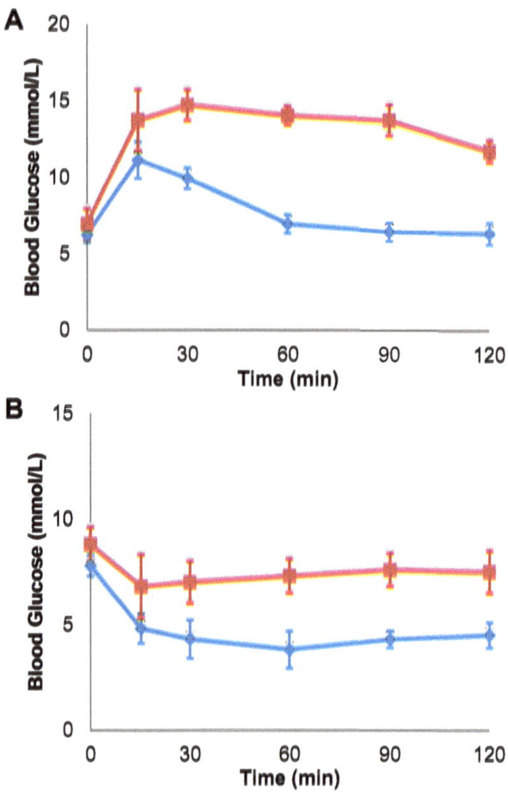

Fig. 1 Data presentation for the GTT and ITT. Sample line graphs showing substantial differences in intraperitoneal glucose tolerance test (**a**) and insulin tolerance test (**b**) in 6-month-old wild-type (blue circle) and transgenic (dark red rectangle) male mice

5. Overnight (14–18 h) fasting is recommended for the GTT; otherwise, a 5–6-h fast is sufficient to assess insulin action within a more physiological state [3].
6. When studying the effect of diets, the effects on body composition should be noted because they not only affect glucose metabolism but also influence the manner in which metabolic tests are performed and how results are normalized [2].
7. Sampling from the tail tip is appropriate when only small volumes of blood are needed. Larger samples often require prolonged massaging and hence cause more stress to the mouse, resulting in increased catecholamine levels, endogenous glucose production, and whole-body glucose disappearance [2].
8. A fasting baseline value greater than 7 mmol/L or 126 mg/dL indicates diabetes mellitus.
9. Remove any air bubbles prior to injection. Perform injection slowly at an angle of 30–45° into the posterior quadrant of the abdomen.

10. The administration of glucose via intraperitoneal injection means there is no incretin response that is known to significantly potentiate the glucose-mediated insulin response [3]. Thus, a difference in glucose level and insulin response is expected based on the route of its administration.
11. Animals should not eat during the test; water is acceptable and can be left in the cage.
12. If measuring glucose fluxes using isotopic techniques, it is recommended to measure plasma glucose levels instead of whole-blood glucose. Plasma glucose levels may be measured using an enzymatic spectrophotometric or fluorometric assay, or alternatively using a commercially available plasma glucose analyzer.
13. For insulin-resistant mice, increase the dosage of insulin injection to 2–2.5 U/kg.
14. When an event occurs, inject 10% glucose intraperitoneally and allow the mouse to recover. No further experiments should be conducted on the animal.
15. Select the same time of the day to perform all of the GTT and ITT trials as the mice display variations in hormone levels and associated metabolism at different times during the 24-h cycle.
16. If fasting glucose levels differ in mice from different experimental groups undergoing the GTT, then data should be presented along with a calculation of the area under the curve.

Acknowledgments

Research in the laboratory of SM is supported by Natural Sciences and Engineering Research Council of Canada (RGPIN-2017-04962), Research Manitoba, Health Sciences Centre Foundation, and URGP-University of Manitoba. YXZX is a recipient of the Alexander Graham Bell Canada Graduate Fellowship-Doctoral (NSERC).

References

1. Lee YS, Wollam J, Olefsky JM (2018) An integrated vies of immunometabolism. Cell 172:22–40
2. Ayala JE, Samuel VT, Morton GJ et al (2010) Standard operating procedures for describing and performing metabolic tests of glucose homeostasis in mice. Dis Model Mech 3:525–534
3. Andrikopoulos S, Blair AR, Deluca N, Fam BC, Proietto J (2008) Evaluating the glucose tolerance test in mice. Am J Physiol Endocrinol Metab 295:E1323–E1332
4. Nguyen KH, Yao X-H, Moulik S, Mishra S, Nyomba BLG (2011) Human IGF binding protein-3 overexpression impairs glucose regulation in mice via an inhibition of insulin secretion. Endocrinology 152:2184–2196
5. Nguyen KH, Yao XH, Erickson AG, Mishra S, Nyomba BL (2015) Glucose intolerance in aging male IGFBP-3 transgenic mice: differential effects of human IGFBP-3 and its mutant IGFBP-3 devoid of IGF binding ability. Endocrinology 156:462–474

Chapter 17

Immunometabolism and Its Potential to Improve the Current Limitations of Immunotherapy

Andrew D. Sheppard and Joanne Lysaght

Abstract

The last century of research in tumor immunology has culminated in the advent of immunotherapy, most notably immune checkpoint inhibitors. These drugs have shown encouraging results across a multitude of malignancies and have shifted the paradigm of cancer treatment. However, no more than 40% of patients treated with these immune checkpoint blockade inhibitors respond. Thus, resistance is a barrier to therapy that remains poorly understood. All cells require energy and biosynthetic precursors for survival, growth, and functioning, where multiple metabolic pathways allow for flexibility in how nutrients are utilized. A defining hallmark of many cancers is altered cellular metabolism, creating an imbalanced demand for nutrients within the tumor microenvironment. Immunometabolism is increasingly understood to be vital to the functions and phenotypes of a myriad of immune cell subsets. In tumors, the high demand for nutrients by the tumor drives competition between tumor cells and infiltrating immune cells, culminating in dysfunctional immune responses. This chapter discusses the recent successes in cancer immunotherapy and highlights challenges to therapy. We also outline the major metabolic processes involved in the generation of an immune response, how this can become dysregulated in the context of the tumor microenvironment, and how this contributes to resistance to immunotherapy. Finally, we explore the potential for targeting immunometabolic pathways to improve immunotherapy, and examine current trials targeting various aspects of metabolism in an attempt to improve the outcomes from immunotherapy.

Key words Immunometabolism, Immunotherapy, Metabolism, Tumor microenvironment

1 Introduction

1.1 The History of Immunotherapy

At the beginning of the twentieth century, scientists and physicians began to realize that the immune system was a potent and specific tool that could identify and eradicate infectious agents. The idea that cancers could be treated with the immune system was novel, but largely dismissed. A number of determined scientists and clinicians pushed the idea forward to varying degrees of success, with notable examples including William Coley's toxins [1] (demonstrating that an antitumor response can be elicited in response to bacterial components), and some 16 years later, the theory that nascent cancer can be recognized by the immune system,

Suresh Mishra (ed.), *Immunometabolism: Methods and Protocols*, Methods in Molecular Biology, vol. 2184, https://doi.org/10.1007/978-1-0716-0802-9_17,

postulated by Paul Ehrlich [2]. Despite this, the field of cancer immunology lay dormant largely for another 50 years. However, discoveries of how soluble factors modulate immune responses, of pathogen- and damage-associated molecular pattern receptors, and even the development of modern immune checkpoint blockade drugs have been built on the foundation of these pioneers' works.

The field awoke once again in 1957 with the discovery of interferon by Lindenmann and Isaacs, and the foundation of the New York Cancer Research Foundation (established by Coley's daughter) [3]. In 1959, the first ever "cancer vaccine" was trialed in the state of New York with a 22% remission rate achieved in gynecological cancers treated with tumor lysates. Despite these encouraging results, the trials went largely unnoticed by the scientific community [3]. In the 1960s and 1970s, a series of landmark discoveries resulted in an increase in our understanding of the cellular basis for immune responses: the T cell in 1967 by Jacques Miller, the dendritic cell in 1973 by Ralph Steinman, major histocompatibility complex (MHC) restriction in 1974 by Zinkernagel and Doherty, and natural killer (NK) cells in 1975 by Eva Klein [4–8]. Most modern-day immunotherapies are based around these discoveries from this time period. Despite these discoveries, clinicians were not yet convinced by the idea that the immune system could be used to fight cancer; this was the era of cytotoxic compounds, where new and more potent drug combinations were being discovered and trialed constantly. Ironically, at the same time, immune-based treatments were being employed for treatments of some cancers, although the immune system's involvement was not well known, namely in bone marrow transplants for hematological cancers and the eventual cloning and subsequent mass production of IFN-α used to treat chronic myeloid leukemia [9].

The first concrete evidence that the immune system could play a role in treating malignancies came in the late 1990s, when Schreiber et al. described how lymphocyte-derived IFN-γ worked to protect against carcinogen-induced cancers, and that over time the immune system placed a selective pressure on cancer to select for immune evasive clones [10, 11]. This proved that the immune system provides benefits and drawbacks when dealing with cancers; on the one hand, it is capable of destroying cancer cells with sufficient immunogenicity, and on the other it can cause the selection of more aggressive clones capable of avoiding immune destruction.

1.2 Immune Checkpoint Inhibitors and Modern Immunotherapy

In the decades that followed these landmark discoveries, extensive research was carried out to understand the mechanisms of lymphocyte activation, destruction, and resolution. It was during these decades that the molecules CTLA-4 and CD28 were shown to have opposing effects, and how preventing signaling through CTLA-4 could boost antitumor activity (work for which Jim

Allison would later win the 2018 Nobel Prize in Physiology or Medicine [12, 13]). The programmed death receptor 1 (PD-1) was also discovered by Tasuku Honjo, who shared the 2018 Nobel Prize in Physiology or Medicine, and was later shown to be involved in the inhibition of the immune response [14]. However, it was not until 2010 that the anti-CTLA-4 monoclonal antibody ***ipilimumab*** was reported to increase overall survival in metastatic melanoma—an event which put immunology and immunotherapy at the forefront of cancer research [15].

Immunotherapy has arguably become the most important advance in cancer treatment since the use of cytotoxic mustard gas, being named the breakthrough of the year in 2013 by *Science* journal and the subject of the Nobel Prize in Physiology or Medicine in 2018, as previously stated. Currently, drugs targeting CTLA-4, PD-1, and PD-L1 have been approved, including in 2017 when ***pembrolizumab*** was approved for use based on a biomarker (MSI-hi or dMMR), and not the tumor type. These so-called immune checkpoint inhibitors (ICIs) have revolutionized cancer treatment, inducing durable responses and improvements in overall survival, and progression-free survival for cancers with the most negative prognoses. While most trials continue to focus on combinations of anti-PD-1/PD-L1 and anti-CTLA4, trials are now underway to investigate new checkpoint pathways for blockades (LAG-3, TIGIT, TIM-3, and VISTA), as well as the immune stimulatory pathways ICOS, GITR, 4-1BB, and CD40 [16]. As more progress is made, it is likely that new combinations of inhibitors and stimulators will be devised, extracting the maximum possible benefit from these pathways. A series of landmark clinical trials demonstrated the potential of immunotherapy in various malignancies, including melanoma, NSCLC, and head and neck cancer. In Checkmate 037, a phase III RCT Nivolumab was compared to chemotherapy for late-stage melanoma patients; the overall response rate (ORR) was 30.7% in the nivolumab group compared to 10.6% in the chemotherapy group [17]. This result was improved further with PD-L1 positivity (>5% staining for PD-L1 as per IHC), where the ORR was 43.6% [17]. Combining the CTLA-4 and PD-1 blockade also produced encouraging results in the melanoma Checkmate 069 trial, with a response rate of 60% following a combination treatment [17]. For NSCLC patients, the KEYNOTE trials challenged the dogma that lung cancer would not benefit from immunotherapy. In metastatic disease, pembrolizumab was significantly better at increasing both overall survival (OS) and progression-free survival (PFS) compared with docetaxel for patients with >1% PD-L1 positivity [18]. In 2016 pembrolizumab received first-line approval for the treatment of metastatic NSCLC, as it had significantly better PFS than chemotherapy 10.3 months vs. 6 months [19]. In the years since these trials, a number of follow-up trials have been conducted testing combinations of anti-CTLA-4 and anti-PD-1 with the standard of care. These trials have resulted in immunotherapy drugs being approved for first-line therapy in many malignancies, including NSCLC, melanoma, and triple-negative breast cancer.

Despite these remarkable results, the overall response to PD-1/PD-L1 and CTLA-4 antibodies remains, at best, between 10 and 40% [20]. There are many reasons for which overall efficacy remains low; exploring these reasons remains at the center of immuno-oncology research and focuses on the interactions occurring in the tumor microenvironment (TME). Patients are now being screened for several predictive biomarkers, such as the expression of PD-1, PD-L1 markers of genomic instability, and DNA repair, as well as immune markers, including total TIL count, $CD8^+$ specific infiltration, and even expression of cytotoxic genes [21]. These biomarkers could potentially predict responders and also indicate whether patients could benefit from combination therapies. Most significantly, microsatellite instability-high (MSI-hi) or DNA mismatch repair (MMR)-deficient tumors have been approved for treatment with pembrolizumab, highlighting the importance of understanding how tumor and TME biology interact with the immune system [22]. Of equal importance is understanding how the immune cell infiltrate will respond to the immune checkpoint blockade, as studies have shown that tumors with increased cytotoxic T lymphocyte (CTL) infiltration respond better to therapy (although this has proven controversial, perhaps due to the heterogeneity of tumor-infiltrating lymphocyte (TIL) populations, and their ability to respond to immunotherapy), while others have shown that gene expression of key cytotoxic effector molecules, such as IFN-γ, correlates with a response to the blockade [23, 24]. We are now also beginning to understand the importance of clonal diversity in immunotherapy—with less TCR diversity being associated with a poorer response, it is likely that a high TCR diversity will allow T cells to respond to multiple tumor-associated antigens and neoantigens and increase the efficacy of immunotherapy [25]. Improving the therapeutic delivery of the immune checkpoint blockade drugs, specifically with the scheduling of conventional therapy, is of vital importance. Understanding how the tumor and the surrounding microenvironment respond to chemotherapy and radiotherapy, via the release of immune-stimulatory damage-associated molecular patterns (DAMPs), and the activation of immunogenic signaling pathways, which likely contribute to the efficacy of immunotherapy, will help delineate the most effective time to deliver immunotherapy [26, 27]. Crucially, understanding how radiotherapy and chemotherapy combine to modulate the immune response will guide which types of immunotherapy will be most efficacious and support the combinations of various treatment modalities. Conventionally, chemotherapy has been viewed as immunosuppressive due to its common side effect of lymphodepletion, while radiotherapy has been viewed as immunostimulatory due to its effect on the local immune response, including the release of DAMPs, which engage and activate CD8$^+$ T cells, as well as increase the visibility of antigens [28]. New evidence, however, suggests that not all chemotherapeutic agents induce immune suppression; indeed, some classes such as the anthracycline base compounds work by inducing an influx of immune cells to the tumor [29, 30]. Other cytotoxic chemotherapies are also known to induce DAMP expression, thereby activating the immune system [30, 31]. Therefore, there is an impetus on researchers to discover combinations of conventional therapies that will induce maximum immune-stimulatory conditions, which will synergize with immunotherapy.

1.3 Mechanisms of Resistance

What is clear from the current body of research is that antitumor immune responses are often dysfunctional. De novo mutations can elicit responses from the immune system early in tumorigenesis, in what is known as the elimination phase, as mutated proteins and peptides are identified by the immune system, culminating in the destruction of the mutated clone. Following elimination, an equilibrium phase is reached, whereby mutated clones are kept in check by the immune system or lie in a dormant state. Eventually, the selective pressure placed on neoplastic cells by the immune system results in the emergence of non-immunogenic tumorigenic cells and the resultant formation of a tumor [32]. In this model of cancer immune evasion, a gradual shift toward tolerance of the tumor occurs, whereby at the final stage the immune cell population of the tumor is skewed toward a pro-tumor phenotype, with antitumor immune cells being largely suppressed and exhibiting exhaustion. While TILs may be a positive prognostic marker in some cases, these lymphocytes are largely unable to mount an appropriate antitumor response, perhaps explaining the discrepancies in the use of TILs as a prognostic marker [33]. T cells in the tumor are subject to a vast network of suppressive signaling, either directly from tumor cells or from other cells in the TME [34]. This suppressive microenvironment is a major barrier to immunotherapy.

The aforementioned immune checkpoint molecules, CTLA-4 and PD-1, are expressed on the surface of T cells, and serve to homeostatically blunt activation signals and provide negative feedback for immune responses [16]. Given enough co-inhibitory stimulation, T-cell activation may be attenuated or indeed be forced to switch off. Their ligands are expressed on a number of cells in the TME, including cancer cells and other infiltrating immune cells. The expression of these ligands is therefore an important mechanism of immunosuppression and tolerance in the TME [35]. It is now also established that PD-L1, the ligand for PD-1, is shed in exosomes, which travel to the draining lymph node and influence the activation of local T cells [36].

A key mediator of TME-suppressive signaling is TGF-β, with transformed cells secreting larger amounts of TGF-β than normal cells [37]. TGF-β has pleiotropic effects on many aspects of tumor immunity, as well as autocrine effects on cancer cells, including the regulation of stemness and EMT/metastasis [38]. Mice with T cells lacking TGF-β display spontaneous differentiation and autoimmune disease [39]. TGF-β also regulates the immune exclusion of cytotoxic cells from tumors, explaining the low levels of TIL infiltration into the TME, where a higher expression of *TGFB1* in patient samples is associated with the immune-excluded phenotype, and reduced overall survival [40, 41]. TGF-β also directly affects the generation of a suppressive T-cell response by promoting the generation of T_{regs} and downregulating the key cytotoxic genes perforin, GzmA, GzmB, and FasL; thus, TGF-β encourages an immunosuppressive TME [42]. TGF-β also regulates the recruitment and induction of suppressive innate immune cells, including tumor-associated macrophages (TAMs) and myeloid-derived suppressor cells (MDSCs), which in turn secrete TGF-β and exert other suppressive effects in the TME [43]. There are a number of early-phase clinical trials targeting TGF-β in combination with checkpoint inhibitors, with some encouraging preliminary results; however as these are mostly phase I trials no conclusion regarding OS or PFS can be reached as of yet [44, 45]. TGF-β may also prove to be a challenging therapeutic target due to its wide expression and roles in many biological processes.

Part of the suppressive signaling within the TME is derived from an aberrant accumulation of T_{reg} cells [46], which function as a suppressive cell of the adaptive immune system, and are characterized by the expression of the master transcription factor Foxp3 [47]. Naturally occurring T_{reg} cells develop in the thymus (tT_{reg}) and generally possess TCRs with an affinity for self-antigens [48]. Induced T_{reg} (iT_{reg}) cells develop from native $CD4^+$ T cells in the presence of suppressive cytokines such as TGF-β. They function by inducing peripheral tolerance, acting as a sink for IL-2 (a key survival cytokine for other T cells such as Th1 cells, important for inflammation), via their high expression of CD25 and IL-2Rα, as well as via the expression of IL-10 and TGF-β, which contribute to their direct suppression of cytotoxic responses [49, 50]. Interestingly, T_{reg} cells within the TME are more suppressive than their counterparts in healthy tissue; this is thought to be a consequence of the secretion of suppressive factors by tumors [51]. The net effect of this increase of T_{reg} cells, both in abundance and suppressive capability, is immunosuppression both directly on effector T cells and indirectly via the induction of MDSCs and M2-like macrophages [52, 53]. T_{reg} cells also suppress cytotoxic responses in a CTLA-4-dependent manner, as T_{reg} CTLA-4 downregulates the co-stimulatory ligands' CD80/86 expression by antigen-presenting cells (APCs) [54], thus reducing co-stimulatory signals for effector T cells. T_{reg} cells are now being targeted with depletion strategies such as anti-CTLA-4 and agonists for other stimulatory molecules such as GITR [55].

As previously noted, around 40% of patients respond to checkpoint blockade immunotherapy [21, 56]. It is likely that no single-agent immunotherapy will be effective against all branches of the immunosuppressive TME, and that a combination of agents will need to target multiple pathways to improve the efficacy for the majority of patients.

A recent focus on the cellular metabolism of both cancer and immune cells has demonstrated the vital importance of metabolic competition and nutrient availability in the generation of an effective antitumor response. This field of immunometabolism has offered new insights into the function of leukocytes and has implications for traditional anticancer therapy as well as novel immunotherapies.

2 Immunometabolism

Lymphocytes are necessary for successful human existence; they are the main branch of the adaptive immune system, a diverse and powerful defense force. Once activated, T cells undergo complex intracellular reprogramming resulting in a rapid increase in cellular metabolism, and eventually the appropriate immune response to resolve the issue at hand. Eventually, the immune response must be resolved to prevent tissue damage and autoimmunity. The removal of inflammatory cells is coupled with the persistence of memory cells to allow for a rapid response to a recurring challenge.

Metabolism has long been seen as a bystander effect of cellular processes, a passenger that follows the function of cells. In recent years studies have begun to show that metabolism plays a more active role in cellular function than previously thought, guiding translational programs by providing metabolic intermediates that feed into signaling pathways, as well as allowing for the production of vital components for the immune response, such as fatty acids for cell membranes, amino acids for protein production, and ribose sugars and nucleotides for DNA synthesis [57]. An overview of cellular metabolism is given in Fig. 1.

Metabolism is a product of the microenvironment in which the cell finds itself, i.e. nutrient availability, and the sum of other inputs such as cytokines, chemokines and growth factors. In the early days of immunology, very little consideration was given to how metabolic pathways could guide functions. There was considerable interest in the mechanistic target of rapamycin (mTOR) as it was shown to be a potent immune-regulatory agent [58], and in AMPK, which seemed to be active during nutrient deprivation to limit immune cell activation [59]. However, this area was not explored thoroughly for a number of years.

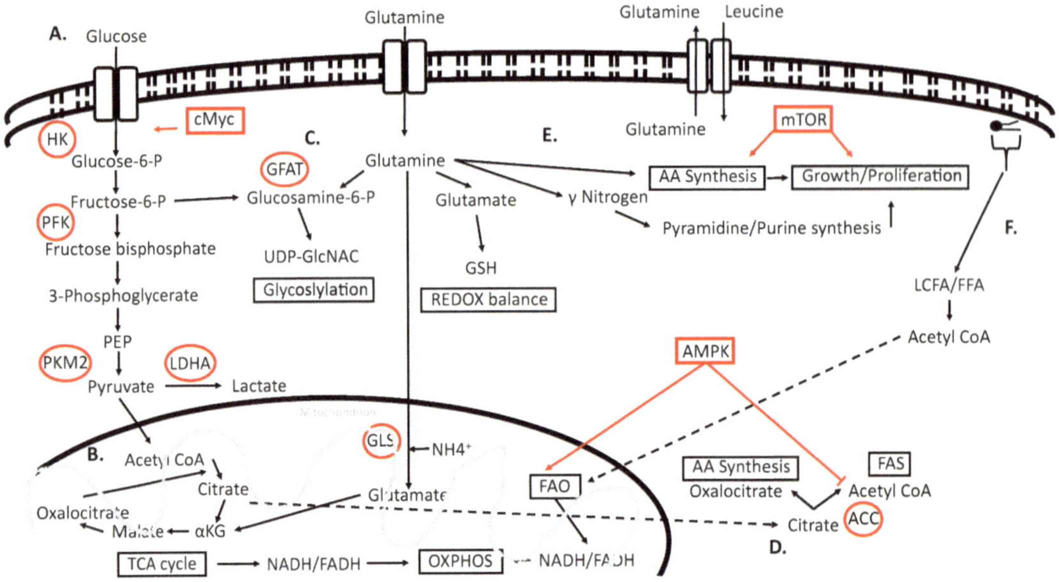

Fig. 1 Overview of cellular metabolism. (**a**) Glucose is the primary fuel for cellular metabolism and is metabolized through two main pathways: glycolysis and oxidative phosphorylation (OXPHOS). Glucose enters the cell via glucose transporters and is subsequently broken down to pyruvate. (**b**) Pyruvate is taken into the mitochondria and converted to acetyl coenzyme A via the addition of coenzyme A, which subsequently enters the tricarboxylic acid cycle (TCA cycle). In the TCA cycle the sequential oxidation of acetyl CoA products yields $NADH^+$ and CO_2. $NADH^+$ is then processed by the mitochondrial enzyme complexes I–IV which create a proton gradient in the intermembrane space. This electrochemical gradient is then used to generate ATP via the ATP synthase enzyme. Various breakpoints exist where carbon products from glucose can be shuttled away from this pathway and utilized in other cellular products. (**c**) Glutamine imported via the glutamine transporters can be converted to glucosamine-6-phosphate, via the conversion of fructose-6-phosphate with the enzyme GFAT. UDP-GlcNAC participates in protein glycosylation and regulates the activity of many proteins involved in T-cell activation. (**d**) Citrate derived from the TCA cycle can be exported from the mitochondria and via the enzyme acetyl coenzyme carboxylase (ACC) is converted to acetyl CoA which is then used to synthesize new fatty acids, important for cell growth and proliferation. (**e**) Glutamine participates in the synthesis of new amino acids via a transamination reaction; similarly glutamine taken up from the microenvironment possesses a gamma amide group which is a crucial donor of nitrogen for nucleotide synthesis. (**f**) Phospholipids from the cell membrane are converted to acetyl CoA and imported to the mitochondria, whereby they are oxidized (fatty acid oxidation), which yields $NADH^+$ and $FADH^+$ which are also used to generate ATP via ATP synthase and the electrochemical proton gradient in the mitochondria

As our technology has advanced substantially, the real-time monitoring of metabolism has made it possible to probe various metabolic processes and their effects on immune cell function. High-throughput metabolomics has enabled the discovery of metabolites that are necessary for various subsets of immune cells. Consequently, we are beginning to understand how metabolic pathways impact immune cells' functions from glycolysis through oxidative phosphorylation (OXPHOS) and fatty acid oxidation (FAO). We can now measure how differences in amino acid metabolism result in the emergence of functionally distinct subtypes of immune cells.

At the core of immunometabolism, particularly in the context of cancer immunotherapy, it is important to familiarize oneself with the following paradigm: quiescent or inactive immune cells are metabolically dormant, producing low amounts of energy primarily through catabolic metabolic pathways, mainly OXPHOS [60]. An immunological challenge results in a cascade of signals, inducing rapid metabolic reprogramming. Without this initial burst of energy, there would be no immune response, and as we begin to deepen our understanding of immunometabolism, we are discovering that metabolism is fundamental to the initiation, guidance, and resolution of immune responses. Another fundamental of immunometabolism (in the specific context of cancer) is that cancer cells and immune cells share nutritional requirements; this results in the establishment of competition for nutrients (a competition which the lymphocytes generally lose). Alterations in the levels of metabolites at activation, or indeed, upon infiltration to peripheral tumors, can have deleterious effects on the functions of immune cells, including T cells [61].

In this chapter, we look deeply into T-lymphocyte metabolism; however, many of the pathways discussed hold true for other cell types, such as macrophages and NK cells.

A naïve T cell lies in wait for its cognate antigen, which when presented with the relevant co-stimulation and cytokine signaling will undergo clonal expansion and take up an effector phenotype [62]. Naïve $CD4^+$ and $CD8^+$ T cells are long-lived resting cells with low bioenergetic demands, needing only to maintain homeostasis [63]. Thus, naïve T cells mainly use OXPHOS as their primary source of ATP; this incorporates low levels of glycolysis-derived pyruvate, as well as glutaminolysis-derived α-ketoglutarate (α-Ket), and some FAO to run electrons along the electron transport chain [64, 65]. This reliance on OXPHOS is best illustrated by the fact that naïve $CD4^+$ and $CD8^+$ T cells survive in conditions of glucose deprivation, but not in hypoxia [57, 66].

2.1 T-Cell Activation: A Major Metabolic Event

When presented with a pathogen or other immunogenic challenge, such as malignant neo-antigens, T cells activate and perform their effector functions, which involve clonal expansion, expression of chemokines and chemokine receptors to facilitate migration, and production and release of soluble mediators of immunity. The process of T-cell activation involves signaling from the T-cell receptor (TCR), as well as appropriate co-stimulation; without both of these signals, T cells fail to activate and a state of anergy is induced in the T cell. T-cell activation is metabolically demanding and requires increased nutrient uptake from the environment. While OXPHOS is the most efficient process by which cells make ATP, Otto Warburg observed that activated lymphocytes primarily use glycolysis, even in the presence of oxygen (termed aerobic glycolysis), as their main metabolic pathway [67, 68]. In a striking similarity to cancer cells, this cellular "decision" to create ATP via glycolysis seemed to make little sense, with only 2 ATP molecules being created per molecule of glucose, compared to 32 per glucose molecule in OXPHOS [57, 69]. By utilizing glycolysis, ATP is produced at a higher rate and the end products (pyruvate and NADH) are available for a number of further cellular processes, including shuttling into the TCA cycle and OXPHOS, enzymatic cofactors, and biosynthesis of products involved in growth proliferation and cytokine release [70].

T-cell activation induces a metabolic switch from OXPHOS to glycolysis as the major metabolic pathway being utilized by the cell; this allows for the production of large amounts of biosynthetic intermediates to be produced, which are then available for shuttling to other important anabolic pathways in the cell in preparation for effector functions [70]. Signaling through the TCR induces the transcription factor c-myc, which directly results in the translation of glycolytic enzymes [63], as well as the expression of key nutrient transporters, such as GLUT1 and amino acid transporters [71]. In addition, signaling through CD28 also regulates glycolytic metabolism via activation of the phosphatidylinositol 3′-kinase (PI3 K)-Akt pathway, and this amplifies the signal from the TCR via activation of the mechanistic target of rapamycin (mTOR) [72, 73]. Interestingly, naïve T cells have the mRNA of glycolytic proteins expressed at steady levels, but these mRNAs are prevented from being translated; activation of the TCR and co-stimulatory pathways induces the rapid translation of these mRNAs, suggesting that naïve T cells sit ready for rapid activation, and that this metabolic switch lies at the heart of early T-cell activation [74]. Glycolytic reprogramming is important not only for the availability of precursor metabolites to be shunted into biosynthetic and growth pathways, but also for the activity of glycolysis and glycolytic enzymes to regulate the release of cytokines (important for the differentiation and emergence of effector subsets). Specifically, GAPDH is an enzyme involved in glycolysis which binds to the 3′ UTR AU-rich sequences in IL-2 and IFN-γ mRNA, preventing them from being translated [74]. On initiation of glycolysis during activation, GAPDH engages its enzymatic activity, losing its ability to prevent the translation of IL-2 or IFN-γ [75].

This may be in part due to the fact that glutamine is involved in the de novo synthesis of purine and pyrimidine synthesis, which is necessary for cell growth and proliferation. Glutamine metabolism also fuels the enzyme O-GlcNAc transferase, which regulates O-GlcNAcylation, a process by which changes in glutamine availability in the T cell, modifies T-cell biology. T cells deficient in glutamine display dysfunctional O-GlcNAcylation and fail to activate MYC, and subsequently fail to activate [76, 77]. Therefore, glutamine is an essential metabolite for the activation of lymphocytes.

Mitochondrial metabolism is also increased during the activation of lymphocytes, although this does not reach the level of glycolytic metabolism. Interestingly, the mitochondria are found closer to the immune synapse following T-cell activation [78]. Rather than generating ATP via OXPHOS, the mitochondria in activating T cells play a role in one carbon metabolism via the TCA cycle and generate a number of intermediates available for lipid biosynthesis [79]. Mitochondria also produce reactive oxygen species (ROS), which play a signaling role in T-cell activation via their regulation of NFAT and IL-2 [80]. In addition, ROS are required for activation, demonstrated by the fact that T cells without ROS lose their ability to activate [81]. The role of ROS in T-cell activation is complicated, as chronic or high levels of ROS result in the failure to activate; similarly the inhibition of glutamate-cysteine ligase (the rate-limiting enzyme in glutathione synthesis) increases ROS and reduces mTORC1 activity [82]; therefore the status of cellular REDOX is important in activation [83].

Fatty acid (FA) metabolism plays a central role in cells that are rapidly proliferating by synthesizing new FAs to be incorporated into daughter cell membranes. Upon activation, T cells upregulate metabolic pathways associated with de novo fatty acid synthesis (FAS), as well as with a reduction in pathways relating to FAO [63]. This FAS is related to the myc transcription factor, which is downstream of the mTORC1, with constitutive mTORC1 activity also driving FAS [84]. Inhibition of the FAS pathway reduces the proliferation of both $CD8^+$ and $CD4^+$ T cells, and inhibits the generation of both human and murine effector T cells [85, 86]. An overview of how extracellular signals and T-cell activation influence metabolic pathways is given in Fig. 2.

2.2 T-Cell Differentiation: Metabolism Acts as a Guide

Metabolism is thought to be not only important in the initial reprogramming during lymphocyte activation, but it also plays a role in the differentiation of distinct effector subtypes in both $CD4^+$ and $CD8^+$ T cells. In general, effector subsets tend to display a higher dependency on glycolytic metabolism and suppressive subtypes (T_{regs}) rely less on glycolysis post-activation, favoring the use of OXPHOS and FAO instead [87].

As previously mentioned, the activity of glycolytic enzymes intrinsically links glycolysis with effector functions via the modification of mRNA encoding IFN-γ and IL-2 [88]. In murine Th1 effector cells, the glycolytic intermediate 2-phosphoenolpyruvate (PEP) prevents Ca^{2+} uptake into the endoplasmic reticulum, maintaining TCR-mediated Ca^{2+} signaling, and thereby sustaining the activation of NFAT, aiding in the generation of an effector's function (the overexpression of PEP enhances the function of $CD4^+$ and $CD8^+$ T cells) [89]. Similarly, pyruvate, the final metabolite of glycolysis, is converted to acetyl-coenzyme A (A-CoA) via the enzyme pyruvate dehydrogenase (PDH). This step promotes the

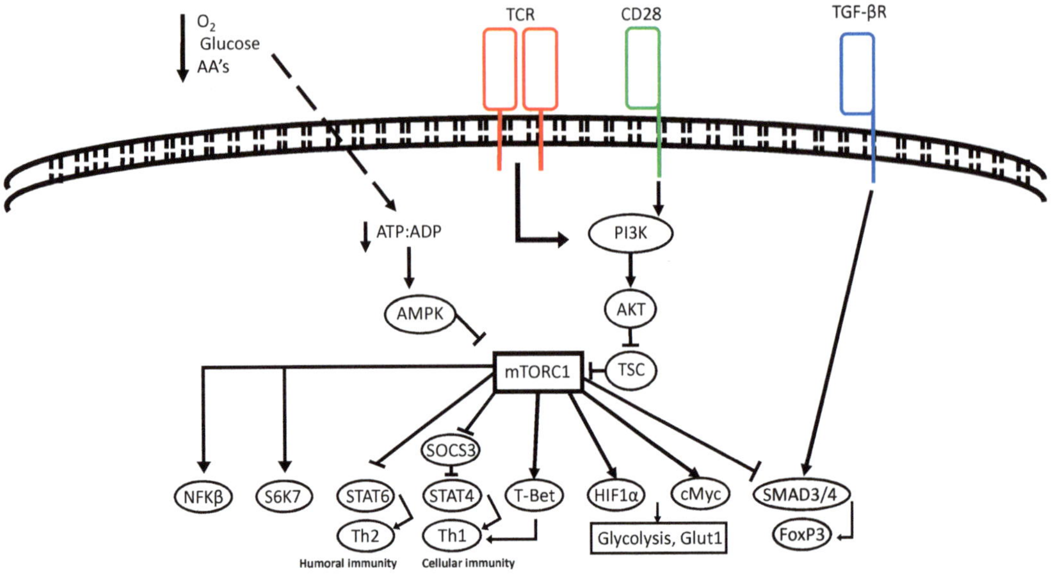

Fig. 2 T-cell activation, metabolism, and feedback from nutrient availability. When the T-cell receptor (TCR) is provided with its cognate antigen in the presence of sufficient co-stimulation (signals 1 and 2) the signaling cascade ultimately results in the stabilization of mechanistic target of rapamycin complex 1 (mTORC1). mTORC1 has a number of key downstream effector functions including engaging the glycolytic machinery via activation of HIF1α and cMyc. Other targets are associated with a pro-inflammatory immune response such as T-bet, the master transcription factor of cellular mediated immunity in $CD4^+$ and $CD8^+$ T cells. SMAD3/4 is blocked by the activation of mTORC1 resulting in the abrogation of FoxP3 (the transcription factor guiding T_{reg} induction). Similarly STAT6, involved in humoral or Th2-like immunity, is also blocked by mTOCR1 activity. Environmental factors and exogenous signaling can feed into the TCR activation signal; for example TGF-β, frequently found within the TME, results in the activation of SMAD3/4 which can overcome mTORC1 signaling and skew a T_{reg}-like response. Similarly, lack of key nutrients such as glucose or amino acids like glutamine results in the reduction of the ATP:ADP ratio; this reduced intracellular concentration of ATP activates AMPK, which blocks mTORC1 signals and programs its own distinct set of metabolic processes. Thus upon activation of the TCR the microenvironment plays a pivotal role in determining which immunity program is activated

oxidative metabolism of glucose and is itself controlled by the PDH kinase 1 (PDHK1), which inhibits PDH and promotes the glycolytic pathway [90]. The inhibition of PDHK1 promotes the skewing of effector responses toward a T_{reg}-like response [90, 91]. The importance of glycolysis and glucose availability in the generation of robust antitumor responses is demonstrated in studies that have shown that reduced glucose availability suppresses Ca^{2+} signaling, IFN-γ production, cytotoxicity, and motility in T cells—all hallmarks of the anticancer immune response [92, 93].

Similarly, amino acid metabolism plays a role in the determination of effector subtypes. Glutamine is not only necessary for the activation of T cells; glutamine metabolism also plays a distinct role in the differentiation of different effector subtypes [57]. In an elegant demonstration of glutamine's role in differentiation, Metzler et al. demonstrated that as glutamine concentrations decrease, the proportion of T cells expressing FoxP3 (the master transcription factor for T_{reg} cells) increases, as too does their suppressive ability [94]. This phenomenon may help to understand how T_{reg} populations may emerge in tumors with glutamine addiction, or when antimetabolite chemotherapeutics such as 5-FU are used. A recent study by Johnson et al. revealed that where glutamine metabolism is necessary for Th1 and Th17 differentiation, the amount of T_{regs} was significantly higher when glutamine was absent, even in the presence of cytokines that would normally guide the differentiation of Th1 and Th17 T cells [95]. The amino acids L-arginine and L-tryptophan are also known to be of vital importance to T-cell differentiation, particularly in the context of antitumor immunity. Enzymes catabolizing both L-arginine and L-tryptophan (ARG1 and IDO, respectively) are frequently expressed in tumors [96]; however, it is now recognized that the elevated activity of these enzymes is derived largely from polarized suppressive cells within the stroma, namely TAMs, MDSCs and cancer-associated fibroblasts (CAFs) [97]. L-arginine promotes T-cell function by inducing a central memory-like phenotype, which ensures longevity and stronger antitumor responses [98]. Conversely, the absence of L-arginine prevents T-cell proliferation and promotes the downregulation of the TCR ζ chain [99]. Tryptophan is another amino acid which is important in generating immune responses; it is the rarest of the essential amino acids, and 90% of the tryptophan consumed in humans is metabolized by the kynurenine pathway, which is essential for producing NAD^+. NAD^+ is an important cofactor in many cellular reactions including DNA damage and cellular metabolism [100]. The depletion of tryptophan in the context of T-cell responses induces regulatory responses via the polarization toward a T_{reg} phenotype, as well as the downregulation of the TCR ζ chain [101], thus blunting antigen-driven signaling.

The role of mitochondrial respiration in driving T-cell responses has conflicting reports, with some studies suggesting that mitochondrial metabolism is necessary for driving antitumor immunity, and others proposing that it plays a role in the generation of a suppressive immune response [102]. T cells with a knockout for Tfam (a mitochondrial transcription factor controlling mitochondrial biogenesis) possess dysfunctional mitochondria; this phenotype promotes a more glycolytic profile associated with a higher production of Th1-related cytokine IFN-γ [103]. This would suggest that Th1-like responses depend less on mitochondrial metabolism; this is supported by the fact that T cells deficient in HIF1-α rely on OXPHOS, which promotes a T_{reg} phenotype [104]. Furthermore, T cells expressing FoxP3 suppress myc and concurrently upregulate mitochondrial complexes associated with OXPHOS [105]. Interestingly, functionally competent mitochondria are required for the generation of both regulatory and inflammatory T-cell responses; persistent mitochondrial dysfunction leads to a reduction in PGC1α (another key transcription factor in mitochondrial biogenesis) [106]. This results in a gradual reduction of proliferation and the production of IFN-γ [107]. These studies, when taken together with the necessity for mitochondrial activity during T-cell activation, suggest that mitochondrial metabolism is important for the function of T cells; however the role of OXPHOS is more complex. It is likely that the traditional view of glycolysis being pro-inflammatory and OXPHOS being anti-inflammatory is oversimplified and that the balance of each pathway is more important in the differentiation of effector functions. Therefore, therapeutic targeting of mitochondrial metabolism could inadvertently induce immune dysfunction via other mechanisms.

FA metabolism also plays a role in the generation of effector subsets, subsequent to the upregulation of FAS; during activation, different effector T cells depend on varying amounts and types of FA metabolism. In $CD4^+$ T cells, the balance of FAO and FAS is a critical determinant of T_{eff} versus T_{reg} phenotype [108]. Th1 and Th17 T cells direct glucose to the TCA cycle, shunting biosynthetic intermediates toward FAS, which promotes the effector function and proliferation of those cells. When inhibited, these cells had a marked decrease in Th1, Th17 and to a lesser extent Th2, while T_{reg} development was unaltered [86]. The differentiation of T_{regs}, on the other hand, was abrogated by the addition of etomoxir, a selective inhibitor of CPT1 (an enzyme involved in FAO), which suggests that they depend on the oxidation of lipids to develop after activation [109]. FAO is also essential in the generation of $CD8^+$ memory cells [110]; the progression from the $CD8^+$ T_{eff} cell toward a T_{mem} phenotype involves metabolic reprogramming from FAS to FAO, a process which involves AMPK signaling [111]. Similarly, in $CD4^+$ T cells, the shift toward a memory phenotype is influenced by FA metabolism, the inhibition of ACC1, and the rate-limiting enzyme of FAS, which promotes a $CD4^+$ T_{mem} phenotype [112]. These data suggest a critical role for FA metabolism (in particular the oxidation of lipids) in the generation of immunological memory. An overview of metabolic pathways active during different stages of T-cell activity is given in Fig. 3.

The cellular "choice" of metabolic pathways plays a critical role in the determination of what type of immune response is elicited from T-cell activation. The traditional view of glycolytic metabolism being inflammatory and oxidative metabolism being anti-inflammatory is outdated. It is perhaps more accurate to say that anabolic metabolism, as elicited by the activation of mTOR, is pro-inflammatory, while catabolic metabolism elicited by AMPK drives a more anti-inflammatory response (Fig. 4). The activation of mTOR versus AMPK is dependent on the environment in which

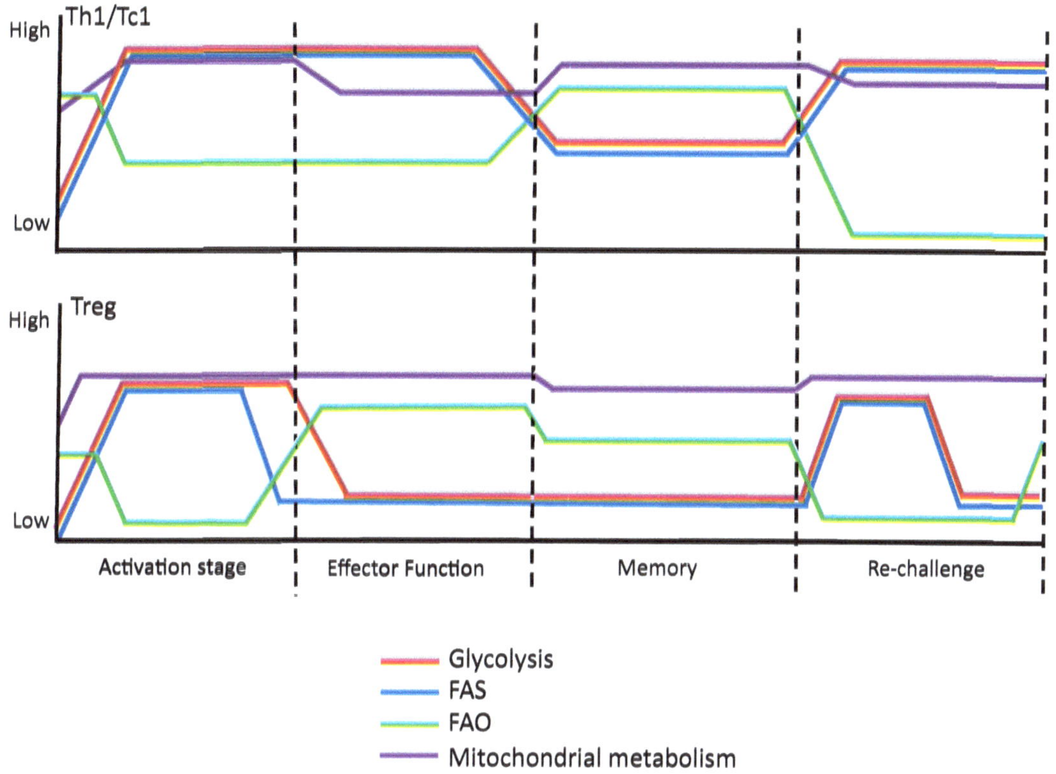

Fig. 3 Metabolic pathways active during different stages of T-cell activity. Quiescent or naïve T cells exhibit low levels of glycolysis and fatty acid synthesis (FAS); their ATP is generally generated by oxidative phosphorylation (OXPHOS) and fatty acid oxidation (FAO). Upon activation and co-stimulation all T cells must undergo rapid metabolic reprograming, whereby glycolysis becomes the predominant pathway; however increased mitochondrial metabolism is also necessary for activation which assists in activation by locating close to the immune synapse and producing reactive oxygen species (ROS). FAS is also increased upon activation and is driven by mTOR; FAS is necessary for both $CD4^+$ and $CD8^+$ activation. During effector functions, metabolic programs differ between pro-inflammatory cells and regulatory or suppressive cells. mTOR-driven programs such as glycolysis and FAS are lower while AMPK-driven programs such as OXPHOS and FAO are higher. This prevents cells from producing pro-inflammatory mediators such as IL-2 or IFN-γ. mTOR-driven metabolism allows pro-inflammatory subsets to undergo aggressive production of amino acids for protein production as well as de novo synthesis necessary for the production of new membrane lipids; these processes allow for the clonal expansion associated with T-cell activation. Upon resolution of an immune response some cells persist as memory cells. Memory T-cell differentiation is associated with the return to a more quiescent-like metabolic program, driven mainly by FAO and OXPHOS; as memory cells are long lived and have little bioenergetics requirements, ATP can be efficiently generated through these pathways. Upon reactivation in response to antigen the memory T cells exit this quiescent metabolic state and require the metabolic programming of their previous effector state; however this process is still poorly understood

the activating and differentiating T cell finds itself. Our immune system evolved to respond to infectious pathogens and to limit damage to itself, and as such has many layers of suppressive tolerance to prevent autoimmunity. It is possible that immunometabolism is an additional fundamental mechanism for tolerance,

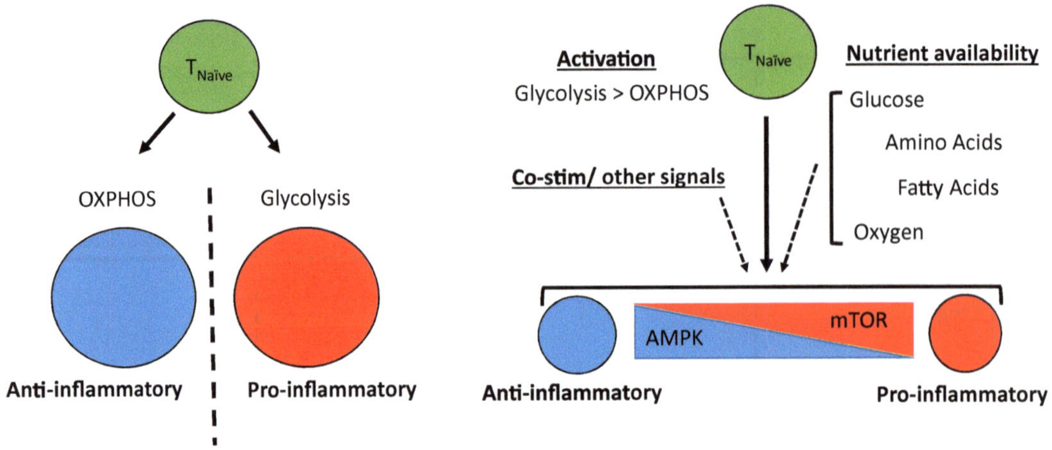

Fig. 4 Classical versus emerging model of how cellular metabolism drives immunity. Historically, metabolism of immune cell subsets was seen as a consequence of a particular set of transcriptional programs. The general observation that glycolysis appeared higher in pro-inflammatory subsets when compared to anti-inflammatory subsets resulted in the hypothesis that these metabolic pathways were associated with that phenotype. However, it is now known that metabolism is more intrinsically connected with immune responses, feedback from co-stimulation, and cytokine signaling as well as the microenvironment along with classical activation signals drives specific metabolic programs, often involving AMPK and mTOR pathways, which in turn allow subset-specific responses to develop. It is likely that immune responses exist in a spectrum and are subject to change given a particular set of environmental factors

whereby progressing inflammation depletes metabolites in the microenvironment, resulting in the gradual generation of regulatory and suppressive cells. In cancer, the unfortunate shared nutritional needs between cancer and immune cells drive an accelerated tolerance response ultimately to the detriment of the host.

3 Targeting Immunometabolism in Cancer Therapy

Given the essential role metabolism plays in the generation of a robust immune response, many research groups are targeting immunometabolism as a novel therapeutic approach in cancer treatment. Given the limitations of conventional immune checkpoint blockades, new approaches are required to overcome suppressive mechanisms, which result in the lack of therapeutic efficacy.

Some of the most common mutations driving cancer are genes regulating metabolic pathways; metabolic dysregulation is a hallmark of cancer [113]. These oncogenic driver mutations include MYC, PTEN, and PI3K, and confer metabolic advantages to cancer cells, allowing for rapid proliferation and resulting in a subsequent depletion of metabolites in the TME, which varies in different tumors (and sometimes even within the same tumor owing to inherent heterogeneity) [114]. The net result is the creation of a metabolically challenging TME for newly activated T cells; when coupled with existing suppressive mechanisms, this undermines the dysfunctional immune response characteristic of TILs.

Ironically, since the advent of the ICI drugs, novel signaling roles have emerged, highlighting that the checkpoint receptors and their ligands themselves often regulate cellular metabolism, both in T cells and in cancer cells themselves. Both CTLA-4 and PD-1 receptors interfere with the signaling of CD28 co-stimulation, which promotes effector functions via the promotion of a mTOR-driven metabolic profile [73]. It would appear that the link between ICI therapy and cellular metabolism is an important mechanism of action, as treatment with these ICI inhibitors restores T-cell metabolism to promote an antitumor response, while anti-PD-1 therapy promotes glycolytic metabolism and FAS [115]. Similarly with other negative checkpoint molecules new metabolic roles are emerging, with CTLA-4 inhibiting PI3K, thereby blocking glycolytic metabolism [116]; LAG-3 maintaining quiescence in T cells by blocking glycolytic and mitochondrial metabolism, potentially via elevated PTEN signaling [117]; and TIGIT blocking glycolysis and effector functions in $CD8^+$ T cells by downregulating pS6K, a downstream target of mTOR [118]. Similarly the stimulatory immune checkpoint molecules, such as GITR, have shown to promote T-cell glycolysis and mitochondrial metabolism by signaling through pS6K and phosphor-4EBP1, another downstream target of mTOR, thereby promoting antitumor effector functions [119]. Thus, by blocking inhibitory checkpoints, or antagonizing stimulatory checkpoints, it is likely that this is modulating T-cell immunometabolism to promote effector functions. Investigating these metabolic effects closely will help in the development of novel combination therapies to ensure that the right metabolic profile is enhanced to promote an appropriate and robust antitumor response. Interestingly, PD-L1 expression in cancer cells has been associated with the promotion of glycolysis via activation of PI3K/AKT [120]. This suggests that inhibition of both PD-1 and its ligand could synergize by reducing cancer cell glycolytic metabolism, increasing T-cell metabolism and potentially freeing up glucose in the TME to be utilized by TILs; this has been supported in one study to date, suggesting that further studies are warranted in this area [121]. The ongoing clinical trials in checkpoint inhibitors were not designed on the basis of their metabolic reprogramming ability and thus the design of new trials will need to take this into account.

There are ongoing clinical trials targeting various metabolic pathways which are altered and have opposing effects in both tumor and immune biology. Targeting shared pathways may result in inadvertent reductions in efficacy; for example, broadly targeting the promotion of glycolysis, while mechanistically promoting anti-tumor immunity, may instead result in an increase in tumor glycolysis and promote growth and proliferation of the tumor. Instead, groups are beginning to look at lactate, the product of aerobic glycolysis, in both tumor and immune cells. Accumulation of lactate in the TME is another mechanism of immunosuppression utilized by tumors; the lactate can be used as an alternative fuel for cancer cells and can also suppress the TILs' ability to respond to the tumor [122, 123]. Inhibition of the enzyme controlling lactate production, lactate dehydrogenase (LDHA), or transporter associated with the uptake into cells (MCT1) is therefore a promising therapeutic target currently under review in clinical trials. Preclinical evidence supports the targeting of this metabolic pathway, as low levels of LDH are associated with a better response to PD-1 inhibition in metastatic melanoma patients and treating melanoma cells with LDHA inhibitors significantly increases tumor cell destruction by T cells [121, 124].

As discussed previously, glutamine metabolism is vital in the activation and generation of effective antitumor responses. Glutamine is also an important fuel for tumor metabolism given its role in the production of metabolic intermediates involved in cell proliferation. The enzyme glutaminase (GLS) is involved in the metabolism of glutamine and has shown promise in vitro in various cancers, including breast cancer, pancreatic cancer, and leukemia [125, 126]. GLS plays a role in guiding the differentiation of effector T cells, as the ablation of GLS promotes Th1-like responses over the generation of Th17/T_{reg} [95]. GLS, therefore, was a logical target in the manipulation of metabolism; CB-839 is a selective inhibitor of GLS that synergizes with anti-PD-1 therapy in several malignancies (such as triple-negative breast and clear-cell renal carcinoma) [127], although an exact mechanism for this synergism has yet to be established. It is likely that the antitumor effects of GLS inhibition work in harmony with the promotion of CTL responses and allow further cytotoxicity when combined with the inhibition of PD-1.

There is some evidence for targeting T-cell fatty acid metabolism, as different subsets preferentially use FAS over FAO. However, as FAO is important for memory cell differentiation and T_{mem} cells are important in the long-term antitumor response, inhibiting FAO may prove ineffective. Understanding how suppressive subsets utilize fatty acids may provide new therapeutic targets. T_{reg} cells undergoing activation require FAS initially; however, unlike their Th1/Th17 counterparts, T_{regs} do not depend on de novo synthesis; instead, they uptake FAs from the microenvironment [86]. Therefore, targeting FA uptake in T cells may encourage effector subsets to differentiate within the tumor and decrease T_{reg}-mediated suppression.

Previously, we discussed how mitochondrial metabolism is necessary for the generation of a robust immune response to cancer, but as the immune response progresses, so too does mitochondrial dysfunction, promoting an exhaustion phenotype [128]. PD-1 signaling represses PCG1α signaling and reduced mitochondrial biogenesis; TIL's display reduced mitochondrial metabolism and the blockade of PD-1 rescues their mitochondrial function [129, 130]. Therefore, there is potential in promoting mitochondrial metabolism for improving immunotherapy; given the role anti-PD-1 therapy plays in improving mitochondrial metabolism, it is possible that other inhibitory checkpoints may also promote mitochondrial functions and improve T-cell effector and memory formation. There is therefore an impetus to understand the metabolic effects of checkpoint blockades on the various stages of an immune cycle, including differentiation and memory formation.

3.1 Immunometabolism and Adoptive Cell Transfer (ACT)

Additional immunotherapies to ICIs have been trialed with varying degrees of success. Of note is the advent of the chimeric antigen receptor (CAR) T cells. CAR-Ts employ engineered T cells to target a known tumor-associated antigen, resulting in the destruction of cells expressing that antigen. CAR-Ts have shown the most success in dealing with hematological malignancies, where they have proven to be a relatively safe and sometimes curative therapy, particularly for $CD19^+$ B-cell malignancies [131]. In solid malignancies, CARs have struggled to show effectiveness, in part due to the heterogeneity of tumor antigen expression, as well as challenges associated with recruitment and persistence within the tumor [132]. As CAR-Ts are genetically engineered, newer generations of CARs are beginning to target the TME in an attempt to overcome suppressive signaling. A recent report highlighted a new CAR T cell that expressed the extracellular domain of the TGF-β receptor fused to the intracellular domain of the 4-1BB stimulatory receptor, which produces potent antitumor responses to stimuli that physiologically would induce immunosuppression [133]. This concept can be expanded to a wide variety of potential uses; for example, engineering the overexpression of GLUT1 may improve metabolic competition-driving effector responses against cancer. Similarly, by joining the extracellular adenosine receptor domain to the intracellular domain of GITR, this may drive favorable immunometabolism to produce an immune-stimulatory response to a suppressive signal. The applications of this technology are widespread and increasing our understanding of immunometabolism and suppression within the TME will guide new applications.

Another form of adoptive cell transfer (ACT) involves the isolation of TILs from patients and a subsequent ex vivo expansion and reinfusion. TIL ACT has shown some encouraging results in melanoma, with response rates between 40% and 70% [134]. This is being trialed in other malignancies such as breast cancer, ovarian cancer and renal cell carcinoma [135, 136]. Traditionally, TILs are expanded via high-dose IL-2 treatments, followed by CD3 stimulation. IL-2 is known to be a driver of T_{eff} subtypes via the promotion of a metabolic phenotype favoring glycolysis and mTOR metabolism [137]. This effector type of T cell is known to be suboptimal for ACT, where a central memory-like phenotype produces better antitumor results [138]. Therefore, culturing T cells isolated from tumors in IL-21 and IL-15, which are known to improve mitochondrial metabolism in T cells and promote a less differentiated T_{CM}-like phenotype, may prove to be superior for a graft versus tumor response [139]. Likewise, it is important to consider the levels of nutrients in which TILs are expanded; cell culture media often contains 10–25 mM of glucose, whereas in blood, glucose levels are around 5 mM. This may result in the generation of T cells that are dependent on glucose and may be unable to function in the TME that is nutrient deprived.

4 Discussion and Future Prospects

The field of immunotherapy is still relatively young, and yet has demonstrated remarkable anticancer properties. As our knowledge of how the immune system develops and responds to tumors deepens, so too does our ability to modulate it in order to promote more potent and robust antitumor responses. The advent of immune checkpoint inhibitors revolutionized cancer therapy by serving as a platform to demonstrate the power of the immune system. However, it is estimated that around 40% of patients will respond to ICI therapy, suggesting that resistance is a major barrier [19, 140]. With the development of new and improved combinations of both antagonizing inhibitory checkpoints and antagonizing stimulatory receptors, we will likely continue to see improved results.

However, the fact remains that the TME is an unfavorable environment for immune cells, characterized by various stressors such as hypoxia, nutrient deprivation, and cancer/stromal derived suppressive signaling. Over time, this challenging milieu recruits immune cells that themselves become aiders and abettors of tumor growth and metastasis—MDSCs, TAMs, and T_{regs} working together to abrogate immunity and promote local and distant metastasis by inducing the breakdown of the extracellular matrix, angiogenesis, and epithelial to mesenchymal transition (EMT).

Further complicating matters, nutritional needs are shared in proliferating cells—including tumor cells and activated immune cells. The resulting competition places additional selective pressures on any developing immune response, which inevitably results in the generation of metabolically dysfunctional TILs, often forced to generate ATP from whatever nutrients are available. This lack of nutrients results in different metabolic states of TILs, which tend to prefer catabolic metabolism due to a lack of exogenous uptake, and is primarily AMPK driven and results in metabolic programs such as

FAO and OXPHOS that ultimately induce suppressive or dysfunctional cells. Thus, altered immunometabolism is a driving factor in resistance to immunotherapy.

Targeting TIL metabolism can boost their ability to use nutrients within the TME, resulting in better anticancer immune responses. The mechanisms of the ICI therapies themselves have been demonstrated to act through metabolic pathways, demonstrating the potency of improving metabolic fitness. Glycolysis is a prerequisite for anticancer immunity by T cells as well as M1 macrophages; therefore, strategies that increase immune cell competition for glucose or decrease cancer cells' affinity for glucose may promote stronger antitumor responses. Simultaneously, it is not as simple as promoting glycolysis over mitochondrial metabolism; functional and healthy mitochondria are necessary, not only for the activation of lymphocytes but also for the differentiation of robust effector cells. Ensuring that T cells engage in mitochondrial biogenesis and prevent the degradation of mitochondria (mitophagy) is becoming an important area of research; a recent study has demonstrated that by knocking out *atg5* in mice, the $CD8^+$ T cells improve antitumor immunity via an increase in glycolytic metabolism and increase the production of IFN-γ [141]. Combining ICI drugs that block metabolic pathways associated with suppressive immune responses with antagonists for the stimulatory checkpoints that promote metabolism linked to antitumor immunity will ensure that the most potent antitumor response is elicited from therapy (Fig. 5).

A number of early-stage clinical trials are underway to investigate the potential of combining conventional immunotherapy with agents targeting metabolism. Glutamine metabolism is being targeted with CB-839 (a GLS inhibitor) in combination with nivolumab in advanced renal cell carcinoma, melanoma, and NSCLC (NCT02771626). Metformin is also currently under investigation in combination with both pembrolizumab and nivolumab for use in NSCLC and melanoma (NCT03048500, NCT03311308). Targeting glucose metabolism with metformin may prove to have both favorable and unfavorable consequences; by reducing glucose consumption in cancer cells, it may free up glucose for use by TILs; however, as a potent activator of AMPK, metformin may inadvertently promote a suppressive metabolic profile in immune cells. The immunosuppressive IDO and ARG1 enzymes are frequently overexpressed in cancer and utilized by TAMs and MDSCs, and are also the target of extensive clinical trials. A recent phase III trial of epacadostat (a selective IDO1 inhibitor) in combination with pembrolizumab showed little benefit despite significant supporting preclinical data across multiple cancer types, with no significant difference between the treated group and the control [142]. Multiple hypotheses have been suggested to explain these discouraging results, with the dosage and intra-tumoral pharmacodynamics

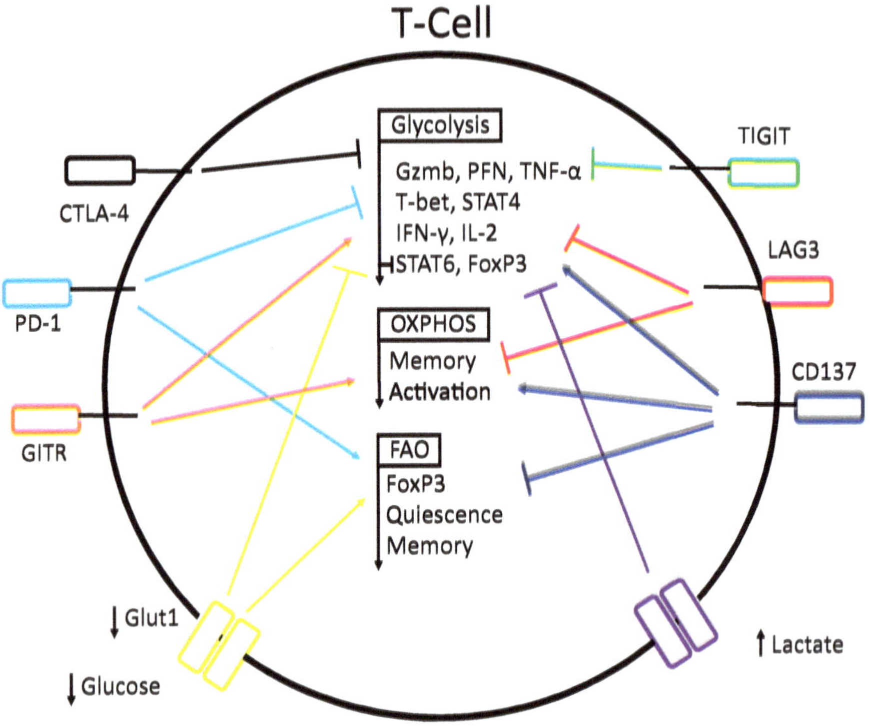

Fig. 5 Activatory and inhibitory feedback from the T-cell microenvironment. After T-cell activation and migration from the lymph node to the tumor, the tissue microenvironment changes. The tumor microenvironment (TME) is not conducive to appropriate metabolic function, in part due to the altered availability of key nutrients such as reduced glucose and increased lactate. Altered nutritional status impacts the metabolic pathways that are active within tumor-infiltrating lymphocytes (TILs) and can result in functional changes which impair antitumor immunity. Similarly, due to chronic antigen stimulation, the TILs undergo a state of differentiation that is known as exhaustion, characterized by high expression of suppressive checkpoint receptors including CTLA-4, PD-1, TIGIT, and LAG3. The ligands for these suppressive receptors are expressed in the TME and subsequent signaling downstream of these receptor-ligand interactions is now being shown to act in part by promoting metabolic pathways associated with dysfunction and immunosuppression. Conversely, the stimulatory checkpoint receptors GITR and CD137 are known to act by promoting metabolic pathways that are conducive to antitumor inflammation. Targeting these pathways alone and in combination may provide synergistic benefits to TILs and increase antitumor function. Arrowheads indicate promotion; capped lines indicate inhibition

being among the most relevant issues to address [143]. ARG1 inhibitors are also undergoing early trials in combination with immunotherapy. These trials, however, have not yet reported any effective results but demonstrate significantly elevated arginine levels in plasma [144].

In cancer, the unfortunate similarities in metabolic demands between immune cells and proliferating tumor cells mimic the conditions of tissue in need of immunosuppression, hence cancer's moniker, "the wound that never heals." In the context of infection, the resolution of the immune response is characterized by the further differentiation of some clones to a memory phenotype and awaiting antigen restimulation in order to induce a quicker, more effective adaptive response. The overarching goal of immunotherapy is to induce a durable memory response in humans, whereby tumor rejection is mediated by T_{eff} cells and long-term, progression-free survival is maintained by T_{mem}. Memory formation is at least in part metabolically governed, with certain metabolites and pathways being imperative. L-arginine is implicated in memory formation as well as effector functions, but increasing intracellular levels of L-arginine result in the promotion of a T_{cm} phenotype associated with increased OXPHOS and decreased effector cytokine production [98]. This implies that memory formation is dysfunctional within the TME. Inhibitors of ARG1 may assist in the development of memory T cells resulting in better responses to immunotherapies; there are trials in progress combining ARG1 inhibitors with ICI drugs.

In summary, there is much to still discover in the field of immuno-oncology; understanding the development of exhaustion from activation through to memory formation in the context of the TME will provide us with new therapeutic targets and opportunities for modulation. Likewise, understanding the mechanisms of existing immunotherapies on T-cell responses within the TME, as well as understanding if existing therapies such as chemotherapy and radiotherapy can be leveraged to complement the immune system, will help us to develop new regimens that have maximal impact on antitumor immunity.

It is clear now that metabolism lies at the center of the generation of an immune response. Although still poorly understood, we are beginning to identify novel targets that will bolster the therapeutic repertoire. It is dubious, however, that targeting immunometabolism alone will bring about the complete mobilization of the immune system against cancer. Inter- and intratumor heterogeneity are significant barriers to any novel therapy in cancer. It is to be expected that different tumor types will be characterized by varying levels of metabolic perturbation. Similarly, as regions within a tumor often harbor multiple genetic phenotypes, it is likely that one area within the TME will lack particular nutrients distinct from a neighboring region. For these reasons it will be important to attempt to target only the most fundamental aspects of immunometabolism. The therapeutic modulation of immunometabolism should be seen as a way of tailoring the response to maximize antitumor activity, as well as assisting in the generation of immunological memory—the overarching goal of tumor immunology. Through understanding of the correct targets and optimal combinations, the immune system can be more effectively deployed in the fight against malignancy.

References

1. Coley WB (1891) II. Contribution to the knowledge of sarcoma. Ann Surg 14 (3):199–220
2. Ehrlich P (1909) Ueber den jetzigen Stand der Karzinomforschung. Ned Tijdschr Geneeskd 5:273–290
3. Isaacs A, Lindenmann J (1957) Virus interference. I. The interferon. Proc R Soc Lond Ser B Biol Sci 147(927):258–267
4. Graham JB, Graham RM (1959) The effect of vaccine on cancer patients. Surg Gynecol Obstet 109(2):131–138
5. Miller JF, Mitchell GF, Weiss NS (1967) Cellular basis of the immunological defects in thymectomized mice. Nature 214 (5092):992–997
6. Steinman RM, Cohn ZA (1973) Identification of a novel cell type in peripheral lymphoid organs of mice. I. Morphology, quantitation, tissue distribution. J Exp Med 137 (5):1142–1162
7. Zinkernagel RM, Doherty PC (1974) Restriction of in vitro T cell-mediated cytotoxicity in lymphocytic choriomeningitis within a syngeneic or semiallogeneic system. Nature 248 (5450):701–702
8. Kiessling R, Klein E, Pross H, Wigzell H (1975) "Natural" killer cells in the mouse. II. Cytotoxic cells with specificity for mouse Moloney leukemia cells characteristics of the killer cell. Eur J Immunol 5(2):117–121. https://doi.org/10.1002/eji.1830050209
9. Anguille S, Lion E, Willemen Y, Van Tendeloo V, Berneman Z, Smits E (2011) Interferon-α in acute myeloid leukemia: an old drug revisited. Leukemia 25(5):739
10. Kaplan DH, Shankaran V, Dighe AS, Stockert E, Aguet M, Old LJ et al (1998) Demonstration of an interferon gamma-dependent tumor surveillance system in immunocompetent mice. Proc Natl Acad Sci U S A 95(13):7556–7561
11. Shankaran V, Ikeda H, Bruce AT, White JM, Swanson PE, Old LJ et al (2001) IFNgamma and lymphocytes prevent primary tumour development and shape tumour immunogenicity. Nature 410(6832):1107–1111. https://doi.org/10.1038/35074122
12. Leach DR, Krummel MF, Allison JP (1996) Enhancement of antitumor immunity by CTLA-4 blockade. Science 271 (5256):1734–1736
13. Krummel MF, Allison JP (1995) CD28 and CTLA-4 have opposing effects on the response of T cells to stimulation. J Exp Med 182(2):459–465
14. Ishida Y, Agata Y, Shibahara K, Honjo T (1992) Induced expression of PD-1, a novel member of the immunoglobulin gene superfamily, upon programmed cell death. EMBO J 11(11):3887–3895
15. Hodi FS, O'Day SJ, McDermott DF, Weber RW, Sosman JA, Haanen JB et al (2010) Improved survival with ipilimumab in patients with metastatic melanoma. N Engl J Med 363 (8):711–723. https://doi.org/10.1056/NEJMoa1003466
16. Marin-Acevedo JA, Dholaria B, Soyano AE, Knutson KL, Chumsri S, Lou Y (2018) Next generation of immune checkpoint therapy in cancer: new developments and challenges. J Hematol Oncol 11(1):39. https://doi.org/10.1186/s13045-018-0582-8
17. Weber JS, D'Angelo SP, Minor D, Hodi FS, Gutzmer R, Neyns B et al (2015) Nivolumab versus chemotherapy in patients with advanced melanoma who progressed after anti-CTLA-4 treatment (CheckMate 037): a randomised, controlled, open-label, phase 3 trial. Lancet Oncol 16(4):375–384. https://doi.org/10.1016/S1470-2045(15)70076-8
18. Hodi FS, Chesney J, Pavlick AC, Robert C, Grossmann KF, McDermott DF et al (2016) Combined nivolumab and ipilimumab versus ipilimumab alone in patients with advanced melanoma: 2-year overall survival outcomes in a multicentre, randomised, controlled, phase 2 trial. Lancet Oncol 17 (11):1558–1568. https://doi.org/10.1016/S1470-2045(16)30366-7
19. Chen C, de Alwis DP, Lubiniecki GM, Gergich K, Tice MAB, Kang SP et al (2017) Pembrolizumab KEYNOTE-001: an adaptive study leading to accelerated approval for two indications and a companion diagnostic. Ann Oncol 28(6):1388–1398. https://doi.org/10.1093/annonc/mdx076
20. Darvin P, Toor SM, Sasidharan Nair V, Elkord E (2018) Immune checkpoint inhibitors: recent progress and potential biomarkers. Exp Mol Med 50(12):165–165. https://doi.org/10.1038/s12276-018-0191-1
21. Duan J, Wang Y, Jiao S (2018) Checkpoint blockade-based immunotherapy in the

context of tumor microenvironment: opportunities and challenges. Cancer Med 7 (9):4517–4529. https://doi.org/10.1002/cam4.1722
22. Haratani K, Hayashi H, Tanaka T, Kaneda H, Togashi Y, Sakai K et al (2017) Tumor immune microenvironment and nivolumab efficacy in EGFR mutation-positive non-small-cell lung cancer based on T790M status after disease progression during EGFR-TKI treatment. Ann Oncol 28(7):1532–1539. https://doi.org/10.1093/annonc/mdx183
23. Ayers M, Lunceford J, Nebozhyn M, Murphy E, Loboda A, Kaufman DR et al (2017) IFN-gamma-related mRNA profile predicts clinical response to PD-1 blockade. J Clin Invest 127(8):2930–2940. https://doi.org/10.1172/jci91190
24. Herbst RS, Soria JC, Kowanetz M, Fine GD, Hamid O, Gordon MS et al (2014) Predictive correlates of response to the anti-PD-L1 antibody MPDL3280A in cancer patients. Nature 515(7528):563–567. https://doi.org/10.1038/nature14011
25. McNeel DG (2016) TCR diversity—a universal cancer immunotherapy biomarker? J Immunother Cancer 4:69–69. https://doi.org/10.1186/s40425-016-0175-4
26. Tumeh PC, Harview CL, Yearley JH, Shintaku IP, Taylor EJ, Robert L et al (2014) PD-1 blockade induces responses by inhibiting adaptive immune resistance. Nature 515 (7528):568–571. https://doi.org/10.1038/nature13954
27. Song W, Shen L, Wang Y, Liu Q, Goodwin TJ, Li J et al (2018) Synergistic and low adverse effect cancer immunotherapy by immunogenic chemotherapy and locally expressed PD-L1 trap. Nat Commun 9 (1):2237. https://doi.org/10.1038/s41467-018-04605-x
28. Walle T, Martinez Monge R, Cerwenka A, Ajona D et al (2018) Radiation effects on antitumor immune responses: current perspectives and challenges. Ther Adv Med Oncol 10:1758834017742575–17588340177425-75. https://doi.org/10.1177/1758834017742575
29. Wang Y, Deng W, Li N, Neri S, Sharma A, Jiang W et al (2018) Combining immunotherapy and radiotherapy for cancer treatment: current challenges and future directions. Front Pharmacol 9:185–185. https://doi.org/10.3389/fphar.2018.00185
30. Galluzzi L, Buqué A, Kepp O, Zitvogel L, Kroemer G (2015) Immunological effects of conventional chemotherapy and targeted anticancer agents. Cancer Cell 28 (6):690–714. https://doi.org/10.1016/j.ccell.2015.10.012
31. Kroemer G, Galluzzi L, Kepp O, Zitvogel L (2013) Immunogenic cell death in cancer therapy. Annu Rev Immunol 31:51–72
32. Matsushita H, Vesely MD, Koboldt DC, Rickert CG, Uppaluri R, Magrini VJ et al (2012) Cancer exome analysis reveals a T-cell-dependent mechanism of cancer immunoediting. Nature 482(7385):400
33. Boon T, Coulie PG, BJVD E, PVD B (2006) Human T cell responses against melanoma. Annu Rev Immunol 24:175–208
34. Schlosser HA, Theurich S, Shimabukuro-Vornhagen A, Holtick U, Stippel DL, von Bergwelt-Baildon M (2014) Overcoming tumor-mediated immunosuppression. Immunotherapy 6(9):973–988. https://doi.org/10.2217/imt.14.58
35. Pardoll DM (2012) The blockade of immune checkpoints in cancer immunotherapy. Nat Rev Cancer 12:252. https://doi.org/10.1038/nrc3239
36. Poggio M, Hu T, Pai C-C, Chu B, Belair CD, Chang A et al (2019) Suppression of exosomal PD-L1 induces systemic anti-tumor immunity and memory. Cell 177 (2):414–427.e413. https://doi.org/10.1016/j.cell.2019.02.016
37. Pasche B (2001) Role of transforming growth factor beta in cancer. J Cell Physiol 186 (2):153–168
38. Bellomo C, Caja L, Moustakas A (2016) Transforming growth factor β as regulator of cancer stemness and metastasis. Br J Cancer 115:761. https://doi.org/10.1038/bjc.2016.255
39. Gorelik L, Flavell RA (2002) Transforming growth factor-beta in T-cell biology. Nat Rev Immunol 2(1):46–53. https://doi.org/10.1038/nri704
40. Wahl SM, Wen J, Moutsopoulos N (2006) TGF-beta: a mobile purveyor of immune privilege. Immunol Rev 213:213–227. https://doi.org/10.1111/j.1600-065X.2006.00437.x
41. Mariathasan S, Turley SJ, Nickles D, Castiglioni A, Yuen K, Wang Y et al (2018) TGFbeta attenuates tumour response to PD-L1 blockade by contributing to exclusion of T cells. Nature 554(7693):544–548. https://doi.org/10.1038/nature25501
42. Yang L, Pang Y, Moses HL (2010) TGF-beta and immune cells: an important regulatory

axis in the tumor microenvironment and progression. Trends Immunol 31(6):220–227. https://doi.org/10.1016/j.it.2010.04.002

43. Movahedi K, Guilliams M, Van den Bossche J, Van den Bergh R, Gysemans C, Beschin A et al (2008) Identification of discrete tumor-induced myeloid-derived suppressor cell subpopulations with distinct T cell-suppressive activity. Blood 111(8):4233–4244. https://doi.org/10.1182/blood-2007-07-099226
44. Condeelis J, Pollard JW (2006) Macrophages: obligate partners for tumor cell migration, invasion, and metastasis. Cell 124(2):263–266. https://doi.org/10.1016/j.cell.2006.01.007
45. de Gramont A, Faivre S, Raymond E (2016) Novel TGF-β inhibitors ready for prime time in onco-immunology. Onco Targets Ther 6(1):e1257453–e1257453. https://doi.org/10.1080/2162402X.2016.1257453
46. Jacobs JF, Nierkens S, Figdor CG, de Vries IJM, Adema GJ (2012) Regulatory T cells in melanoma: the final hurdle towards effective immunotherapy? Lancet Oncol 13(1):e32–e42
47. Lu L, Barbi J, Pan F (2017) The regulation of immune tolerance by FOXP3. Nat Rev Immunol 17(11):703–717. https://doi.org/10.1038/nri.2017.75
48. Hsieh C-S, Liang Y, Tyznik AJ, Self SG, Liggitt D, Rudensky AY (2004) Recognition of the peripheral self by naturally arising CD25+ CD4+ T cell receptors. Immunity 21(2):267–277
49. Sharma A, Rudra D (2018) Emerging functions of regulatory T cells in tissue homeostasis. Front Immunol 9:883. https://doi.org/10.3389/fimmu.2018.00883
50. Prieto-Martin P, Yamaguchi T, Sakaguchi S, Wing K, Onishi Y (2009) Regulatory T cells: how do they suppress immune responses? Int Immunol 21(10):1105–1111. https://doi.org/10.1093/intimm/dxp095
51. Yokokawa J, Cereda V, Remondo C, Gulley JL, Arlen PM, Schlom J et al (2008) Enhanced functionality of CD4+ CD25high-FoxP3+ regulatory T cells in the peripheral blood of patients with prostate cancer. Clin Cancer Res 14(4):1032–1040
52. Gasparoto TH, de Souza Malaspina TS, Benevides L, de Melo EJF, Costa MRSN, Damante JH et al (2010) Patients with oral squamous cell carcinoma are characterized by increased frequency of suppressive regulatory T cells in the blood and tumor microenvironment. Cancer Immunol Immunother 59(6):819–828
53. Vinay DS, Ryan EP, Pawelec G, Talib WH, Stagg J, Elkord E et al (2015) Immune evasion in cancer: mechanistic basis and therapeutic strategies. Semin Cancer Biol 35:S185–S198. https://doi.org/10.1016/j.semcancer.2015.03.004
54. Qureshi OS, Zheng Y, Nakamura K, Attridge K, Manzotti C, Schmidt EM et al (2011) Trans-endocytosis of CD80 and CD86: a molecular basis for the cell-extrinsic function of CTLA-4. Science 332(6029):600–603. https://doi.org/10.1126/science.1202947
55. Shimizu J, Yamazaki S, Takahashi T, Ishida Y, Sakaguchi S (2002) Stimulation of CD25(+) CD4(+) regulatory T cells through GITR breaks immunological self-tolerance. Nat Immunol 3(2):135–142. https://doi.org/10.1038/ni759
56. Tanaka A, Sakaguchi S (2017) Regulatory T cells in cancer immunotherapy. Cell Res 27(1):109–118. https://doi.org/10.1038/cr.2016.151
57. O'Neill LA, Kishton RJ, Rathmell J (2016) A guide to immunometabolism for immunologists. Nat Rev Immunol 16(9):553–565. https://doi.org/10.1038/nri.2016.70
58. Liu Y, Zhang DT, Liu XG (2015) mTOR signaling in T cell immunity and autoimmunity. Int Rev Immunol 34(1):50–66. https://doi.org/10.3109/08830185.2014.933957
59. O'Neill LA, Hardie DG (2013) Metabolism of inflammation limited by AMPK and pseudo-starvation. Nature 493(7432):346–355. https://doi.org/10.1038/nature11862
60. Fox CJ, Hammerman PS, Thompson CB (2005) Fuel feeds function: energy metabolism and the T-cell response. Nat Rev Immunol 5(11):844–852. https://doi.org/10.1038/nri1710
61. Kedia-Mehta N, Finlay DK (2019) Competition for nutrients and its role in controlling immune responses. Nat Commun 10(1):2123. https://doi.org/10.1038/s41467-019-10015-4
62. Janeway (2012) Janeway's immunobiology, 8th edn. Garland Science
63. Wang R, Dillon CP, Shi LZ, Milasta S, Carter R, Finkelstein D, McCormick LL, Fitzgerald P, Chi H, Munger J, Green DR (2011) The transcription factor Myc controls metabolic reprogramming upon T lymphocyte activation. Immunity 35(6):871–882. https://doi.org/10.1016/j.immuni.2011.09.021

64. Loftus RM, Finlay DK (2016) Immunometabolism: cellular metabolism turns immune regulator. J Biol Chem 291(1):1–10. https://doi.org/10.1074/jbc.R115.693903
65. Ghesquiere B, Wong BW, Kuchnio A, Carmeliet P (2014) Metabolism of stromal and immune cells in health and disease. Nature 511(7508):167–176. https://doi.org/10.1038/nature13312
66. Dimeloe S, Mehling M, Frick C, Loeliger J, Bantug GR, Sauder U et al (2016) The immune-metabolic basis of effector memory CD4+ T cell function under hypoxic conditions. J Immunol 196(1):106. https://doi.org/10.4049/jimmunol.1501766
67. Warburg O, Gawehn K, Geissler AW (1958) [Metabolism of leukocytes]. Z Naturforsch B 13B(8):515–516
68. Herbel C, Patsoukis N, Bardhan K, Seth P, Weaver JD, Boussiotis VA (2016) Clinical significance of T cell metabolic reprogramming in cancer. Clin Transl Med 5(1):29. https://doi.org/10.1186/s40169-016-0110-9
69. Koppenol WH, Bounds PL, Dang CV (2011) Otto Warburg's contributions to current concepts of cancer metabolism. Nat Rev Cancer 11(5):325–337. https://doi.org/10.1038/nrc3038
70. Pearce EL (2010) Metabolism in T cell activation and differentiation. Curr Opin Immunol 22(3):314–320. https://doi.org/10.1016/j.coi.2010.01.018
71. Macintyre Andrew N, Gerriets Valerie A, Nichols Amanda G et al (2014) The glucose transporter Glut1 is selectively essential for CD4 T cell activation and effector function. Cell Metab 20(1):61–72. https://doi.org/10.1016/j.cmet.2014.05.004
72. Sinclair LV, Rolf J, Emslie E, Shi Y-B, Taylor PM, Cantrell DA (2013) Control of amino-acid transport by antigen receptors coordinates the metabolic reprogramming essential for T cell differentiation. Nat Immunol 14:500. https://doi.org/10.1038/ni.2556. https://www.nature.com/articles/ni.2556-supplementary-information
73. Frauwirth KA, Riley JL, Harris MH, Parry RV, Rathmell JC, Plas DR et al (2002) The CD28 signaling pathway regulates glucose metabolism. Immunity 16(6):769–777. https://doi.org/10.1016/S1074-7613(02)00323-0
74. Ricciardi S, Manfrini N, Alfieri R, Calamita P, Crosti MC, Gallo S et al (2018) The translational machinery of human CD4(+) T cells is poised for activation and controls the switch from quiescence to metabolic remodeling. Cell Metab 28(6):895–906.e895. https://doi.org/10.1016/j.cmet.2018.08.009
75. Nagy E, Rigby WFC (1995) Glyceraldehyde-3-phosphate dehydrogenase selectively binds AU-rich RNA in the NAD+-binding region (Rossmann fold). J Biol Chem 270(6):2755–2763. https://doi.org/10.1074/jbc.270.6.2755
76. Carr EL, Kelman A, Wu GS, Gopaul R, Senkevitch E, Aghvanyan A, Turay AM, Frauwirth KA (2010) Glutamine uptake and metabolism are coordinately regulated by ERK/MAPK during T lymphocyte activation. J Immunol 185(2):1037–1044. https://doi.org/10.4049/jimmunol.0903586
77. Crawford J, Cohen HJ (1985) The essential role of L-glutamine in lymphocyte differentiation in vitro. J Cell Physiol 124(2):275–282. https://doi.org/10.1002/jcp.1041240216
78. Franchina DG, Dostert C, Brenner D (2018) Reactive oxygen species: involvement in T cell signaling and metabolism. Trends Immunol 39(6):489–502
79. Bantug GR, Galluzzi L, Kroemer G, Hess C (2018) The spectrum of T cell metabolism in health and disease. Nat Rev Immunol 18(1):19
80. Sena Laura A, Li S, Jairaman A, Prakriya M, Ezponda T, Hildeman David A et al (2013) Mitochondria are required for antigen-specific T cell activation through reactive oxygen species signaling. Immunity 38(2):225–236. https://doi.org/10.1016/j.immuni.2012.10.020
81. Previte DM, O'Connor EC, Novak EA, Martins CP, Mollen KP, Piganelli JD (2017) Reactive oxygen species are required for driving efficient and sustained aerobic glycolysis during CD4+ T cell activation. PLoS One 12(4):e0175549. https://doi.org/10.1371/journal.pone.0175549
82. Mak TW, Grusdat M, Duncan GS, Dostert C, Nonnenmacher Y, Cox M et al (2017) Glutathione primes T cell metabolism for inflammation. Immunity 46(4):675–689. https://doi.org/10.1016/j.immuni.2017.03.019
83. Flescher E, Ledbetter JA, Schieven GL, Vela-Roch N, Fossum D, Dang H et al (1994) Longitudinal exposure of human T lymphocytes to weak oxidative stress suppresses transmembrane and nuclear signal transduction. J Immunol 153(11):4880–4889
84. Düvel K, Yecies JL, Menon S, Raman P, Lipovsky AI, Souza AL et al (2010) Activation of a metabolic gene regulatory network downstream of mTOR complex 1. Mol Cell 39(2):171–183

85. Kidani Y, Elsaesser H, Hock MB, Vergnes L, Williams KJ, Argus JP et al (2013) Sterol regulatory element-binding proteins are essential for the metabolic programming of effector T cells and adaptive immunity. Nat Immunol 14(5):489
86. Berod L, Friedrich C, Nandan A, Freitag J, Hagemann S, Harmrolfs K et al (2014) De novo fatty acid synthesis controls the fate between regulatory T and T helper 17 cells. Nat Med 20(11):1327
87. Galgani M, De Rosa V, La Cava A, Matarese G (2016) Role of metabolism in the immunobiology of regulatory T cells. J Immunol 197 (7):2567–2575. https://doi.org/10.4049/jimmunol.1600242
88. Chang CH, Curtis JD, Maggi LB Jr, Faubert B, Villarino AV, O'Sullivan D et al (2013) Posttranscriptional control of T cell effector function by aerobic glycolysis. Cell 153(6):1239–1251. https://doi.org/10.1016/j.cell.2013.05.016
89. Ho PC, Bihuniak JD, Macintyre AN, Staron M, Liu X, Amezquita R et al (2015) Phosphoenolpyruvate is a metabolic checkpoint of anti-tumor T cell responses. Cell 162(6):1217–1228. https://doi.org/10.1016/j.cell.2015.08.012
90. Gerriets VA, Kishton RJ, Nichols AG, Macintyre AN, Inoue M, Ilkayeva O et al (2015) Metabolic programming and PDHK1 control CD4+ T cell subsets and inflammation. J Clin Invest 125(1):194–207. https://doi.org/10.1172/jci76012
91. Weisshaar N, Madi A, Cui G (2018) Early TCR signaling sweetens effector function through PDHK1. Trends Endocrinol Metab 29(9):595–597. https://doi.org/10.1016/j.tem.2018.03.016
92. Cham CM, Driessens G, O'Keefe JP, Gajewski TF (2008) Glucose deprivation inhibits multiple key gene expression events and effector functions in CD8+ T cells. Eur J Immunol 38(9):2438–2450. https://doi.org/10.1002/eji.200838289
93. Chang C-H, Qiu J, O'Sullivan D, Buck Michael D, Noguchi T et al (2015) Metabolic competition in the tumor microenvironment is a driver of cancer progression. Cell 162 (6):1229–1241. https://doi.org/10.1016/j.cell.2015.08.016
94. Metzler B, Gfeller P, Guinet E (2016) Restricting glutamine or glutamine-dependent purine and pyrimidine syntheses promotes human T cells with high FOXP3 expression and regulatory properties. J Immunol 196(9):3618–3630. https://doi.org/10.4049/jimmunol.1501756
95. Johnson MO, Wolf MM, Madden MZ, Andrejeva G, Sugiura A, Contreras DC et al (2018) Distinct regulation of Th17 and Th1 cell differentiation by glutaminase-dependent metabolism. Cell 175:1780. https://doi.org/10.1016/j.cell.2018.10.001
96. Huang L, Mellor AL (2014) Metabolic control of tumour progression and antitumour immunity. Curr Opin Oncol 26(1):92–99. https://doi.org/10.1097/CCO.0000000000000035
97. Mondanelli G, Ugel S, Grohmann U, Bronte V (2017) The immune regulation in cancer by the amino acid metabolizing enzymes ARG and IDO. Curr Opin Pharmacol 35:30–39. https://doi.org/10.1016/j.coph.2017.05.002
98. Geiger R, Rieckmann JC, Wolf T, Basso C, Feng Y, Fuhrer T et al (2016) L-arginine modulates T cell metabolism and enhances survival and anti-tumor activity. Cell 167 (3):829–842.e813
99. Rodriguez PC, Quiceno DG, Ochoa AC (2007) L-arginine availability regulates T-lymphocyte cell-cycle progression. Blood 109(4):1568–1573
100. Kennedy BE, Sharif T, Martell E, Dai C, Kim Y, Lee PW et al (2016) NAD+ salvage pathway in cancer metabolism and therapy. Pharmacol Res 114:274–283
101. Fallarino F, Grohmann U, You S, McGrath BC, Cavener DR, Vacca C et al (2006) The combined effects of tryptophan starvation and tryptophan catabolites down-regulate T cell receptor ζ-chain and induce a regulatory phenotype in naive T cells. J Immunol 176 (11):6752–6761
102. Desdín-Micó G, Soto-Heredero G, Mittelbrunn M (2018) Mitochondrial activity in T cells. Mitochondrion 41:51–57. https://doi.org/10.1016/j.mito.2017.10.006
103. Baixauli F, Acín-Pérez R, Villarroya-Beltrí C, Mazzeo C, Nuñez-Andrade N, Gabandé-Rodriguez E et al (2015) Mitochondrial respiration controls lysosomal function during inflammatory T cell responses. Cell Metab 22(3):485–498
104. Karmaus PWF, Chen X, Lim SA, Herrada AA, Nguyen T-LM et al (2019) Metabolic heterogeneity underlies reciprocal fates of TH17 cell stemness and plasticity. Nature 565 (7737):101–105. https://doi.org/10.1038/s41586-018-0806-7
105. Procaccini C, Carbone F, Di Silvestre D, Brambilla F, De Rosa V, Galgani M et al (2016) The proteomic landscape of human ex vivo regulatory and conventional T cells

reveals specific metabolic requirements. Immunity 44(2):406–421

106. Ventura-Clapier R, Garnier A, Veksler V (2008) Transcriptional control of mitochondrial biogenesis: the central role of PGC-1-alpha. Cardiovasc Res 79(2):208–217. https://doi.org/10.1093/cvr/cvn098
107. Hatfield S, Kjaergaard J, Lukashev D, Schreiber T, Belikoff B, Abbott R et al (2015) Immunological mechanisms of the antitumor effects of supplemental oxygenation. Sci Transl Med 7(277):277ra230
108. Howie D, Ten Bokum A, Necula AS, Cobbold SP, Waldmann H (2018) The role of lipid metabolism in T lymphocyte differentiation and survival. Front Immunol 8:1949–1949. https://doi.org/10.3389/fimmu.2017.01949
109. Michalek RD, Gerriets VA, Jacobs SR, Macintyre AN, MacIver NJ, Mason EF et al (2011) Cutting edge: distinct glycolytic and lipid oxidative metabolic programs are essential for effector and regulatory CD4(+) T cell subsets. J Immunol 186(6):3299–3303. https://doi.org/10.4049/jimmunol.1003613
110. Zhang Y, Kurupati R, Liu L, Zhou XY, Zhang G, Hudaihed A et al (2017) Enhancing CD8+ T cell fatty acid catabolism within a metabolically challenging tumor microenvironment increases the efficacy of melanoma immunotherapy. Cancer Cell 32(3):377–391.e379. https://doi.org/10.1016/j.ccell.2017.08.004
111. Rolf J, Zarrouk M, Finlay DK, Foretz M, Viollet B, Cantrell DA (2013) AMPK α1: a glucose sensor that controls CD 8 T-cell memory. Eur J Immunol 43(4):889–896
112. Endo Y, Onodera A, Obata-Ninomiya K, Koyama-Nasu R, Asou HK, Ito T et al (2019) ACC1 determines memory potential of individual CD4+ T cells by regulating de novo fatty acid biosynthesis. Nat Metab 1(2):261–275. https://doi.org/10.1038/s42255-018-0025-4
113. Hanahan D, Weinberg RA (2011) Hallmarks of cancer: the next generation. Cell 144(5):646–674
114. O'Sullivan D, Sanin DE, Pearce EJ, Pearce EL (2019) Metabolic interventions in the immune response to cancer. Nat Rev Immunol 19:324. https://doi.org/10.1038/s41577-019-0140-9
115. Patsoukis N, Bardhan K, Chatterjee P, Sari D, Liu B, Bell LN et al (2015) PD-1 alters T-cell metabolic reprogramming by inhibiting glycolysis and promoting lipolysis and fatty acid oxidation. Nat Commun 6:6692
116. Parry RV, Chemnitz JM, Frauwirth KA, Lanfranco AR, Braunstein I, Kobayashi SV et al (2005) CTLA-4 and PD-1 receptors inhibit T-cell activation by distinct mechanisms. Mol Cell Biol 25(21):9543
117. Previte DM, Martins CP, O'Connor EC, Marre ML, Coudriet GM et al (2019) Lymphocyte activation Gene-3 maintains mitochondrial and metabolic quiescence in naive CD4(+) T cells. Cell Rep 27(1):129–141.e124. https://doi.org/10.1016/j.celrep.2019.03.004
118. He W, Zhang H, Han F, Chen X, Lin R, Wang W et al (2017) CD155T/TIGIT signaling regulates CD8+ T-cell metabolism and promotes tumor progression in human gastric cancer. Cancer Res 77(22):6375. https://doi.org/10.1158/0008-5472.CAN-17-0381
119. Sabharwal SS, Rosen DB, Grein J, Tedesco D, Joyce-Shaikh B, Ueda R et al (2018) GITR agonism enhances cellular metabolism to support CD8(+) T-cell proliferation and effector cytokine production in a mouse tumor model. Cancer Immunol Res 6(10):1199–1211. https://doi.org/10.1158/2326-6066.cir-17-0632
120. Shi T, Ma Y, Cao L, Zhan S, Xu Y, Fu F et al (2019) B7-H3 promotes aerobic glycolysis and chemoresistance in colorectal cancer cells by regulating HK2. Cell Death Dis 10(4):308. https://doi.org/10.1038/s41419-019-1549-6
121. Cascone T, McKenzie JA, Mbofung RM, Punt S, Wang Z, Xu C et al (2018) Increased tumor glycolysis characterizes immune resistance to adoptive T cell therapy. Cell Metab 27(5):977–987.e974. https://doi.org/10.1016/j.cmet.2018.02.024
122. Brand A, Singer K, Koehl GE, Kolitzus M, Schoenhammer G, Thiel A et al (2016) LDHA-associated lactic acid production blunts tumor immunosurveillance by T and NK cells. Cell Metab 24(5):657–671. https://doi.org/10.1016/j.cmet.2016.08.011
123. Sonveaux P, Végran F, Schroeder T, Wergin MC, Verrax J, Rabbani ZN et al (2008) Targeting lactate-fueled respiration selectively kills hypoxic tumor cells in mice. J Clin Invest 118(12):3930–3942. https://doi.org/10.1172/JCI36843
124. Weide B, Martens A, Hassel JC, Berking C, Postow MA, Bisschop K et al (2016) Baseline biomarkers for outcome of melanoma patients treated with pembrolizumab. Clin Cancer Res 22(22):5487–5496. https://doi.org/10.1158/1078-0432.ccr-16-0127

125. Gross MI, Demo SD, Dennison JB, Chen L, Chernov-Rogan T, Goyal B et al (2014) Antitumor activity of the glutaminase inhibitor CB-839 in triple-negative breast cancer. Mol Cancer Ther 13(4):890–901. https://doi.org/10.1158/1535-7163.mct-13-0870
126. DeLaBarre B, Hurov J, Cianchetta G, Murray S, Dang L (2014) Action at a distance: allostery and the development of drugs to target cancer cell metabolism. Cell Chem Biol 21(9):1143–1161. https://doi.org/10.1016/j.chembiol.2014.08.007
127. Tannir NM, Motzer RJ, Agarwal N, Liu P-Y, Whiting SH, O'Keeffe B et al (2018) CANTATA: a randomized phase 2 study of CB-839 in combination with cabozantinib vs. placebo with cabozantinib in patients with advanced/metastatic renal cell carcinoma. J Clin Oncol 36(Suppl 15): TPS4601. https://doi.org/10.1200/JCO.2018.36.15_suppl.TPS4601
128. Bengsch B, Johnson AL, Kurachi M, Odorizzi PM, Pauken KE, Attanasio J et al (2016) Bioenergetic insufficiencies due to metabolic alterations regulated by the inhibitory receptor PD-1 are an early driver of CD8+ T cell exhaustion. Immunity 45(2):358–373
129. Balmer Maria L, Ma Eric H, Bantug Glenn R, Grählert J, Pfister S et al (2016) Memory CD8+ T cells require increased concentrations of acetate induced by stress for optimal function. Immunity 44(6):1312–1324. https://doi.org/10.1016/j.immuni.2016.03.016
130. Gubin MM, Zhang X, Schuster H, Caron E, Ward JP, Noguchi T et al (2014) Checkpoint blockade cancer immunotherapy targets tumour-specific mutant antigens. Nature 515(7528):577
131. Pettitt D, Arshad Z, Smith J, Stanic T, Holländer G, Brindley D (2018) CAR-T cells: a systematic review and mixed methods analysis of the clinical trial landscape. Mol Ther 26(2):342–353. https://doi.org/10.1016/j.ymthe.2017.10.019
132. Chen N, Li X, Chintala NK, Tano ZE, Adusumilli PS (2018) Driving CARs on the uneven road of antigen heterogeneity in solid tumors. Curr Opin Immunol 51:103–110. https://doi.org/10.1016/j.coi.2018.03.002
133. Sukumaran S, Watanabe N, Bajgain P, Raja K, Mohammed S, Fisher WE et al (2018) Enhancing the potency and specificity of engineered T cells for cancer treatment. Cancer Discov 8(8):972–987. https://doi.org/10.1158/2159-8290.cd-17-1298
134. Rosenberg SA, Yang JC, Sherry RM, Kammula US, Hughes MS, Phan GQ et al (2011) Durable complete responses in heavily pretreated patients with metastatic melanoma using T-cell transfer immunotherapy. Clin Cancer Res 17(13):4550–4557
135. Lee HJ, Kim YA, Sim CK, Heo SH, Song IH, Park HS et al (2017) Expansion of tumor-infiltrating lymphocytes and their potential for application as adoptive cell transfer therapy in human breast cancer. Oncotarget 8 (69):113345–113359. https://doi.org/10.18632/oncotarget.23007
136. Stevanovic S, Draper L, Langhan M (2016) Complete regression of metastatic cervical cancer after treatment with human papillomavirus-targeted tumor-infiltrating T cells (vol 33, p 1543, 2015). J Clin Oncol 34(5):519–519
137. Pipkin ME, Sacks JA, Cruz-Guilloty F, Lichtenheld MG, Bevan MJ, Rao A (2010) Interleukin-2 and inflammation induce distinct transcriptional programs that promote the differentiation of effector cytolytic T cells. Immunity 32(1):79–90
138. Contreras A, Beems MV, Tatar AJ, Sen S, Srinand P, Suresh M, Luther TK, Cho CS (2018) Co-transfer of tumor-specific effector and memory CD8+ T cells enhances the efficacy of adoptive melanoma immunotherapy in a mouse model. J Immunother Cancer 6 (1):41. https://doi.org/10.1186/s40425-018-0358-2
139. van der Windt GJ, Everts B, Chang C-H, Curtis JD, Freitas TC, Amiel E et al (2012) Mitochondrial respiratory capacity is a critical regulator of CD8+ T cell memory development. Immunity 36(1):68–78
140. Haslam A, Prasad V (2019) Estimation of the percentage of US patients with cancer who are eligible for and respond to checkpoint inhibitor immunotherapy drugs. JAMA Netw Open 2(5):e192535. https://doi.org/10.1001/jamanetworkopen.2019.2535
141. DeVorkin L, Pavey N, Carleton G, Comber A, Ho C, Lim J et al (2019) Autophagy regulation of metabolism is required for CD8+ T cell anti-tumor immunity. Cell Rep 27(2):502–513.e505. https://doi.org/10.1016/j.celrep.2019.03.037
142. Long GV, Dummer R, Hamid O, Gajewski T, Caglevic C, Dalle S et al (2018) Epacadostat plus pembrolizumab versus pembrolizumab alone in patients with unresectable or metastatic melanoma: results of the phase 3 ECHO-301/KEYNOTE-252 study. J Clin Oncol 36(Suppl 15):108

143. Labadie BW, Bao R, Luke JJ (2019) Reimagining IDO pathway inhibition in cancer immunotherapy via downstream focus on the tryptophan–kynurenine–aryl hydrocarbon axis. Clin Cancer Res 25(5):1462. https://doi.org/10.1158/1078-0432.CCR-18-2882
144. Papadopoulos KP, Tsai FY-C, Bauer TM, Muigai L, Liang Y, Bennett MK et al (2017) CX-1158-101: a first-in-human phase 1 study of CB-1158, a small molecule inhibitor of arginase, as monotherapy and in combination with an anti-PD-1 checkpoint inhibitor in patients (pts) with solid tumors. J Clin Oncol 35(Suppl 15):3005. https://doi.org/10.1200/JCO.2017.35.15_suppl.3005

Chapter 18

Sex Differences in Immunometabolism: An Unexplored Area

Suresh Mishra, Geetika Bassi, and Yang Xin Zi Xu

Abstract

The last three decades have seen a growing interest in research in the field of immunometabolism, likely because of promising discoveries made in this field. This includes demonstration of the crucial roles of cellular metabolism in the regulation of functional plasticity of various immune cells, their cross talk with major metabolic tissues (and consequently in the regulation of metabolic homeostasis) at the systemic level, and their potential in improving the efficacy of current immunotherapy or developing new therapeutics for a variety of metabolic and immune diseases (Lee YS, Wollam J, Olefsky JM, Cell 172:22–40, 2018). Surprisingly, sex differences, which are integral to metabolic and immune health and disease, have received a short shrift from researchers in this field. The purpose of this chapter in this protocols book in the *Immunometabolism: Methods in Molecular Biology* series is to bring attention to this understudied, but crucial, feature of immunometabolism within the scientific community. Sex differences in adipose (and by extension, metabolic) and immune functions are pervasive in metabolic and immune health and disease; it is likely that a better insight into them may open new research directions to better capitalize on the promising discoveries made in this field, and thereby contribute to the development of sex-based precision medicine. It is counterintuitive to ignore a fundamental aspect of immunometabolism, and thereby limit our ability to capitalize on its promising features in improving or maintaining health, and for the therapeutic targeting of associated diseases. Here we briefly discuss the potential drivers and touch upon some unanswered questions in sex differences in immunometabolism, especially those that require attention from the scientific community.

Key words Sex steroids, X chromosome inactivation (XCI), Mitochondria, Adipose-immune cross talk, Metabolic-immune cross talk

1 Metabolic and Immune Functions Are Interlinked and Display Sex Differences

Immune cells play crucial roles in metabolic homeostasis and have a reciprocal relationship with adipose tissue [1]. Consequently, they often undergo parallel changes in normal physiology (e.g., puberty and pregnancy), during aging, and in various metabolic and immune diseases [2]. For instance, meta-inflammation in adipose tissue negatively affects metabolic homeostasis in the body, whereas obesity adversely affects the body's immune system and responses [3]. The exact cellular and molecular mediators responsible for integrating metabolic and immune functions are beginning to be

Suresh Mishra (ed.), *Immunometabolism: Methods and Protocols*, Methods in Molecular Biology, vol. 2184,
https://doi.org/10.1007/978-1-0716-0802-9_18,

discovered. An essential component of this close relationship between the two is sex differences. For instance, sex differences are apparent in adipose tissue distribution and functions, as well as in immune functions in the body, and sex steroid hormones are integral to them. In general, females possess a higher percentage of body fat, but display a resistance to obesity-related metabolic dysregulation compared to males. This difference in metabolic function between females and males is attributed to sex differences in adipose tissue distribution in different adipose depots and their functions [4]. A parallel sex difference also exists in immune responses. In general, males experience a greater risk of developing various infections and cancers than females, whereas females exhibit a greater response to antigenic challenges, such as infection and vaccination, and are more prone to developing autoimmune diseases [5]. Thus, there are fundamental aspects of metabolic homeostasis and immune functions that are regulated differently in males and females. It is likely that these differences influence both the development of metabolic and immune diseases and the response to different therapeutics. However, this basic tenet of metabolic and immune functions has not yet been capitalized on for the development of more effective sex-based interventions, or therapeutics. A major hurdle in developing sex-based precision medicine is our lack of knowledge in this field. For example, the identities of the effector molecules that mediate these effects and determine sex differences in metabolic (adipose) and immune functions are largely unknown. It is also unclear whether these effector molecules in different metabolic tissues and immune cell types are distinct, or if there are common effector molecules with cell-type-specific functions, and whether the sex differences in metabolic and immune functions influence each other. Identifying these factors and illuminating the mechanisms involved in the sex differences in metabolic and immune functions are crucial, because of their importance in immunometabolism, and their dysregulation in metabolic and immune diseases.

2 Putative Factors for Sex Differences in Immunometabolism

The major factors that could theoretically contribute to sex differences in immune cell functions are sex steroid hormones, sex chromosomes, epigenetics, and environmental factors. Thus, sex differences in immunometabolism may be a consequence of hormone-induced and cell-intrinsic properties, or a combination of both. Among these factors, sex steroids appear to play a prominent role. For instance, changes in sex steroid levels and immune-related alterations occur in parallel during puberty, pregnancy, and aging [6]. Notably, sex differences in immune responsiveness are substantially altered during puberty and pregnancy when sex hormone levels change significantly [7, 8], suggesting that sex hormones play a pivotal role. Similarly, sex differences in the distribution of adipose tissue become apparent during puberty when sex steroid levels rise, and are altered in transgender people during sex hormone replacement therapy [3, 9]. An alteration in adipose tissue distribution also occurs after menopause when estradiol levels decline [3, 4]. The ways in which sex steroids influence sex-dependent immune cell phenotypes and adipose tissue distribution and functions are an emerging topic of study, and accumulating evidence is providing new insights [5, 10]. However, our current knowledge of the factors that mediate such effects, and the underlying mechanisms involved, remains limited. Of note, co-regulators of sex steroid receptors play an important role in the regulation of steroid hormone actions [11]. Therefore, steroid hormone co-regulators potentially play a role in mediating sex differences in metabolic and immune functions. In addition, sex hormone-binding globulins (SHBG) that regulate the tissue availability of sex steroids may contribute to sex-dimorphic functions. In addition to sex steroids, corticosteroids, which play a role in sex-dimorphic functions [12], also have potent immunomodulatory roles [13]. Different factors that can regulate sex steroids' actions (as described above) also apply to the regulation of corticoids' actions. Thus, the regulation of sex-dimorphic functions is not limited to sex steroid hormones, and may include corticoids and various factors that regulate the biology of steroid hormone actions.

Moreover, the X chromosome may contribute to sex differences in metabolic and immune functions, for two reasons. Firstly, the X chromosome contains the largest number of immune-related genes of the human genome [5, 10]. Secondly, the majority of these genes escape from X chromosome inactivation (XCI) [3, 14], contributing to differences in gene expression profiles in male and females. Approximately 15% of X-linked genes that are known to escape from XCI also contain nuclear coded mitochondrial genes [14]. As mitochondria play an important role in the regulation of different immune cell functions [15], and in metabolic tissues [17], it is likely that mitochondrial genes that escape from XCI may contribute to sex differences in immunometabolism. In addition, the X chromosome also contains the highest number of microRNAs, including those with roles in the immune system [2, 16]. Thus, the combined effect of X chromosome-associated mitochondrial and immune-related genes, escape from XCI, as well as sex steroids and their co-regulators may contribute substantially to sex differences in metabolic and immune functions in health and disease. An overview of various factors that can potentially contribute to sex differences in immunometabolic functions is given in Fig. 1.

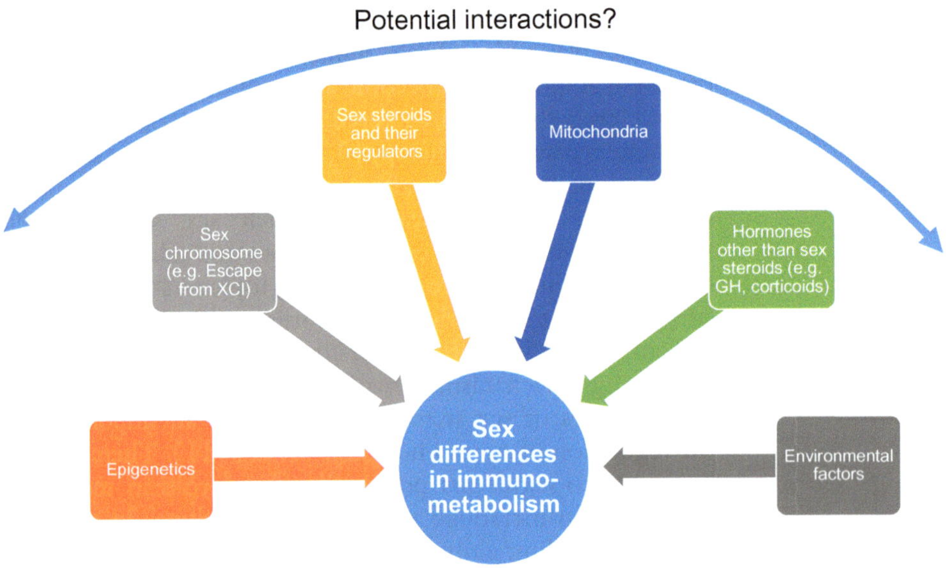

Fig. 1 The putative drivers for sex-dimorphic functions. A schematic showing putative factors that contribute to sex differences in metabolic and immune functions

3 The Interplay Between Sex Differences in Metabolic and Immune-Related Diseases

As metabolic and immune functions closely regulate each other at the systemic level, it is likely that their cross talk plays a role in the regulation of their sex-dimorphic functions, and may involve an inverse or a direct relationship in a context-dependent manner. Such a relationship, in turn, may lead to sex differences in various aspects of metabolic and immune diseases, including incidence, onset, progression, and outcomes. For example, sex differences are known to exist in the incidence of a variety of cancer types worldwide, which generally occur more frequently in males [17, 18]. Similar sex differences exist in different types of autoimmune diseases; however, unlike cancers, autoimmune diseases are more common in females [5, 10]. This would suggest that the known sex differences in immunity might be responsible for this dichotomy, because immune checkpoints that play a role in cancer surveillance are also crucial for self-tolerance [19]. However, the precise relationship between the sex differences in different cancer types and autoimmune diseases remains largely unexplored.

Another example of the interplay in sex differences in different metabolic tissues (and possibly their resident immune components) is hepatocellular carcinoma (HCC). HCC is sexually dimorphic worldwide, with 2–3 times higher incidence in males, an effect that is dependent on sex steroids [20, 21]. The sex differences in the development of HCC have also been found in rodent models, suggesting that they are biological [22]. However, the mechanisms involved remain poorly understood. In the past, the focus has been on the direct effect of sex steroids on hepatocytes [20, 21]. New evidence suggests that sex differences in adipose tissue functions contribute to sex differences in HCC incidence [22, 23], and it is plausible to prevent HCC development by targeting adipose tissue dysregulation [23]. This follows because adipose tissue and liver work in a highly coordinated manner in the regulation of metabolic homeostasis in the body. Thus, sex differences in extra hepatic tissue and in HCC development provide a new paradigm in sex differences in metabolic health and disease. It is possible that such a relationship between different tissues/organs also exist in other related diseases that display sexual dimorphism, such as age-related neurodegenerative and immune diseases. Thus, a better understanding of the similarities and dissimilarities in sex-specific disease incidences (and their interrelationship) is expected to have major implications in maintaining metabolic and immune health, and in developing sex-specific precision medicine. In addition, such information may provide new insights into the sex differences in incidence and mortality rates in different types of cancers, and in predicting the outcome of cancer immunotherapy under conditions that are known to affect immunity (such as obesity, a risk factor for different types of cancer) [24, 25].

In conclusion, the sex differences in various aspects of metabolic and immune health and disease (and their cross talk) suggest that the prevention and treatment of immunometabolic diseases should be sex specific. Furthermore, studies of the mechanisms contributing to context-specific relative protection in one sex and increased susceptibility in another are needed to help develop better approaches to treat immunometabolic diseases in the future. The mechanisms behind this fundamental aspect of metabolic and immune functions are not clear yet and require further research. The identification of the molecular pathways that regulate sex differences in metabolic and immune cells could offer potentially novel approaches to manipulating immunometabolism in an effort to maintain health and prevent the onset of diseases, as well as improving the efficacy of promising immunotherapy.

4 Outstanding Questions

1. Are factors that mediate sex differences in different metabolic and immune cell types distinct, or are there pleiotropic effectors with cell-type-specific functions, or is there a combination of both?

2. What are the relative contributions of cell-intrinsic and hormone-induced attributes in sex differences in different metabolic and immune cell types?
3. What are the relative contributions of X-linked genes that escape from XCI in sex differences in different metabolic and immune cell types?
4. Do cell-intrinsic and hormone-induced factors interact with each other, or do they operate independently in defining a sex-dimorphic phenotype of a particular cell type?
5. Do sex differences in metabolic and immune functions influence each other?
6. What is the role of X-linked mitochondrial genes that escape from XCI in sex differences in mitochondrial phenotypes? Do they interact with sex steroids in defining sex-dimorphic mitochondrial phenotypes in metabolic and immune cells?
7. Does mitochondrial retrograde signaling and epigenetics play a role in sex differences in metabolic and immune functions?
8. Do components of immune checkpoints display sex differences in their regulation and functions?
9. Do co-regulators of sex hormone receptors and SHBG play a role in sex differences in different metabolic and immune cell types?
10. What are the impacts of sex hormone replacement therapy on metabolic and immune functions in transgender people and their potential health consequences?

Acknowledgments

Research in the laboratory of SM is supported by Natural Sciences and Engineering Research Council of Canada (RGPIN-2017-04962), Research Manitoba, Health Sciences Centre Foundation, and URGP-University of Manitoba.

References

1. Lee YS, Wollam J, Olefsky JM (2018) An integrated view of immunometabolism. Cell 172:22–40
2. Mishra S, Nyomba BG (2017 Jun) Prohibitin—at the crossroads of obesity-linked diabetes and cancer. Exp Biol Med (Maywood) 242 (11):1170–1177
3. Mishra S, Nyomba BG (2019 Feb) Prohibitin: a hypothetical target for sex-based new therapeutics for metabolic and immune diseases. Exp Biol Med (Maywood) 244(2):157–170
4. Mauvais-Jarvis F (2015) Sex differences in metabolic homeostasis, diabetes, and obesity. Biol Sex Differ 6:14
5. Markle JG, Fish EN (2014) SeXX matters in immunity. Trends Immunol 35:97–104
6. Nair RR, Verma P, Singh K (2017) Immune-endocrine crosstalk during pregnancy. Gen Comp Endocrinol 242:18–23
7. Lin JH, Zhang SM, Rexrode KM et al (2013) Association between hormones and colorectal

cancer risk in men and women. Clin Gastroenterol Hepatol 11:419–424
8. Butterworth M et al (1967) Influence of sex in immunoglobulin levels. Nature 214:1224–1225
9. Xu YXZ, Ande SR, Mishra S (2018) Gonadectomy in Mito-Ob mice revealed a sex-dimorphic relationship between prohibitin and sex steroids in adipose tissue biology and glucose homeostasis. Biol Sex Differ 9(1):37
10. Fischer J, Jung N, Robinson N et al (2015) Sex differences in immune responses to infectious diseases. Infection 43:399–403
11. González-Arenas A, Neri-Gómez T, Guerra-Araiza C, Camacho-Arroyo I (2004) Sexual dimorphism in the content of progesterone and estrogen receptors, and their cofactors in the lung of adult rats. Steroids 69(5):351–356
12. Liu S, Sun Q (2018) Sex differences, endogenous sex-hormone hormones, sex-hormone binding globulin, and exogenous disruptors in diabetes and related metabolic outcomes. J Diabetes 10(6):428–441
13. Cao J, Yu L, Zhao J, Ma H (2019) Effect of dehydroepiandrosterone on the immune function of mice in vivo and in vitro. Mol Immunol 112:283–290
14. Balaton BP, Cotton AM, Brown CJ (2015) Derivation of consensus inactivation status for X-linked genes from genome-wide studies. Biol Sex Differ 6:35
15. Weinberg SE, Sena LA, Chandel NS (2015) Mitochondria in the regulation of innate and adaptive immunity. Immunity 42(3):406–417
16. Matarrese P, Tieri P, Anticoli S, Ascione B, Conte M, Franceschi C, Malorni W, Salvioli S, Ruggieri A (2019) X-chromosome-linked miR548am-5p is a key regulator of sex disparity in the susceptibility to mitochondria-mediated apoptosis. Cell Death Dis 10(9):673
17. Xu YXZ, Ande SR, Mishra S (2018) Prohibitin: a new player in immunometabolism. Cancer Lett 415:208–216
18. Ande SR, Nguyen KH, Nyomba BLG, Mishra S (2016) Prohibitin in adipose and immune functions. Trends Endocrinol Metab 27(8):531–541
19. Topalian SL, Taube JM, Anders RA, Pardoll DM (2016 May) Mechanism-driven biomarkers to guide immune checkpoint blockade in cancer therapy. Nat Rev Cancer 16(5):275–287
20. Li Z et al (2012) Foxa1 and Foxa2 are essential for sexual dimorphism in liver cancer. Cell 148:72–83
21. Bojkowska K et al (2012) Liver-specific ablation of Krüppel-associated box-associated protein 1 in mice leads to male-predominant hepatosteatosis and development of liver adenoma. Hepatology 56:1279–1290
22. Ande SR, Nguyen KH, Grégoire Nyomba BL, Mishra S (2016) Prohibitin-induced, obesity-associated insulin resistance and accompanying low-grade inflammation causes NASH and HCC. Sci Rep 6:23608
23. Corbit KC, Wilson CG, Lowe D, Tran JL, Vera NB, Clasquin M, Mattis AN, Weiss EJ (2019) Adipocyte JAK2 mediates spontaneous metabolic liver disease and hepatocellular carcinoma. JCI Insight 5. pii: 131310
24. Xu H, Cao D, He A, Ge W (2019) The prognostic role of obesity is independent of sex in cancer patients treated with immune checkpoint inhibitors: a pooled analysis of 4090 cancer patients. Int Immunopharmacol 74:105745
25. Cortellini A, Bersanelli M, Buti S et al (2019) A multicenter study of body mass index in cancer patients treated with anti-PD-1/PD-L1 immune checkpoint inhibitors: when overweight becomes favorable. J Immunother Cancer 7(1):57

Chapter 19

Isolation and Preparation of Bone Marrow-Derived Immune Cells for Metabolic Analysis

Nnamdi M. Ikeogu, Chidalu A. Edechi, Gloria N. Akaluka, Aida Feiz-Barazandeh, and Jude E. Uzonna

Abstract

The isolation of immune cells from the bone marrow is important for obtaining sufficient numbers for downstream analysis. Immune cells derived from the bone marrow may be subjected to metabolic assays for analysis or used to test the effect of infectious agents on immune cells. Here, we describe a process for the isolation of macrophages, dendritic cells, and neutrophils from mice. Using the methods described herein, specific immune cells with purity above 85–90% can be obtained from the bone marrow of mice.

Key words Bone marrow, Immune cells, Macrophages, Dendritic cells, Neutrophils, Metabolic analysis, Mice, Flow cytometry

1 Introduction

The bone marrow, which is the major site of hematopoiesis (a process which leads to the formation of all blood cells) [1] is made up of the parenchyma and stroma components. Immune cells are derived from the hematopoietic stem cells originating from the parenchyma [2]. In studying certain infectious diseases and other disease conditions, it is sometimes difficult to obtain the required number of immune cells for experimental purposes. Therefore, it becomes necessary to derive immune cells in vitro from laboratory animals of interest in order to obtain adequate cell numbers for analyses. For instance, bone marrow-derived macrophages (BMDMs) and bone marrow-derived dendritic cells (BMDCs) are routinely used to study infectivity of *Leishmania major* parasites in mice because the parasite infects these cells in vivo and their response to the infection significantly impacts the outcome of the disease [3, 4]. Neutrophils isolated from the bone marrow can be

The authors contributed equally to this manuscript.

Suresh Mishra (ed.), *Immunometabolism: Methods and Protocols*, Methods in Molecular Biology, vol. 2184,
https://doi.org/10.1007/978-1-0716-0802-9_19,

2 M

MIX
Papier aus verantwortungsvollen Quellen
Paper from responsible sources
FSC® C105338

If you have any concerns about our products,
you can contact us on
ProductSafety@springernature.com

In case Publisher is established outside the EU,
the EU authorized representative is:
Springer Nature Customer Service Center GmbH
Europaplatz 3, 69115 Heidelberg, Germany

Printed by Libri Plureos GmbH
in Hamburg, Germany